AF633533

*INTRODUCTION TO ELECTRIC POWER AND DRIVE SYSTEMS*

**IEEE Press**
445 Hoes Lane
Piscataway, NJ 08854

**IEEE Press Editorial Board**
Tariq Samad, *Editor in Chief*

Giancarlo Fortino
Dmitry Goldgof
Don Heirman
Ekram Hossain
Xiaoou Li
Andreas Molisch
Saeid Nahavandi
Jeffrey Nanzer
Ray Perez
Linda Shafer
Mohammad Shahidehpour
Zidong Wang

# INTRODUCTION TO ELECTRIC POWER AND DRIVE SYSTEMS

PAUL KRAUSE
OLEG WASYNCZUK
TIMOTHY O'CONNELL
MAHER HASAN

IEEE PRESS SERIES ON POWER ENGINEERING

IEEE PRESS

WILEY

Copyright © 2017 by The Institute of Electrical and Electronics Engineers, Inc.

Published by John Wiley & Sons, Inc., Hoboken, New Jersey. All rights reserved
Published simultaneously in Canada

No part of this publication may be reproduced, stored in a retrieval system, or transmitted in any form or by any means, electronic, mechanical, photocopying, recording, scanning, or otherwise, except as permitted under Section 107 or 108 of the 1976 United States Copyright Act, without either the prior written permission of the Publisher, or authorization through payment of the appropriate per-copy fee to the Copyright Clearance Center, Inc., 222 Rosewood Drive, Danvers, MA 01923, (978) 750-8400, fax (978) 750-4470, or on the web at www.copyright.com. Requests to the Publisher for permission should be addressed to the Permissions Department, John Wiley & Sons, Inc., 111 River Street, Hoboken, NJ 07030, (201) 748-6011, fax (201) 748-6008, or online at http://www.wiley.com/go/permission.

Limit of Liability/Disclaimer of Warranty: While the publisher and author have used their best efforts in preparing this book, they make no representations or warranties with respect to the accuracy or completeness of the contents of this book and specifically disclaim any implied warranties of merchantability or fitness for a particular purpose. No warranty may be created or extended by sales representatives or written sales materials. The advice and strategies contained herein may not be suitable for your situation. You should consult with a professional where appropriate. Neither the publisher nor author shall be liable for any loss of profit or any other commercial damages, including but not limited to special, incidental, consequential, or other damages.

For general information on our other products and services or for technical support, please contact our Customer Care Department within the United States at (800) 762-2974, outside the United States at (317) 572-3993 or fax (317) 572-4002.

Wiley also publishes its books in a variety of electronic formats. Some content that appears in print may not be available in electronic formats. For more information about Wiley products, visit our web site at www.wiley.com.

*Library of Congress Cataloging-in-Publication Data is available.*

ISBN: 978-1-119-21425-0

Printed in the United States of America

10 9 8 7 6 5 4 3 2 1

# CONTENTS


*ABOUT THE AUTHORS* ix

*PREFACE AND ACKNOWLEDGMENT* xi

**CHAPTER 1** *BASIC CONCEPTS* 1

1.1 Introduction 1
1.2 Phasor Analysis and Power Calculations 1
Power and Reactive Power 5
1.3 Elementary Magnetic Circuits 8
Field Energy and Coenergy 13
1.4 Stationary Coupled Circuits 16
Magnetically Linear Transformer 16
Field Energy 20
1.5 Coupled Circuits in Relative Motion 22
Field Energy 25
1.6 Electromagnetic Force and Torque 26
1.7 Elementary Electromechanical Device 32
1.8 Two- and Three-Phase Systems 35
Two-Phase Systems 35
Three-Phase Systems 37
References 40
Problems 40

**CHAPTER 2** *ELECTRIC MACHINES* 43

2.1 Introduction 43
2.2 Fundamentals of Electric Machine Analysis 44
Concentrated Winding 44
Distributed Windings 47
Rotating Air-Gap MMF – Tesla's Rotating Magnetic Field 52
Two-Pole Two-Phase Stator 52
Three-Phase Stator 56
P-Pole Machines 58
Machine Inductances 62
2.3 Two-Phase Permanent-Magnet AC Machine 63
2.4 Analysis of a Two-Phase Permanent-Magnet AC Machine 69
Transformation 70
Steady-State Analysis 77
2.5 Three-Phase Permanent-Magnet AC Machine 81
Voltage Equations and Winding Inductances 82

Torque 83
The $qs^r$-, $ds^r$-, and $0s$- Equations 84
References 89
Problems 89

CHAPTER 3 *POWER ELECTRONICS* 91

3.1 Introduction 91
3.2 Switching-Circuit Fundamentals 91
Power Conversion Principles 92
Fourier Analysis 95
Switches and Switching Functions 97
Energy Storage Elements 101
3.3 DC-DC Conversion 103
Buck Converter 104
Boost Converter 113
Advanced Circuit Topologies 118
3.4 AC-DC Conversion 118
Half-Wave Rectifier 118
Full-Wave Rectifier 125
3.5 DC-AC Conversion 133
Single-Phase Inverter 133
Three-Phase Inverter 136
3.6 Harmonics and Distortion 144
References 147
Problems 147

CHAPTER 4 *PERFORMANCE AND SIMULATION OF AN ELECTRIC DRIVE* 149

4.1 Introduction 149
4.2 Operating Modes of a Brushless DC Motor 149
Brushless DC Motor Operation with $\phi_v = 0$ 150
Maximum Torque Per Volt Operation of a Brushless DC Motor ($\phi_v = \phi_{vMT/V}$) 155
Maximum Torque Per Ampere Operation of a Brushless DC Motor ($\phi_v = \phi_{vMT/A}$) 160
4.3 Operation of a Brushless DC Drive 164
Operation of Brushless DC Drive with $\phi_v = 0$ 166
Operation of Brushless DC Drive with $\phi_v = \phi_{vMT/V}$ 170
Operation of Brushless DC Drive with $\phi_v = \phi_{vMT/A}$ 172
Modes of Control of a Brushless DC Drive 174
4.4 Simulation of a Brushless DC Drive 180
Simulation of Coupled Circuits 180
Simulation of Drive Inverter and Transformation 181
Simulation of Permanent-Magnet AC Machine 183
References 186
Problems 186

CHAPTER 5 *POWER SYSTEMS* 187

5.1 Introduction 187

5.2 Three-Phase Transformer Connections 187
Wye-Wye Connection 188
Delta-Delta Connection 190
Wye-Delta or Delta-Wye Connection 191
Ideal Transformers 192
5.3 Synchronous Generator 193
Damper Windings 198
Torque 198
Steady-State Operation and Rotor Angle 198
5.4 Reactive Power and Power Factor Correction 204
5.5 Per Unit System 209
Per Unitizing the Synchronous Generator 210
5.6 Discussion of Transient Stability 215
Three-Phase Fault 215
References 220
Problems 220

APPENDIX A *TRIGONOMETRIC RELATIONS, CONSTANTS AND CONVERSION FACTORS, AND ABBREVIATIONS* 221

APPENDIX B *WINDING INDUCTANCES* 225

APPENDIX C *ANIMATIONS* 229

*INDEX* 231

# ABOUT THE AUTHORS

**Paul Krause** retired from Purdue University School of Electrical and Computer Engineering in 2009 after serving 39 years as a Professor. He also held teaching positions at the University of Kansas and the University of Wisconsin–Madison. He is a life fellow of IEEE and has authored or co-authored more than 100 technical papers and three textbooks, and he was the 2010 recipient of the IEEE Nikola Tesla Award. He received B.S. degrees in electrical and mechanical engineering, and an M.S. in electrical engineering from the University of Nebraska, and a Ph.D. in electrical engineering from the University of Kansas. He is presently Board Chairman of PC Krause and Associates, Inc. (PCKA) with offices in West Lafayette, IN and Dayton, OH.

**Oleg Wasynczuk** is a Professor of Electrical and Computer Engineering at Purdue University and also serves as Chief Technical Officer of PC Krause and Associates, Inc. He has authored or co-authored more than 100 technical papers and two textbooks on electric machines. He is a fellow of the IEEE and was the 2008 recipient of the IEEE PES Cyril Veinott Electromechanical Energy Conversion Award. He received a B.S.E.E. degree from Bradley University in 1976, and the M.S.E.E. and Ph.D. degrees from Purdue University in 1977 and 1979, respectively.

**Timothy O'Connell** is a Senior Lead Engineer with PC Krause and Associates, Inc. (PCKA), in West Lafayette, IN. He jointly holds an Adjunct Research Assistant Professorship in the Department of Electrical and Computer Engineering (ECE) at the University of Illinois at Urbana-Champaign (UIUC). He is a senior member of IEEE, has served as Chair of the IEEE Central Illinois Section since 2015, and is the Department of Defense aerospace industry liaison for the IEEE Transportation Electrification Community (TEC). He has authored or co-authored more than 20 technical papers on a wide variety of topics, including electric machine analysis and design, aerospace power systems, and modeling and simulation. He received a B.A. degree, *summa cum laude*, in physics from Carleton College in Northfield, MN and M.S. and Ph.D. degrees in ECE from UIUC. He has been teaching a course on Green Electric Energy at UIUC since 2012.

**Maher Hasan** is currently a Senior Lead Engineer with PC Krause and Associates, Inc., which he joined in 2005. He received a B.S. in electrical engineering, an M.S. in electrical and computer engineering, and a Ph.D. degree from Purdue University in 2004, 2007, and 2009, respectively. He has authored or co-authored several technical papers in the fields of modeling and simulation techniques for circuits and electromechanical systems.

# PREFACE AND ACKNOWLEDGMENT

**TRADITIONALLY**, the lead-in to the study of electric power engineering has been a junior-level course on electric machines. However, due to the advent of power electronics and the computer, the present-day electrical engineering student interested in the power area has several areas from which to choose, including electric machines, power electronics, electric drives, and power systems. This has caused some concern as to the most appropriate lead-in course to the power discipline. Although the analysis of electric machinery is fundamental to power systems and electric drives, it may not be the most effective way to introduce the power area to a student who is trying to decide on a career path. As an alternative, some schools have instituted a survey course; however, this often takes the form of a broad and general review lacking the sufficient depth to give the student an indication of the physical concepts and analyses common to the areas or to provide an analytical foundation on which to build in follow-on courses. From the standpoint of the faculty, who must make full use of every credit hour allotted to their area of study, an introductory course that establishes a useful foundation seems appropriate.

This book is an attempt to provide students with an analytical base on which they can build in follow-on courses in electric machines, power electronics, electric drives, and power systems. This is accomplished with only sophomore-level calculus, physics, and basic electric circuits as pre- or co-requisites, allowing the course to be taught early in most engineering disciplines by a professor with either an engineering or physics background. The text is suggested for use in a 3-hour course in electrical engineering at the second-semester sophomore-level or first-semester junior-level in schools with a viable power program. Alternatively, it supports a junior/senior technical elective course in schools without a power program, or in non-electrical engineering disciplines such as mechanical engineering. It is written so that the professor can select the material to fit the school's power program and interests. Although most of the material in Chapter 1 is common to all areas of power and should be covered in its entirety, each of the subsequent chapters contains material fundamental to the areas of electric machines, power electronics, electric drives, and power systems, respectively, along with material that is somewhat more advanced in each of these disciplines. This allows the instructor to choose between a brief and an in-depth coverage of each of the areas. It is not intended that the entirety of the text be covered in a three-credit course.

Electric machines are covered in Chapter 2, focusing on the analysis of the round-rotor permanent-magnet ac machine. This device, when used as an inverter-driven brushless dc motor, has become the low-power electric drive of choice. Since this is a synchronous machine, the analysis applies, in part, to the synchronous generator that is covered in Chapter 5 on power systems. Tesla's rotating magnetic field is shown to be the key to machine analysis in that it provides a sound basis for a concise analytical derivation of all known mathematical transformations that are used in machine and power system analyses. Moreover, it provides an analytical means of positioning the stator and rotor poles on the phasor diagram, thereby providing a straightforward and instructive illustration of machine operation that adds credence to the approach of analyzing the two-phase devices first and then the three phase. Tesla not only invented the ac machine, his work is instrumental in analyzing and visualizing its operation. An animation described in Appendix C will help the student to visualize Tesla's rotating magnetic field. All animations can be accessed at www.wiley.com/go/krause/electricpower.

Basic power electronic circuit analysis is covered in Chapter 3, including switching-circuit fundamentals, dc-dc converters, ac-dc rectifiers, dc-ac inverters, and a brief introduction to harmonics and distortion. This includes an analysis of the three-phase six-step inverter, which establishes basic relationships between the transistor switching signals and the input (dc) and output (ac) voltages. This inverter is ubiquitous in the electric drives area; as such, its analysis in Chapter 3 is built upon in Chapter 4 on electric drives.

The performance and simulation of an electric drive are covered in Chapter 4. The drive chosen for analysis is the round-rotor permanent-magnet machine described in Chapter 2 powered by the three-phase six-step inverter introduced in Chapter 3. Although this so-called brushless dc drive represents only a small subset of the electric-drives area, it is widely used and its salient features can be presented without becoming too involved. Three methods of brushless dc motor operation – traditional, maximum torque per volt, and maximum torque per ampere – are demonstrated by changing the phase angle and effective value of the applied stator voltages. A direct comparison is made between the simulated performance of the drive with both sinusoidal and six-step applied voltages. The chapter concludes with a straightforward method for simulating the drive system. An animation described in Appendix C will help the student visualize the operation of a brushless dc drive.

Finally, the power systems area, including transformer connections, the synchronous generator, power factor correction, and the per-unit system, is introduced in Chapter 5. The chapter concludes with a brief discussion of transient stability, provided to illustrate some of the basic challenges faced in the power systems area.

This text covers several power engineering sub-disciplines. It is not surprising, then, that throughout its pages, some differences in notation, variable naming conventions, and analysis methods exist. However, rather than attempt to standardize, we have chosen to recognize and explain any discrepancies as they arise. We have done this so that each chapter may more easily be taught separately from the others and, if the instructor desires it, be incorporated into existing curricula that uses common notation. We hope that this approach will aid the power engineering student to anticipate, understand, and accept these differences.

Obviously, there are other aspects of the power area that would be appropriate for an introductory text; however, the choices we have made seem to be representative at this time. Nevertheless, as the power area continues to evolve through the twenty-first century, so must an introductory text.

We would like to acknowledge Dr. Brett Robbins of PC Krause and Associates for developing the drawings and formatting the computer traces for the text. We would also like to acknowledge the efforts and assistance of the reviewers, in particular Mohamed El-Hawary, the staff of IEEE Press and John Wiley & Sons, especially Mary Hatcher, Danielle Lacourciere, Victoria Bradshaw, Jeanne Audino and finally the Production Editor Suresh Srinivasan of Aptara for the final typesetting of the manuscript.

Paul Krause
Oleg Wasynczuk
Timothy O'Connell
Maher Hasan

CHAPTER 1

# BASIC CONCEPTS

## 1.1 INTRODUCTION

The twentieth century began with the electric power industry in its infancy; Thomas Edison and Nikola Tesla were locked in battle with Edison advocating direct current (dc) and Tesla alternating current (ac). The century ended with the electric power industry expanding rapidly from the traditional power generation, transmission, and utilization into propulsion of air, ground, and sea transportation. The advent of the computer and the silicon-controlled rectifier in the mid-1900s brought about an expansion of the power area to include the smart grid, microgrids, efficient and robust electric drives, more-electric aircraft, ships, and land vehicles. This growth is likely to continue into the foreseeable future.

Before the advent of the computer, engineers were essentially limited to steady-state analysis and therefore unable to conveniently deal with the analytical challenges of the expanding power industry. This chapter sets forth some of the basic concepts and analysis tools that are part of the present-day power and electric drives area. Although not inclusive, the material covered in this chapter is representative and common to all disciplines of the power area.

## 1.2 PHASOR ANALYSIS AND POWER CALCULATIONS

Since the early twentieth century, we have lived in an alternating current (ac) world. Thanks to George Westinghouse and Nikola Tesla, power systems are predominately ac; power is generated by large ac generators, transmitted by high voltage transmission lines, and transformed to a low voltage and distributed to homes and factories. The evolution of the ac power system brought about many engineering challenges and, as we look back, it is difficult to comprehend how these problems were solved without a computer. Even steady-state ac-circuit analysis posed a problem until the early 1900s when Charles Stienmetz, who was a less flamboyant colleague of Edison and Tesla, came up with the concept of what is now known as phasors. Some may consider the phasor a casualty of the computer age along with the slide rule. It is, however, still a very useful means for understanding and portraying the steady-state performance of electric machines, power systems, and electric drives. Moreover, the

*Introduction to Electric Power and Drive Systems*, First Edition.
Paul Krause, Oleg Wasynczuk, Timothy O'Connell and Maher Hasan.
© 2017 by The Institute of Electrical and Electronics Engineers, Inc. Published 2017 by John Wiley & Sons, Inc.
Companion website: www.wiley.com/go/krause/electricpower

phasor concept provides a means of visualizing sinusoidal variations from different frames of reference and in Chapter 2 we will find that the voltage and current phasors combined with Tesla's rotating magnetic field provides a straightforward means of analyzing and portraying the steady-state operation of ac machines.

The phasor can be established by expressing a steady-state sinusoidal variable as

$$F_a(t) = F_p \cos\theta_{ef} \tag{1.2-1}$$

where the $a$ subscript is used here to denote sinusoidal quantities. The sinusoidal variations are expressed as cosines, capital letters are used to denote steady-state quantities, and $F_p$ is the peak value of the sinusoidal variation. Generally, $F$ or $f$ represents voltage ($V$ or $v$) or current ($I$ or $i$) in circuit analysis, but it could be any sinusoidal variable. For steady-state conditions, $\theta_{ef}$ may be written as

$$\theta_{ef}(t) = \omega_e t + \theta_{ef}(0) \tag{1.2-2}$$

where $\omega_e$ is the electrical angular velocity in radians/second ($2\pi$ times the frequency) and $\theta_{ef}(0)$ is the time-zero position of the electrical variable. Substituting (1.2-2) into (1.2-1) yields

$$F_a(t) = F_p \cos[\omega_e t + \theta_{ef}(0)] \tag{1.2-3}$$

Now, Euler's Identity is

$$e^{j\alpha} = \cos\alpha + j\sin\alpha \tag{1.2-4}$$

and since we are expressing the sinusoidal variation as a cosine, (1.2-3) may be written as

$$F_a(t) = \mathrm{Re}\left\{F_p e^{j[\omega_e t + \theta_{ef}(0)]}\right\} \tag{1.2-5}$$

where Re is the shorthand for "real part of." Equations (1.2-3) and (1.2-5) are equivalent. Let us rewrite (1.2-5) as

$$F_a(t) = \mathrm{Re}\left\{F_p e^{j\theta_{ef}(0)} e^{j\omega_e t}\right\} \tag{1.2-6}$$

We need to take a moment to define what is referred to as the root-mean-square (rms) of a sinusoidal variation. In particular, the rms value is defined as

$$F = \left(\frac{1}{T}\int_0^T F_a^2(t)dt\right)^{\frac{1}{2}} \tag{1.2-7}$$

where $F$ is the rms value of $F_a(t)$ and $T$ is the period of the sinusoidal variation. It is left to the reader to show that the rms value of (1.2-3) is $F_p/\sqrt{2}$. Therefore, we can express (1.2-6) as

$$F_a(t) = \mathrm{Re}[\sqrt{2}Fe^{j\theta_{ef}(0)} e^{j\omega_e t}] \tag{1.2-8}$$

By definition, the phasor representing $F_a(t)$, which is denoted with a raised tilde, is

$$\tilde{F}_a = Fe^{j\theta_{ef}(0)} \tag{1.2-9}$$

which is a complex number. The reason for using the rms value as the magnitude of the phasor will be addressed later in this section. Equation (1.2-6) may now be written as

$$F_a(t) = \text{Re}\left[\sqrt{2}\tilde{F}_a e^{j\omega_e t}\right] \tag{1.2-10}$$

A shorthand notation for (1.2-9) is

$$\tilde{F}_a = F\underline{/\theta_{ef}(0)} \tag{1.2-11}$$

Equation (1.2-11) is commonly referred to as the *polar form* of the phasor. The *Cartesian* form is

$$\tilde{F}_a = F\cos\theta_{ef}(0) + jF\sin\theta_{ef}(0) \tag{1.2-12}$$

When using phasors to calculate steady-state voltages and currents, we think of the phasors as being stationary at $t = 0$; however, we know that a phasor is related to the instantaneous value of the sinusoidal quantity it represents. Let us take a moment to consider this aspect of the phasor and thereby, give some physical meaning to it. From (1.2-4), we realize that $e^{j\omega_e t}$ is a line of unity length rotating counterclockwise at an angular velocity of $\omega_e$. Therefore, backing up for a minute

$$\sqrt{2}\,\tilde{F}_a e^{j\omega_e t} = \sqrt{2}F e^{j\theta_{ef}(0)} e^{j\omega_e t} \tag{1.2-13}$$

is a line with a constant amplitude of $\sqrt{2}F$ rotating counterclockwise in the real-imaginary plane at an angular velocity of $\omega_e$ with a time-zero displacement from the positive real axis of $\theta_{ef}(0)$. Since $\sqrt{2}F$ is the peak value of the sinusoidal variation, the instantaneous value of $F_a(t)$ expressed as a cosine is the real part of (1.2-13). In other words, the real projection of the phasor $\tilde{F}_a$ rotating counterclockwise at $\omega_e$ is the instantaneous value of $F_a(t)/\sqrt{2}$. Thus, with $\theta_{ef}(0) = 0$ in (1.2-3)

$$F_a(t) = \sqrt{2}F\cos\omega_e t \tag{1.2-14}$$

the phasor representing (1.2-14) is

$$\tilde{F}_a = Fe^{j0} = F\underline{/0^\circ} = F + j0 \tag{1.2-15}$$

For

$$\begin{aligned} F_a(t) &= \sqrt{2}F\sin\omega_e t \\ &= \sqrt{2}F\cos(\omega_e t - 90^\circ) \end{aligned} \tag{1.2-16}$$

the phasor is

$$\tilde{F}_a = Fe^{-j\pi/2} = F\underline{/-90^\circ} = 0 - jF \tag{1.2-17}$$

We will use degrees and radians interchangeably when expressing phasors. Although there are several ways to arrive at (1.2-17) from (1.2-16), it is helpful to ask yourself where the rotating phasor must be positioned at time-zero so that, when it rotates counterclockwise at $\omega_e$, its real projection is $(1/\sqrt{2})F_p \sin\omega_e t$. It follows that a phasor of amplitude $F$ positioned at 90° represents $-\sqrt{2}F\sin\omega_e t$.

In other words, we are viewing a sinusoidal variation as the real projection in the real-imaginary plane of a rotating line equal in magnitude to the positive peak value ($\sqrt{2}F$) of the variation and rotating at the electrical angular velocity of the sinusoidal variation. Since we are dealing with a steady-state variation, we can stop the rotation at any time and view it as a fixed line, but knowing full well that it, in fact, represents a sinusoidal variation and to represent the sinusoidal variation we must rotate it counterclockwise at $\omega_e$ and take the real projection. Please understand that if we ran at $\omega_e$ in unison with the rotating $\sqrt{2}F$ line it would appear as a constant to us; therefore, in viewing a sinusoidal variation in this manner it would appear to us as a constant. This is no different than stopping the phasor at some arbitrary time-zero; but realizing that it actually represents a sinusoidal variation. We will talk more about this important aspect as we go along; in particular, see Example 1A.

In order to show the facility of the phasor in the analysis of steady-state performance of ac circuits and devices, it is useful to consider a series circuit consisting of a resistance, an inductance $L$, and a capacitance $C$. Thus, using uppercase letters to indicate steady-state variables

$$V_a = RI_a + L\frac{dI_a}{dt} + \frac{1}{C}\int I_a dt \tag{1.2-18}$$

Throughout the text, we will use either $R$ or $r$ to represent resistance. For steady-state operation, let

$$V_a = \sqrt{2}V\cos[\omega_e t + \theta_{ev}(0)] \tag{1.2-19}$$

$$I_a = \sqrt{2}I\cos[\omega_e t + \theta_{ei}(0)] \tag{1.2-20}$$

where we have dropped the functional notation and the subscript $a$ helps to distinguish the instantaneous value from the rms value of the steady-state variables. The steady-state voltage equation may be obtained by substituting (1.2-19) and (1.2-20) into (1.2-18), whereupon we can write

$$\begin{aligned}\sqrt{2}V\cos[\omega_e t + \theta_{ev}(0)] = {} & R\sqrt{2}I\cos\left[\omega_e t + \theta_{ei}(0)\right] \\ & + \omega_e L\sqrt{2}I\cos\left[\omega_e t + \tfrac{1}{2}\pi + \theta_{ei}(0)\right] \\ & + \frac{1}{\omega_e C}\sqrt{2}I\cos\left[\omega_e t - \tfrac{1}{2}\pi + \theta_{ei}(0)\right]\end{aligned} \tag{1.2-21}$$

The second term on the right-hand side of (1.2-21), which is $L\frac{dI_a}{dt}$, can be written

$$\omega_e L\sqrt{2}I\cos\left[\omega_e t + \tfrac{1}{2}\pi + \theta_{ei}(0)\right] = \omega_e L\,\mathrm{Re}\left[\sqrt{2}Ie^{j\frac{1}{2}\pi}e^{j\theta_{ei}(0)}e^{j\omega_e t}\right] \tag{1.2-22}$$

Since $\tilde{I}_a = Ie^{j\theta_{ei}(0)}$, from (1.2-21), we can write

$$L\frac{dI_a}{dt} = \omega_e L e^{j\frac{1}{2}\pi}\tilde{I}_a \tag{1.2-23}$$

Since $e^{j\frac{1}{2}\pi} = j$, (1.2-23) may be written

$$L\frac{dI_a}{dt} = j\omega_e L\tilde{I}_a \qquad (1.2\text{-}24)$$

If we follow a similar procedure, we can show that

$$\frac{1}{C}\int I_a dt = -j\frac{1}{\omega_e C}\tilde{I}_a \qquad (1.2\text{-}25)$$

Differentiation of a steady-state sinusoidal variable rotates the phasor counterclockwise by $\frac{1}{2}\pi$ or $j$; integration rotates the phasor clockwise by $\frac{1}{2}\pi$ or $-j$.

The steady-state voltage equation given by (1.2-21) can now be written in phasor form as

$$\tilde{V}_a = \left[R + j(\omega_e L - \frac{1}{\omega_e C})\right]\tilde{I}_a \qquad (1.2\text{-}26)$$

We can express (1.2-26) compactly as

$$\tilde{V}_a = Z\tilde{I}_a \qquad (1.2\text{-}27)$$

where the impedance, $Z$, is a complex number; it is not a phasor. It may be expressed as

$$Z = R + j(X_L - X_C) \qquad (1.2\text{-}28)$$

where $X_L = \omega_e L$ is the inductive reactance and $X_C = \frac{1}{\omega_e C}$ is the capacitive reactance. We should be careful here. Some prefer to write (1.2-28) as $R + jX$ where $X$ is $X_L + X_C$ and let $X_C$ be negative. This is essentially a matter of choice and does not change the end result. We will deal primarily with $X_L$ and not $X_C$, therefore, this will have little impact on our work; nevertheless, since some authors will use a negative $X_C$ we should make the reader aware of this difference.

It is appropriate to discuss the notation that will be used throughout the text. When an equation is written with the variables in lowercase letters it is valid for transient and steady state. If the variables are written with uppercase letters as in (1.2-18) the equation is a function of time and valid for instantaneous steady-state conditions. Equations (1.2-26) and (1.2-27) are phasor equations representing steady-state sinusoidal variables and are written in uppercase letters with an over tilde.

## Power and Reactive Power

The instantaneous steady-state power is

$$\begin{aligned} P &= V_a I_a \\ &= \sqrt{2}V\cos\left[\omega_e t + \theta_{ev}(0)\right]\sqrt{2}I\cos\left[\omega_e t + \theta_{ei}(0)\right] \end{aligned} \qquad (1.2\text{-}29)$$

where $V$ and $I$ are rms values. After some manipulation, we can write (1.2-29) as

$$P = VI\cos[\theta_{ev}(0) - \theta_{ei}(0)] + VI\cos[2\omega_e t + \theta_{ev}(0) + \theta_{ei}(0)] \qquad (1.2\text{-}30)$$

The instantaneous steady-state power given by (1.2-30) varies about an average value at a frequency of $2\omega_e$. That is, the second term of (1.2-30) has a zero average value and the average power $P_{ave}$ may be written

$$P_{ave} = |\tilde{V}_a|\,|\tilde{I}_a| \cos[\theta_{ev}(0) - \theta_{ei}(0)] \tag{1.2-31}$$

where $|\tilde{V}_a|$ and $|\tilde{I}_a|$ are $V$ and $I$, respectively which are the magnitudes of the phasors (rms value); $\theta_{ev}(0) - \theta_{ei}(0)$ is referred to as the *power factor angle* $\varphi_{pf}$, and $\cos[\theta_{ev}(0) - \theta_{ei}(0)]$ is the *power factor*. Power is in watts. If current is assumed positive in the direction of voltage drop then (1.2-31) is positive if power is consumed and negative if power is generated. It is interesting to point out that in going from (1.2-29) to (1.2-30), the coefficient of the two right-hand terms is $\frac{1}{2}(\sqrt{2}V\sqrt{2}I)$ or one-half the product of the peak values of the sinusoidal variables. Therefore, it was considered more convenient to use the rms values for the phasors, whereupon average steady-state power could be calculated by the product of the magnitude of the voltage and current phasors as given by (1.2-31).

The reactive power is defined as

$$Q = |\tilde{V}_a|\,|\tilde{I}_a| \sin[\theta_{ev}(0) - \theta_{ei}(0)] \tag{1.2-32}$$

The unit of $Q$ is var (volt-ampere reactive). An inductance is said to absorb reactive power where the current lags the voltage by 90° and $Q$ is positive and supplied by a capacitor where the current leads the voltage by 90° and $Q$ is negative. Actually, $Q$ is a measure of the interchange of energy stored in the electric (capacitor) and magnetic (inductor) fields. However, unlike instantaneous real power, the average value of instantaneous reactive power is zero. We will talk more about reactive power when we get to Chapter 5.

**Example 1A. Phasor analysis.** The parameters of a series RLC circuit are $R = 6\ \Omega$, $L = 20$ mH, $C = 1 \times 10^3\ \mu$F. The 60 Hz applied voltage is $V_a = 155.6 \cos \omega_e t$. Calculate $\tilde{I}_a$, $P_{ave}$, $Q$ and draw the phasor diagram and the sinusoidal variations as viewed running at counterclockwise with the phasor representing maximum $V_a$ $(\sqrt{2}V_s\underline{/0°})$. From the expression of $V_a$

$$\tilde{V}_a = 110\underline{/0°}\ \text{V} \tag{1A-1}$$

Now, $\omega_e = 2\pi f = 2\pi \times 60 = 377$ rad/s and

$$\begin{aligned} Z &= R + j(X_L - X_C) \\ &= R + j\left(\omega_e L - \frac{1}{\omega_e C}\right) \\ &= 6 + j\left(377 \times 20 \times 10^{-3} - \frac{1}{377 \times 1 \times 10^{-3}}\right) = 7.73\underline{/39.1°}\ \Omega \end{aligned} \tag{1A-2}$$

$$\tilde{I}_a = \frac{\tilde{V}_a}{Z} = \frac{110\underline{/0°}}{7.73\underline{/39.1°}} = 14.2\underline{/-39.1°}\ \text{A} \tag{1A-3}$$

$$P_{ave} = |\tilde{V}_a|\ |\tilde{I}_a| \cos \phi_{pf} \tag{1A-4}$$

where

$$\begin{aligned}
\phi_{pf} &= \theta_{ev}(0) - \theta_{ei}(0) \\
&= 0 - (-39.1°) = 39.1° \qquad (1A\text{-}5) \\
P_{ave} &= 110 \times 14.2 \cos 39.1° \\
&= 1212.2\,\text{W} \qquad (1A\text{-}6) \\
Q &= |\tilde{V}_a|\,|\tilde{I}_a| \sin \phi_{pf} \\
&= 110 \times 14.2 \sin 39.1° = 985.1 \text{ vars} \qquad (1A\text{-}7)
\end{aligned}$$

The phasor diagram is shown in Fig. 1A-1.

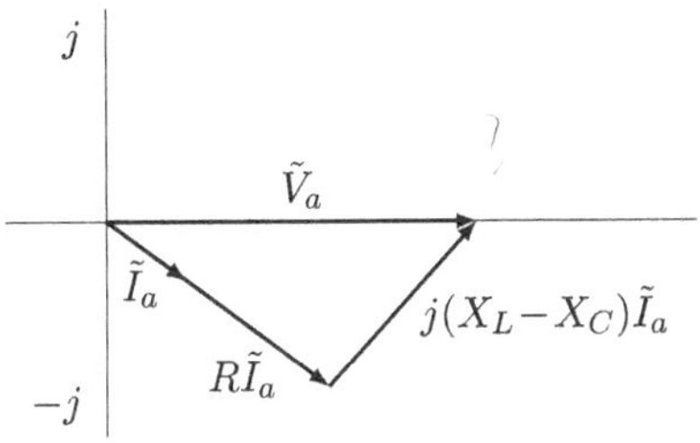


Figure 1A-1 Phasor diagram.

Fig. 1A-2 shows the "stationary" waveforms of the voltages viewed with 360° vision while running at $\omega_e$ with the counterclockwise rotating phasor representing $V_a$. The current $I_a$ is not shown; however, $\sqrt{2}\,|\tilde{I}_a|$ would be in phase with the voltage $R\sqrt{2}\,|\tilde{I}_a|$. Also, the type of diagram shown in Fig. 1A-2, is not convenient to portray that the real part of the rotating phasor is the instantaneous value of the sinusoidal variation. Rotating the phasor diagram given in Fig. 1A-1 best illustrates that feature of a phasor.

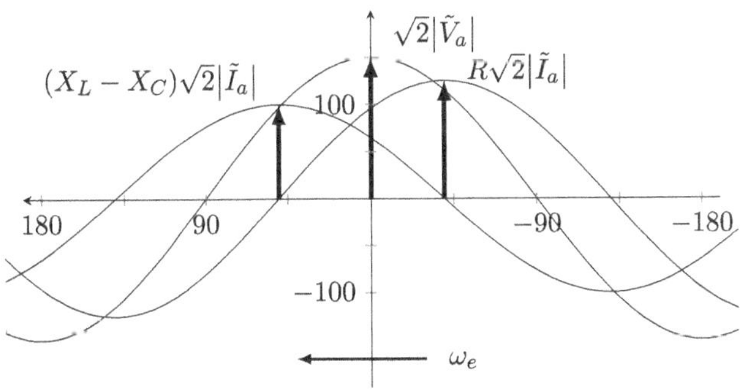


Figure 1A-2 Voltage waveforms of *RLC* circuit viewed with 360° vision while rotating with phasors at $\omega_e$ counterclockwise.

---

**SP1.2-1.** Express the instantaneous steady-state power for Example 1A. [Substitute into (1.2-30)]

**SP1.2-2.** Redraw the phasor diagram shown in Fig. 1A-1 showing $jX_L\tilde{I}_a$ and $-jX_C\tilde{I}_a$ as individual voltages. [Show $jX_L\tilde{I}_a$ and then from the terminus of $jX_L\tilde{I}_a$ show $-jX_C\tilde{I}_a$]

**SP1.2-3.** We know that $P_{\text{ave}} = |\tilde{I}|^2R$, does $Q$ equal $|\tilde{I}_a|^2X_L - |\tilde{I}_a|^2X_C$? [Yes]

**SP1.2-4.** If $\tilde{V} = 1\underline{/0^\circ}$ V and $\tilde{I} = 1\underline{/180^\circ}$ A in the direction of the voltage drop, calculate $Z$ and $P_{\text{ave}}$. Is power generated or consumed? [$(-1 + j0)$ ohms, 1 watt, generated]

**SP1.2-5.** Express the instantaneous power for 60 Hz voltage, $\tilde{V}_a = 1\underline{/0^\circ}$ V, applied to a resistive circuit, $\tilde{I}_a = 1\underline{/0^\circ}$ A. [$1 + \cos 754t$]

**SP1.2-6.** Repeat SP1.2-5 for (a) an inductance, $\tilde{I}_a = I_L\underline{/-90^\circ}$ A and (b) a capacitance, $\tilde{I}_a = I_C\underline{/90^\circ}$ A. [(a) $I_L \cos(754t - 90^\circ)$, (b) $I_C \cos(754t + 90^\circ)$]

## 1.3 ELEMENTARY MAGNETIC CIRCUITS

Electric machines and transformers, which are the backbone of the power industry, are electromagnetic systems. Therefore, magnetic circuits and magnetic coupling play a major role in power and drives systems and it is necessary to establish the principles of magnetic systems sufficiently to convey the basic operation of the electromagnetic devices considered in later chapters. We will attempt to do this without becoming too involved.

An elementary magnetic circuit is shown in Fig. 1.3-1. It consists of a ferromagnetic member (core) with a coil of wire of $N$ turns wound on it and an air gap of length $x$. The ferromagnetic member could be iron, nickel, cobalt, or steel, for example. The voltage equation of the electric circuit may be written

$$v = ri + e \tag{1.3-1}$$

where $r$ is the total resistance of the circuit, $v$ is the source voltage, $e$ is the voltage induced in the coil according to Faraday's Law, and $i$ is the current flowing in the circuit. The current flowing through the coil causes a magnetomotive force (mmf), which produces flux in the magnetic circuit denoted as $\Phi_m$ and $\Phi_l$ in Fig. 1.3-1, much as an electromotive force (emf) or source voltage produces current in an electric circuit.

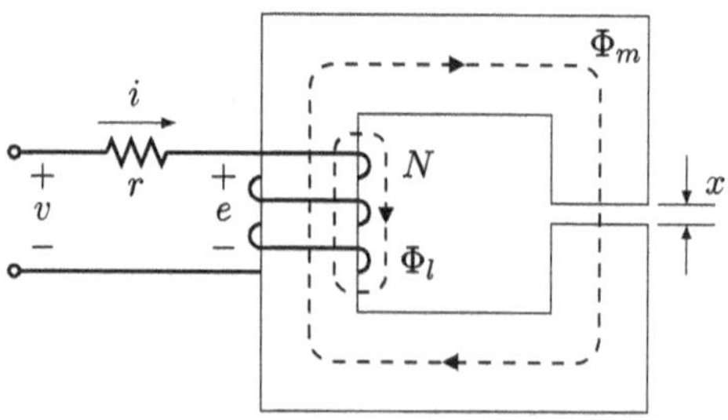


Figure 1.3-1 Elementary magnetic circuit with one air gap.

There are arrows associated with the dashed lines representing the flux paths in Fig. 1.3-1. These arrows indicate the assumed positive direction of flux which is determined from the assumed positive direction of current by the so called "right-hand" rule. If you grasp the coil with your right hand with your fingers in the assumed direction of positive current flow around the coil, your thumb will point in the direction of positive flux. Or, imagine grasping a turn of the coil with thumb in the assumed direction of positive current. If you ungrasp your fingers they will point in the direction of positive flux.

The total flux, $\Phi$, that travels through (links) all of the turns $N$ is

$$\Phi = \Phi_l + \Phi_m \tag{1.3-2}$$

where $\Phi_l$ is the equivalent flux that links all the turns of the coil but does not traverse the ferromagnetic member and $\Phi_m$ is referred to as the magnetizing flux that transverses the ferromagnetic member and links all of the turns of the coil. The leakage flux $\Phi_l$ is shown by one streamline in Fig. 1.3-1, it represents the aggregate of the flux that occurs around each wire of the coil, traveling partially in the ferromagnetic material and partially in air.

The concept of magnetic poles can be used to advantage to explain the operation of electromechanical devices. The poles can be established by considering the magnetic circuit shown in Fig. 1.3-1. In particular, to locate the north and south poles of an electromechanical device for the assumed positive direction of coil current, place yourself on the member that has the coil (windings). Use the right-hand rule to establish the direction of positive flux $\Phi_m$ for the assumed positive direction of coil current; where positive flux issues from the member with the coil, into the air is the assumed north pole. The south pole is where the positive flux returns from the air to the member with the coil. In the case of the magnetic circuit shown in Fig. 1.3-1, with the assumed direction of positive current a north pole exists over the upper face of the air gap and a south pole over the lower face.

A magnetically linear circuit behaves much as a resistive electric circuit. According to Ohm's Law, the current $i$ in a resistive circuit is equal to the applied electromotive force (emf) or applied voltage divided by the resistance $r$. In the case of a magnetic circuit, the total flux $\Phi$ is equal to the magnetomotive force (mmf), which is in ampere turns $Ni$, divided by the equivalent reluctance $\mathcal{R}$ of the magnetic circuit. Thus, in (1.3-2)

$$\Phi_l = \frac{Ni}{\mathcal{R}_l} \tag{1.3-3}$$

$$\Phi_m = \frac{Ni}{\mathcal{R}_i + \mathcal{R}_g} \tag{1.3-4}$$

where $\mathcal{R}_l$ is the reluctance of the leakage flux path, $\mathcal{R}_i$ is the reluctance of the ferromagnetic member, and $\mathcal{R}_g$ is the reluctance of the air gap. Although the leakage reluctance is generally determined by test or an involved calculation, $\mathcal{R}_i$ and $\mathcal{R}_g$ may be calculated from

$$\mathcal{R} = \frac{\ell}{\mu_r \mu_0 A} \tag{1.3-5}$$

where $\ell$ is the length of the flux path, $A$ is the cross sectional area of the flux path, $\mu_0$ is the permeability of free space ($4\pi \times 10^{-7}$ Wb/A · m or H/m), and $\mu_r$ is the permeability relative to free space. The unit of reluctance is (henry$^{-1}$) or H$^{-1}$. In the case of the air gap the relative permeability is considered to be unity ($\mu_{rg} \approx 1$), for the ferromagnetic member typical $\mu_{ri}$ values vary from 500 to 4000 depending upon the type of ferromagnetic material. It follows that the reluctance of the ferromagnetic member is much less than the reluctance of either the air gap or the part of the leakage path that is in air. In fact, the reluctance of the ferromagnetic member is generally neglected when an air gap is present as in the case of an electric machine. The magnetic equivalent circuit shown in Fig. 1.3-2 may be helpful to visualize the flux paths of the magnetic system shown in Fig. 1.3-1.

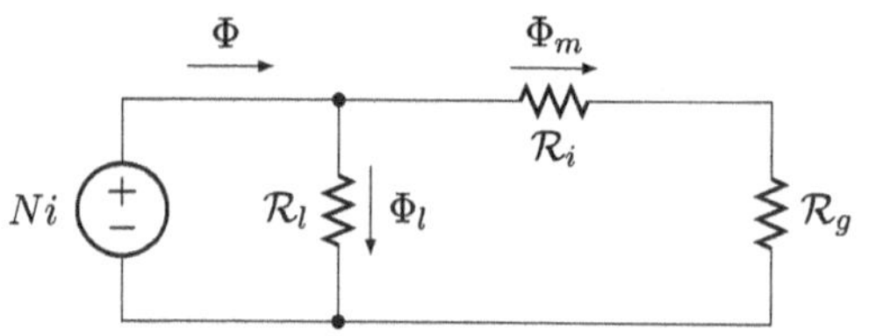


Figure 1.3-2 Magnetic equivalent circuit for magnetic system shown in Fig. 1.3-1.

Before proceeding, there are a couple of things that we should talk about. We have arrived at (1.3-3) and (1.3-4) by exploiting the similarities between a resistive electric circuit and a linear magnetic circuit. Although this approach is straight forward and easy to follow, let us take a minute to apply basic laws of magnetically linear systems that we learned in early physics courses to justify (1.3-3) and (1.3-4). Ampere's Law states that the line integral of the field intensity (field strength), $H$, about a closed path is equal to the net current enclosed within the closed path of integration. Now, for a two dimensional magnetically linear system

$$B = \mu H \tag{1.3-6}$$

where $B$ is the flux density, $\mu$ is the permeability, and $H$ is the field strength. If we assume the flux, $\Phi$, is uniform over the cross-sectional surface area of the magnetic path then

$$B = \frac{\Phi}{A} \tag{1.3-7}$$

where $A$ is the cross-sectional area. Thus, $H$ may be expressed

$$H = \frac{\Phi}{A\mu} \tag{1.3-8}$$

Now, if $H$ is integrated over the closed path that encloses the total turns of the coil and assuming $H$ is the same over this path, then the magnetomotive force, mmf, or $Ni$ is

$$Ni = \int_0^{\ell} \frac{\Phi}{A\mu} d\xi \tag{1.3-9}$$

and

$$Ni = \frac{\Phi}{A\mu}\ell \tag{1.3-10}$$

Thus $\Phi$ may be expressed

$$\begin{aligned} \Phi &= \frac{Ni}{\frac{\ell}{A\mu_r\mu_0}} \\ &= \frac{Ni}{\mathcal{R}} \end{aligned} \tag{1.3-11}$$

where we have substituted $\mu_r\mu_0$ for $\mu$. We now understand the difficulty in determining the reluctance to the leakage flux, however, the reluctance of the iron and air gap can be calculated quite accurately for a magnetically linear system.

We now see that (1.3-3) is obtained by applying Ampere's Law about the leakage flux, $\Phi_\ell$, path and (1.3-4) is obtained by applying Ampere's Law about the magnetizing flux, $\Phi_m$, path. In regard to (1.3-4), since for a magnetically linear system the reluctance of an air gap is much larger than the reluctance of the iron, the mmf drop across the iron is generally neglected and (1.3-4) would become

$$\Phi_m = \frac{Nl}{\mathcal{R}_g} \tag{1.3-12}$$

where the total mmf is dropped across the air gap which is commonly referred to as the air-gap mmf.

Let us modify Fig. 1.3-1 by cutting a second identical air gap as shown in Fig. 1.3-3. We will assume that the positions of the iron members are fixed. If now we neglect the reluctance of the iron, then one-half of the mmf is dropped across each air gap that is, each air-gap mmf is $\frac{Ni}{2}$.

The total flux linking all the turns of the coil may be expressed from (1.3-2) as

$$\begin{aligned} \lambda &= N\Phi \\ &= N(\Phi_l + \Phi_m) \end{aligned} \tag{1.3-13}$$

Substituting (1.3-3) and (1.3-4) into (1.3-13) yields the flux linkage as

$$\lambda = \left(\frac{N^2}{\mathcal{R}_l} + \frac{N^2}{\mathcal{R}_i + \mathcal{R}_g}\right) i \tag{1.3-14}$$

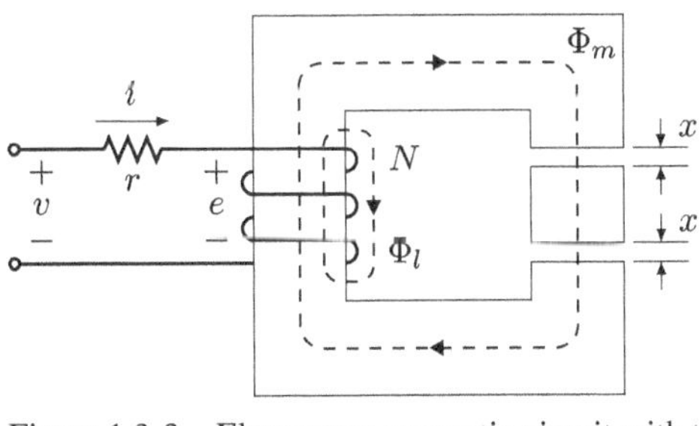


Figure 1.3-3 Elementary magnetic circuit with two air gaps.

It is now time to get back to the electric circuit. Faraday's Law tells us that the voltage induced in the coil, $e$ in Fig. 1.3-1, may be expressed

$$e = \frac{d\lambda}{dt} \tag{1.3-15}$$

Note that the coil resistance has been lumped outside the coil in the circuit shown in Fig. 1.3-1. If we now substitute (1.3-14) into (1.3-15) and then substitute the result into (1.3-1) for $e$ we obtain

$$v = ri + \frac{d}{dt}\left[\left(\frac{N^2}{\mathcal{R}_l} + \frac{N^2}{\mathcal{R}_i + 2\mathcal{R}_g}\right) i\right] \tag{1.3-16}$$

where $\mathcal{R}_g$ is the reluctance of one of the two identical air gaps. Since we are dealing exclusively with a magnetically linear system, the reluctances are assumed to be constant, therefore (1.3-16) becomes

$$v = ri + L\frac{di}{dt} \tag{1.3-17}$$

where $L$ is the self-inductance which is

$$\begin{aligned} L &= \frac{N^2}{\mathcal{R}_l} + \frac{N^2}{\mathcal{R}_i + 2\mathcal{R}_g} \\ &= L_l + L_m \end{aligned} \tag{1.3-18}$$

where $L_l$ is the leakage inductance and $L_m$ is the magnetizing inductance. In particular,

$$L_l = \frac{N^2}{\mathcal{R}_l} \tag{1.3-19}$$

$$L_m = \frac{N^2}{\mathcal{R}_i + 2\mathcal{R}_g} \tag{1.3-20}$$

Since inductance (unit of H) is turns (no unit) squared divided by reluctance, the unit of reluctance is $\text{H}^{-1}$. Moreover, the use of an inductance in the analysis signifies that a magnetically linear system is being assumed.

Let us rewrite (1.3-14) as

$$\lambda = Li \tag{1.3-21}$$

Here we see a linear relationship between flux linkages and current. The $\lambda$ versus $i$ relationship shown in Fig. 1.3-4 is often referred to as the magnetization curve.

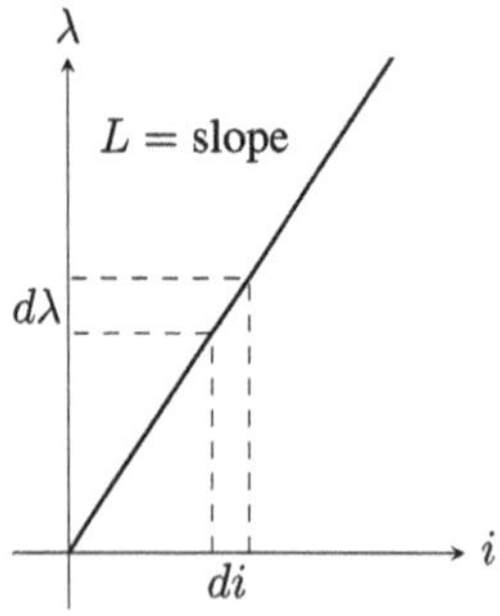


Figure 1.3-4 $\lambda i$ characteristic of a magnetically linear system.

## Field Energy and Coenergy

Since power is the rate of energy transfer, the energy entering the magnetic field from the electrical system (Fig. 1.3-1) may be expressed

$$W_e = \int ei\,dt \tag{1.3-22}$$

As previously mentioned, the resistance of the coil is lumped external to the coil; therefore, $W_e$ is the energy transferred only to the field of the magnetic circuit from the electrical system. Thus, the energy stored in the field which is referred to as the field energy is

$$\begin{aligned} W_f &= W_e \\ &= \int i\frac{d\lambda}{dt}dt = \int i\,d\lambda \\ &= \frac{1}{2}Li^2 \end{aligned} \tag{1.3-23}$$

Therefore, the energy stored in the field ($W_f$) for a given $\lambda$ or $i$ is the area to the left of the magnetization curve shown in Fig. 1.3-4. Also, since, for the magnetic system shown in Fig. 1.3-1, $L_\ell \ll L_m$ and since $\mathcal{R}_i \ll \mathcal{R}_g$, most of the field energy is stored in the air gap.

The area to the right of the magnetization curve is referred to as the coenergy; it is expressed as

$$W_c = \int \lambda\,di \tag{1.3-24}$$

It has no physical meaning; however, it is often used as a convenience in expressing the force and torque, since it is easier to work with $\lambda$ in terms of $i$ rather than $i$ in terms of $\lambda$. Also, we see from Fig. 1.3-4, that

$$\lambda i = W_f + W_c \tag{1.3-25}$$

Clearly, $W_f = W_c$ for a magnetically linear system.

We will not take saturation into account in our analysis in this text. To do so would greatly complicate our work compared to treating all electromechanical devices as magnetically linear and little is sacrificed in portraying their salient operating characteristics. Nevertheless, magnetically nonlinear systems should be mentioned in passing. In this regard, Fig. 1.3-5 is a plot of the flux linkage $\lambda$ versus the current $i$ for a magnetically nonlinear system. The actual characteristic is dependent on the type of ferromagnetic material. The knee of the curve occurs due to the saturation of the ferromagnetic member. We will assume that the $\lambda$ versus $i$ characteristic is a single valued function in that there is only one value of $\lambda$ for a given value of $i$, whereupon hysteresis is neglected. When a ferromagnetic material is subjected to a magnetic field its magnetic characteristics tend to aid in producing flux. This continues at a near linear rate until this rate begins to decrease, causing the knee of the magnetization curve. As the strength of the applied field continues to increase

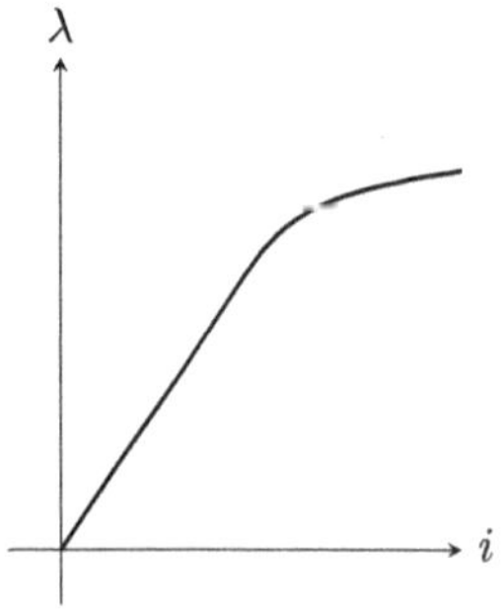


Figure 1.3-5 $\lambda$ versus $i$ for a magnetically nonlinear magnetic system.

(by increasing $i$), the rate will continue to decrease and ultimately the magnetic characteristic of the ferromagnetic material approaches that of air. It is clear that after the knee of the magnetization curve, the field energy $W_f$ is no longer equal to the coenergy $W_c$.

**Example 1B. Inductance calculations of a magnetic circuit.** The parameters of the magnetic circuit shown in Fig. 1.3-1 are: $r = 1\,\Omega$, $N = 100$ turns, $v = 10$ V, $\ell_i = 40$ cm, $x = 3$ mm, $A_i = A_g = 40\,\text{cm}^2$, $\mu_0 = 4\pi \times 10^{-7}$ H/m, $\mu_{ri} = 1000$. Determine (a) the steady-state total flux $\Phi$, assume $\mathcal{R}_l = 6.77 \times 10^7 \text{H}^{-1}$; (b) the leakage inductance $L_l$, the magnetizing inductance $L_m$, and the self-inductance $L$; (c) express the energy from the source to the system following a step applied voltage.

(a) It would be helpful to refer to Fig. 1.3-2. Therein,

$$\mathcal{R}_i = \frac{\ell_i}{\mu_{ri}\mu_0 A_i} = \frac{40 \times 10^{-2}}{(1000)(4\pi \times 10^{-7})(4 \times 10^{-4})} = 7.95 \times 10^5\ \text{H}^{-1} \quad \text{(1B-1)}$$

$$\mathcal{R}_g = \frac{x}{\mu_{rg}\mu_0 A_g} = \frac{3 \times 10^{-3}}{(1)(4\pi \times 10^{-7})(4 \times 10^{-4})} = 5.97 \times 10^6\ \text{H}^{-1} \quad \text{(1B-2)}$$

$$\Phi_m = \frac{Ni}{\mathcal{R}_i + \mathcal{R}_g} = \frac{100\left(\frac{10}{1}\right)}{7.95 \times 10^5 + 5.97 \times 10^6} = 1.48 \times 10^{-4}\ \text{Wb} \quad \text{(1B-3)}$$

$$\Phi_l = \frac{Ni}{\mathcal{R}_l} = \frac{100\left(\frac{10}{1}\right)}{6.77 \times 10^7} = 1.48 \times 10^{-5}\ \text{Wb} \quad \text{(1B-4)}$$

$$\Phi = \Phi_l + \Phi_m = 1.48 \times 10^{-5} + 1.48 \times 10^{-4} = 1.63 \times 10^{-4}\ \text{Wb} \quad \text{(1B-5)}$$

(b)

$$L_l = \frac{N^2}{\mathcal{R}_l} = \frac{(100)^2}{6.77 \times 10^7} = 0.148\,\text{mH} \tag{1B-6}$$

$$L_m = \frac{N^2}{\mathcal{R}_m} = \frac{(100)^2}{(7.96 \times 10^5) + (5.97 \times 10^5)} = 1.48\,\text{mH} \tag{1B-7}$$

$$L = L_l + L_m = (0.148 \times 10^{-3}) + (1.48 \times 10^{-3}) = 1.63\,\text{mH} \tag{1B-8}$$

(c) The expression for the total energy from the source, $W_E$, is

$$W_E = \int vidt \tag{1B-9}$$

From (1.3-1), which can now be written

$$v = ri + L\frac{di}{dt} \tag{1B-10}$$

Substituting (1B-10) into (1B-9) yields

$$\begin{aligned} W_E &= \int ri^2 dt + \int L\frac{di}{dt} idt \\ &= r\int i^2 dt + \int L\,idi \\ &= r\int i^2 dt + \frac{1}{2}Li^2 \\ &= r\int i^2 dt + \frac{1}{2}(L_\ell + L_m)i^2 \end{aligned} \tag{1B-11}$$

The first term on the right side of (1B-11) is the energy dissipated in the resistor while the second term is the energy stored in the inductance (magnetic field).

Since the expression for the current after the step voltage is applied is

$$i = \frac{v}{r}(1 - e^{-t/\tau}) \tag{1B-12}$$

where $\tau$ is the time constant $L/r$, the energy supplied by the source may be expressed

$$\begin{aligned} W_E &= r\int \left[\frac{v}{r}(1 - e^{-t/\tau})\right]^2 dt + \frac{1}{2}Li^2 \\ &= r\left(\frac{v}{r}\right)^2 \int (1 - 2e^{-t/\tau} + e^{-2t/\tau})dt + \frac{1}{2}Li^2 \\ &= \frac{v^2}{r}\left[t + 2\tau e^{-t/\tau} - \frac{\tau}{2}e^{-2t/\tau}\right] \\ &\quad + \frac{1}{2}L\left(\frac{v}{r}\right)^2 (1 - e^{-t/\tau})^2 \end{aligned} \tag{1B-13}$$

For $t \gg \tau$,

$$W_E \cong \frac{v^2}{r}t + \frac{1}{2}L\left(\frac{v}{r}\right)^2 \tag{1B-14}$$

It is interesting to note that the energy stored in the inductance converges to a constant; however, the resistor continues to dissipate energy at a rate of $v^2/r$.

---

**SP1.3-1**. Determine two changes in the magnetic circuit shown in Example 1B that would result in interchanging the position of the poles. [Reverse the direction of positive current or reverse the sense of the winding.]

**SP1.3-2.** Express (a) $W_f$ and (b) $W_c$ if $L = 1$ H and the current is $I = I_p \cos \omega_e t$. (a) $\left[\frac{1}{2}I_p^2\left(\frac{1}{2}+\frac{1}{2}\cos 2\omega_e t\right)\right]$, (b) [same].

**SP1.3-3.** Draw the $\lambda$ versus $i$ plot for Example 1B. Indicate the change in the slope as the length of the air gap increases. [slope decreases]

**SP1.3-4.** Consider Fig. 1.3-1. Assume the reluctance of the iron may be neglected. Express the energy stored in the air gap in terms of its air-gap mmf ($\text{mmf}_{ag}$) and its reluctance $\mathcal{R}_g$. $\left[\frac{(\text{mmf}_{ag})^2}{2\mathcal{R}_g}\right]$

**SP1.3-5.** Consider Fig. 1.3-3. Neglect the reluctance of the iron and express the energy stored in each air gap in terms of its air-gap mmf. Let $\text{mmf}_{ag}$ and $\mathcal{R}_g$ be for one air gap. $\left[\frac{2(\text{mmf}_{ag})^2}{\mathcal{R}_g}\right]$

## 1.4 STATIONARY COUPLED CIRCUITS

Faraday's Law tells us that a voltage is induced in an electric circuit due to a change of flux linkage. In the case of stationary circuits, such as the transformer, a change of flux linkage occurs due to time varying currents. Therefore, in an ac power system we can change the level of the voltage and current. That is, power can be generated at convenient voltage and current levels and transmitted at a high voltage with a low current to minimize transmission losses and then lower the voltage to a safe level at the point of usage.

Flux linkages of an electric circuit can also be changed, and thus voltages induced, due to relative motion between magnetic fields. This allows power to be generated at the source where mechanical motion is converted to electric power (generator action). At the load, electric power is often converted back to mechanical motion (motor action). The work of Faraday and later the inventions and development of the ac power system by Tesla and Westinghouse doomed Edison's dc power system in the very early twentieth century. We will start our analytical journey into the world of Faraday and Tesla with the transformer.

### Magnetically Linear Transformer

Two coupled circuits are shown in Fig. 1.4-1. Actually, all we have done is "closed up" the air gap of Fig. 1.3-1 and added a second coil. About all we have to concern ourselves with is how to handle the coupling between coils since in the previous

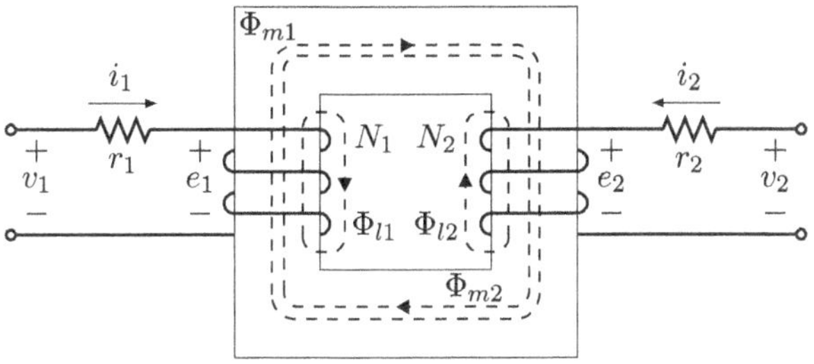


Figure 1.4-1 Magnetically coupled circuits.

section we became familiar with leakage flux and magnetizing flux. The flux linkages are

$$\lambda_1 = N_1 \Phi_1 \tag{1.4-1}$$

$$\lambda_2 = N_2 \Phi_2 \tag{1.4-2}$$

where $\Phi_1$ and $\Phi_2$ are the total flux linking $N_1$ and $N_2$, respectively. From Fig. 1.4-1

$$\begin{aligned} \Phi_1 &= \Phi_{l1} + \Phi_{m1} + \Phi_{m2} \\ &= \frac{N_1 i_1}{\mathcal{R}_{l1}} + \frac{N_1 i_1}{\mathcal{R}_m} + \frac{N_2 i_2}{\mathcal{R}_m} \end{aligned} \tag{1.4-3}$$

$$\begin{aligned} \Phi_2 &= \Phi_{l2} + \Phi_{m2} + \Phi_{m1} \\ &= \frac{N_2 i_2}{\mathcal{R}_{l2}} + \frac{N_2 i_2}{\mathcal{R}_m} + \frac{N_1 i_1}{\mathcal{R}_m} \end{aligned} \tag{1.4-4}$$

Mutual coupling occurs since $\Phi_{m2}$, which is the flux created by current flowing in coil $N_2$, links coil $N_1$. Likewise, $\Phi_{m1}$, which is the flux created by $i_1$ current flowing in coil $N_1$, links coil $N_2$.

Substituting (1.4-3) into (1.4-1) and (1.4-4) into (1.4-2) yields

$$\lambda_1 = \frac{N_1^2}{\mathcal{R}_{l1}} i_1 + \frac{N_1^2}{\mathcal{R}_m} i_1 + \frac{N_1 N_2}{\mathcal{R}_m} i_2 \tag{1.4-5}$$

$$\lambda_2 = \frac{N_2^2}{\mathcal{R}_{l2}} i_2 + \frac{N_2^2}{\mathcal{R}_m} i_2 + \frac{N_2 N_1}{\mathcal{R}_m} i_1 \tag{1.4-6}$$

The coefficients of the first two terms of (1.4-5) and (1.4-6), respectively, are the self-inductances $L_{11}$ and $L_{22}$, that is

$$L_{11} = \frac{N_1^2}{\mathcal{R}_{l1}} + \frac{N_1^2}{\mathcal{R}_m} = L_{l1} + L_{m1} \tag{1.4-7}$$

$$L_{22} = \frac{N_2^2}{\mathcal{R}_{l2}} + \frac{N_2^2}{\mathcal{R}_m} = L_{l2} + L_{m2} \tag{1.4-8}$$

The coefficient of the last term of (1.4-5) and (1.4-6) is the mutual inductance

$$L_{12} = L_{21} = \frac{N_1 N_2}{\mathcal{R}_m} \tag{1.4-9}$$

We can now write $\lambda_1$ and $\lambda_2$ as

$$\lambda_1 = L_{11} i_1 + L_{12} i_2 \tag{1.4-10}$$

$$\lambda_2 = L_{22} i_2 + L_{21} i_1 \tag{1.4-11}$$

It is important to note that $L_{m1}$, $L_{m2}$, and $L_{12}$ have a common term $\mathcal{R}_m$ which is the reluctance of the mutual path around the ferromagnetic core. This fact will allow us to derive a very useful equivalent circuit; however, before getting started on that, let us go back to (1.4-5) and (1.4-6). The mutual inductance can be positive or negative in these equations depending upon the relative directions of $\Phi_{m1}$ and $\Phi_{m2}$. If the assumed positive directions of the currents or the sense of the winding of the coil causes $\Phi_{m1}$ and $\Phi_{m2}$ to aid each other, $L_{12}$ ($L_{21}$) would be positive as in the case here. If they oppose $L_{12}$ ($L_{21}$)would be negative.

Now, $\lambda_1$ may be written

$$\begin{aligned} \lambda_1 &= L_{l1} i_1 + L_{m1} i_1 + L_{12} i_2 \\ &= L_{l1} i_1 + L_{m1} \left( i_1 + \frac{N_2}{N_1} i_2 \right) \end{aligned} \tag{1.4-12}$$

Here we multiplied $L_{12}$, given by (1.4-9), by $\frac{N_1}{N_2}$ whereupon

$$\begin{aligned} \frac{N_1}{N_2} L_{12} &= \left( \frac{N_1}{N_2} \right) \frac{N_1 N_2}{\mathcal{R}_m} \\ &= L_{m1} \end{aligned} \tag{1.4-13}$$

thus,

$$L_{12} = \frac{N_2}{N_1} L_{m1} \tag{1.4-14}$$

From (1.4-12) we see there is a common inductance $L_{m1}$ that carries the currents $i_1$ and $\frac{N_2}{N_1} i_2$. We have referred $\frac{N_2}{N_1} i_2$ to the winding with $N_1$ turns; if we would have expressed $\lambda_2$ rather than $\lambda_1$ we could have referred $i_1$ to the winding with $N_2$ turns by $\frac{N_1}{N_2} i_1$. Let us stay with $\frac{N_2}{N_1} i_2$, that is generally written, for compactness, as

$$i'_2 = \frac{N_2}{N_1} i_2 \tag{1.4-15}$$

or

$$N_1 i'_2 = N_2 i_2 \tag{1.4-16}$$

From (1.4-16), $i'_2$ flowing in $N_1$ produces the same mmf as $i_2$ flowing in $N_2$.

To maintain the same power

$$v'_2 i'_2 = v_2 i_2 \tag{1.4-17}$$

Thus

$$v_2' = \frac{N_1}{N_2} v_2 \tag{1.4-18}$$

Since $\lambda_2$ has units of volt · sec, the same turns ratio used for voltages can be applied to it; thus,

$$\lambda_2' = \frac{N_1}{N_2} \lambda_2 \tag{1.4-19}$$

There is considerable algebraic manipulation to get to the equivalent circuit. We will only set forth the steps. First, let us write the voltage equations in terms of 1- and 2-variables.

$$v_1 = r_1 i_1 + \frac{d\lambda_1}{dt} \tag{1.4-20}$$

$$v_2 = r_2 i_2 + \frac{d\lambda_2}{dt} \tag{1.4-21}$$

where $r_1$ and $r_2$ are the resistances of winding 1 and winding 2, respectively. Since we are referring the 2-variables to the 1-winding, (1.4-20) remains unchanged in form; however, (1.4-21) becomes

$$v_2' = r_2' i_2' + \frac{d\lambda_2'}{dt} \tag{1.4-22}$$

The flux linkage equations become

$$\lambda_1 = L_{l1} i_1 + L_{m1}(i_1 + i_2') \tag{1.4-23}$$

$$\lambda_2' = L_{l2}' i_2' + L_{m1}(i_1 + i_2') \tag{1.4-24}$$

where

$$L_{l2}' = \left(\frac{N_1}{N_2}\right)^2 L_{l2} \tag{1.4-25}$$

$$r_2' = \left(\frac{N_1}{N_2}\right)^2 r_2 \tag{1.4-26}$$

The equivalent circuit is shown in Fig. 1.4-2. This circuit is valid for all modes of operation, transient and steady state; however, it is valid only for a magnetically linear system.

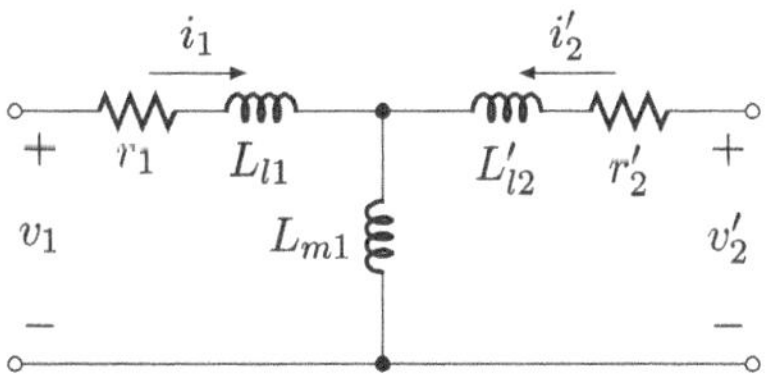


Figure 1.4-2 Transformer equivalent $T$ circuit with winding 1 selected as reference winding.

## Field Energy

Before leaving our work with transformers, let us derive an expression for the energy stored in the field $W_f$. In the case of a single coil the field energy was $\frac{1}{2}Li^2$. Here we have two coils with mutual coupling between them. In this case, the energy from the electrical system is

$$W_e = \int e_1 i_1 dt + e_2 i_2 dt \tag{1.4-27}$$

Since $r_1$ and $r_2$ are external to the respective coils and using unreferred variables

$$e_1 = \frac{d\lambda_1}{dt} \tag{1.4-28}$$

$$e_2 = \frac{d\lambda_2}{dt} \tag{1.4-29}$$

Substituting (1.4-28) and (1.4-29) into (1.4-27) yields

$$W_e = \int i_1 d\lambda_1 + i_2 d\lambda_2 \tag{1.4-30}$$

From (1.4-10) and (1.4-11)

$$d\lambda_1 = L_{11} di_1 + L_{12} di_2 \tag{1.4-31}$$

$$d\lambda_2 = L_{22} di_2 + L_{12} di_1 \tag{1.4-32}$$

Substituting (1.4-31) and (1.4-32) into (1.4-30) and since $W_f = W_e$ we have

$$W_f = \int [i_1(L_{11} di_1 + L_{12} di_2) + i_2(L_{22} di_2 + L_{12} di_1)] \tag{1.4-33}$$

For a magnetically linear system these integrals can be calculated in two steps; first, let the current in winding 1 be the variable of integration and vary it from zero to $i_1$ while holding $i_2$ and $di_2$ at zero. Next, hold the current in winding 1 at $i_1 (di_1 = 0)$ and let the current in winding 2 be the variable of integration varying from zero to $i_2$. Performing the first step we have

$$W_{f(1)} = \int_0^{i_1} L_{11} \xi d\xi = \frac{1}{2} L_{11} i_1^2 \tag{1.4-34}$$

For the second step

$$\begin{aligned} W_{f(2)} &= \int_0^{i_2} L_{12} i_1 d\xi + L_{22} \xi d\xi \\ &= L_{12} i_1 i_2 + \frac{1}{2} L_{22} i_2^2 \end{aligned} \tag{1.4-35}$$

Thus

$$W_f = \frac{1}{2} L_{11} i_1^2 + L_{12} i_1 i_2 + \frac{1}{2} L_{22} i_2^2 \tag{1.4-36}$$

The energy stored in the coupling field is $W_f$ given by (1.4-36) less the energy stored in the leakage inductances. In a problem at the end of the chapter you are asked to express (1.4-36) in terms of $i_1$ and $i'_2$ and to identify the energy stored in the coupling field.

The pattern is clear, the field energy may be expressed for multiple electrical inputs (coils) as $\frac{1}{2}$ the self-inductance of each coil times the square of the current flowing in the coil and the mutual inductance between each set of coils times the product of the current flowing in the mutually coupled coils. For three coupled coils

$$W_f = \frac{1}{2}L_{11}i_1^2 + \frac{1}{2}L_{22}i_2^2 + \frac{1}{2}L_{33}i_3^2 + L_{12}i_1i_2 + L_{13}i_1i_3 + L_{23}i_2i_3 \tag{1.4-37}$$

Now, for multiple electrical inputs (windings), (1.3-25) becomes

$$\sum_{j=1}^{J} \lambda_j i_j = W_f + W_c \tag{1.4-38}$$

where $j$ is the winding. Thus,

$$W_c = \sum_{j=1}^{J} \lambda_j i_j - W_f \tag{1.4-39}$$

In SP1.4-3, you are asked to prove that $W_c$ is $W_f$ using (1.4-37) and (1.4-39).

**Example 1C. Parameters of the transformer equivalent circuit.** It is instructive to illustrate the method of deriving an equivalent $T$ circuit from open- and short-circuit measurements of the transformer. When winding 2 of the two-winding transformer shown in Fig. 1.4-2 is open-circuited and a voltage of 110 V (rms) at 60 Hz is applied to winding 1, the average power supplied to winding 1 is 6.66 W. The measured current in winding 1 is 1.05 A (rms). Next, with winding 2 short-circuited, the current flowing in winding 1 is 2 A when the applied voltage is 30 V at 60 Hz. The average input power is 44 W. If we assume $L_{l1} = L'_{l2}$, an approximate equivalent $T$ circuit can be determined from these measurements with winding 1 selected as the reference winding.

The average power supplied to winding 1 may be expressed from (1.2-31) as

$$P_1 = |\tilde{V}_1|\,|\tilde{I}_1| \cos \varphi_{pf} \tag{1C-1}$$

where

$$\varphi_{pf} = \theta_{ev}(0) - \theta_{ei}(0) \tag{1C-2}$$

Here, $\tilde{V}_1$ and $\tilde{I}_1$ are phasors with the positive direction of $\tilde{I}_1$ taken in the direction of voltage drop, and $\theta_{ev}(0)$ and $\theta_{ei}(0)$ are the phase angles of $\tilde{V}_1$ and $\tilde{I}_1$, respectively. Solving for $\varphi_{pf}$ during the open-circuit test, we have

$$\varphi_{pf} = \cos^{-1}\frac{P_1}{|\tilde{V}_1|\,|\tilde{I}_1|} = \cos^{-1}\frac{6.66}{(110)(1.05)} = 86.7^\circ \tag{1C-3}$$

Although $\varphi_{pf} = -86.7°$ is also a legitimate solution of (1C-3), the positive value is taken since $\tilde{V}_1$ leads $\tilde{I}_1$ in an inductive circuit. With winding 2 open-circuited, the input impedance of winding 1 is

$$Z = \frac{\tilde{V}_1}{\tilde{I}_1} = r_1 + j(X_{l1} + X_{m1}) \tag{1C-4}$$

With $\tilde{V}_1$ as the reference phasor, $\tilde{V}_1 = 110\underline{/0°}$, $\tilde{I}_1 = 1.05\underline{/-86.7°}$. Thus,

$$r_1 + j(X_{l1} + X_{m1}) = \frac{110\underline{/0°}}{1.05\underline{/-86.7°}} = 6 + j104.6\,\Omega \tag{1C-5}$$

From (1C-5), $r_1 = 6\,\Omega$. We also see from (1C-5) that $X_{l1} + X_{m1} = 104.6\,\Omega$.

For the short-circuit test, we will assume that $\tilde{I}_1 = -\tilde{I}'_2$ since transformers are designed so that at rated frequency $X_{m1} \gg |r'_2 + jX'_{l2}|$. Hence, using (1C-1) again,

$$\varphi_{pf} = \cos^{-1}\frac{44}{(30)(2)} = 42.8° \tag{1C-6}$$

In this case, the input impedance is $Z = (r_1 + r'_2) + j(X_{l1} + X'_{l2})$. This may be determined as

$$Z = \frac{30\underline{/0°}}{2\underline{/-42.8°}} = 11 + j10.2\,\Omega \tag{1C-7}$$

Hence, $r'_2 = 11 - r_1 = 5\,\Omega$ and, since it is assumed that $X_{l1} = X'_{l2}$, both are $10.2/2 = 5.1\,\Omega$. Therefore, $X_{m1} = 104.6 - 5.1 = 99.5\,\Omega$. In summary, $r_1 = 6\,\Omega$, $L_{l1} = 13.5$ mH, $L_{m1} = 263.9$ mH, $r'_2 = 5\,\Omega$, $L'_{l2} = 13.5$ mH. It is left to the reader to verify the conversion from $X$'s to $L$'s.

---

**SP1.4-1.** Consider the transformer and parameters calculated in Example 1C. Winding 2 is short-circuited and 12 V (dc) is applied to winding 1. Calculate the steady-state values of $i_1$ and $i_2$. Repeat with winding 2 open-circuited. [$I_1 = 2$ A and $I_2 = 0$ in both cases]

**SP1.4-2.** Calculate the no-load current (winding 2 open-circuited) for the transformer given in Example 1C if $V_1 = \sqrt{2}10\cos 100t$. [$\tilde{I}_1 = 0.35\underline{/-77.8°}$ A]

**SP1.4-3.** Determine $W_c$ from (1.4-37) and (1.4-39). Show that $W_c = W_f$.

**SP1.4-4.** Use the equivalent $T$ circuit given in Fig. 1.4-2 to identify the energy stored in the coupling field. [$\frac{1}{2}L_{m1}(i_1^2 + i'^2_2) + L_{m1}i_1i'_2$]

## 1.5 COUPLED CIRCUITS IN RELATIVE MOTION

In the previous section, we considered the "transformer" coupling that exists between two stationary windings. An induced voltage due to a time varying current (transformer action) is only "half" of Faraday's Law. In the case of electric machines, there can be windings on the stationary member (stator) and windings on the rotating member (rotor). Transformer coupling may exist between only stator windings or between only rotor windings. Transformer coupling may also exist between the stator and rotor windings depending upon their relative positions. In addition, there is a voltage induced in the circuits due to the rate of change of the relative position of the stator and rotor windings; commonly referred to as a speed voltage. In this section, we will take our first look at the magnetic coupling when windings are in relative motion and what Faraday's Law tells us about the voltages that can be induced.

The rotational device shown in Fig. 1.5-1 will be used to illustrate windings in relative motion. This device consists of two coils each containing several turns of a conductor. Winding 1 has $N_1$ turns and it is on the stationary member (stator); winding 2 has $N_2$ turns and it is on the rotating member (rotor). The ⊗ indicates that the assumed direction of positive current flow in the conductors is into the paper, where ⊙ indicates positive current flow in the conductors is out of the paper. In a practical device, the turns of a winding are distributed over an arc (often 30° to 60° of the stator or rotor; however, in this introductory consideration, it is sufficient to assume, for now, that the turns are concentrated in one position, as shown in Fig. 1.5-1. Also, the length of the air gap between the stator and rotor is shown exaggerated relative to the actual inside diameter of the stator.

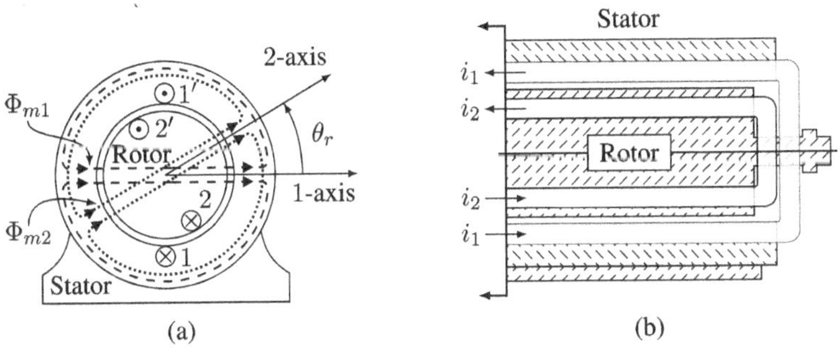


Figure 1.5-1 Elementary rotational electromechanical device: (a) end view; (b) cross-sectional view.

The voltage equations may be written as

$$v_1 = r_1 i_1 + \frac{d\lambda_1}{dt} \tag{1.5-1}$$

$$v_2 = r_2 i_2 + \frac{d\lambda_2}{dt} \tag{1.5-2}$$

where $r_1$ and $r_2$ are the resistances of windings 1 and 2, respectively. The magnetic system is assumed to be linear; therefore, the flux linkages may be expressed as

$$\lambda_1 = L_{11} i_1 + L_{12} i_2 \tag{1.5-3}$$

$$\lambda_2 = L_{21} i_1 + L_{22} i_2 \tag{1.5-4}$$

Since the air gap is uniform, the self-inductances $L_{11}$ and $L_{22}$ are constants and may be expressed in terms of leakage and magnetizing inductances as

$$\begin{aligned} L_{11} &= L_{l1} + L_{m1} \\ &= \frac{N_1^2}{\mathcal{R}_{l1}} + \frac{N_1^2}{\mathcal{R}_m} \end{aligned} \tag{1.5-5}$$

$$\begin{aligned} L_{22} &= L_{l2} + L_{m2} \\ &= \frac{N_2^2}{\mathcal{R}_{l2}} + \frac{N_2^2}{\mathcal{R}_m} \end{aligned} \tag{1.5-6}$$

where $\mathcal{R}_m$ is the reluctance of the complete magnetic path of $\Phi_{m1}$ and $\Phi_{m2}$, which is through the rotor and stator iron and twice across the air gap ($\mathcal{R}_i + 2\mathcal{R}_g$). Clearly, $\mathcal{R}_m$ is the same for the magnetic system established by either winding 1 or 2.

Take a moment to note the designation of axis 1 and axis 2 in Fig. 1.5-1. These axes denote the positive direction of the respective magnetic systems with the assumed positive direction of current flow in the windings (right-hand rule). Now let us consider $L_{12}$. (Is it clear that $L_{12} = L_{21}$?) When $\theta_r$ is zero, the coupling between windings 1 and 2 is maximum. In particular, with $\theta_r = 0$ the magnetic system of winding 1 aids that of winding 2 with positive currents assumed. Hence, the mutual inductance is positive and

$$L_{12}(0) = \frac{N_1 N_2}{\mathcal{R}_m} \tag{1.5-7}$$

When $\theta_r = \frac{1}{2}\pi$, the windings are orthogonal. The mutual coupling is zero. Hence,

$$L_{12}\left(\frac{1}{2}\pi\right) = 0 \tag{1.5-8}$$

When $\theta_r = \pi$, $L_{12}(\pi)$ is the negative of (1.5-7) and at $\theta_r = \frac{3}{2}\pi$, $L_{12}$ is zero; the same as (1.5-8). Let us make it as simple as possible by assuming that as a first approximation the mutual inductance may be written

$$L_{12}(\theta_r) = L_{sr}\cos\theta_r \tag{1.5-9}$$

where $L_{sr}$ is the amplitude of the sinusoidal mutual inductance between the stator and rotor windings as given by (1.5-7).

The flux linkages are given by (1.5-3) and (1.5-4), let us rewrite these equations indicating the functionality

$$\lambda_1(i_1, i_2, \theta_r) = L_{11}i_1 + L_{12}(\theta_r)i_2 \tag{1.5-10}$$

$$\lambda_2(i_1, i_2, \theta_r) = L_{12}(\theta_r)i_1 + L_{22}i_2 \tag{1.5-11}$$

The total derivative of $\lambda_1$ becomes

$$\begin{aligned} \frac{d\lambda_1}{dt} &= \frac{\partial(L_{11}i_1)}{\partial i_1}\frac{di_1}{dt} + \frac{\partial(L_{11}i_1)}{\partial i_2}\frac{di_2}{dt} + \frac{\partial(L_{11}i_1)}{\partial\theta_r}\frac{d\theta_r}{dt} \\ &+ \frac{\partial(L_{12}i_2)}{\partial i_1}\frac{di_1}{dt} + \frac{\partial(L_{12}i_2)}{\partial i_2}\frac{di_2}{dt} + \frac{\partial(L_{12}i_2)}{\partial\theta_r}\frac{d\theta_r}{dt} \end{aligned} \tag{1.5-12}$$

Only the first, fifth, and sixth terms on the right-hand side of (1.5-12) are nonzero, thus $v_1$ can be written

$$v_1 = r_1 i_1 + L_{11}\frac{di_1}{dt} + L_{sr}\cos\theta_r\frac{di_2}{dt} - i_2 L_{sr}\sin\theta_r\frac{d\theta_r}{dt} \tag{1.5-13}$$

Following the same procedure, we can express $v_2$ as

$$v_2 = r_2 i_2 + L_{22}\frac{di_2}{dt} + L_{sr}\cos\theta_r\frac{di_1}{dt} - i_1 L_{sr}\sin\theta_r\frac{d\theta_r}{dt} \tag{1.5-14}$$

Faraday's Law has given us the last three terms of (1.5-13) and (1.5-14). The first of these terms is due to the change of current flowing in the winding; this term would exist independent of the other winding. The second term is due to the "transformer" coupling from one winding to the other. The degree and sense of the coupling would depend upon the relative position or displacement between the windings. The last of these three terms is the voltage induced due to the rate of change of the displacement between windings; commonly referred to as the "speed voltage."

## Field Energy

Regardless, if the electromechanical device is magnetically linear or nonlinear, the $\lambda i$ relationship is a single-valued function with the assumptions we have made. Thus, the energy stored in the field is single-valued state function, meaning that for a given $i_1$, $i_2$, and $\theta_r$ (or $x$ in case of translational motion) there is only one value of $W_f$. Therefore, even though (1.4-33) was derived for magnetically linear, stationary coupled circuits it is also valid for magnetically linear coupled circuits in relative motion. Thus, from (1.4-36)

$$W_f = \frac{1}{2}L_{11}i_1^2 + i_1 i_2 L_{sr}\cos\theta_r + \frac{1}{2}L_{22}i_2^2 \tag{1.5-15}$$

The assumption of a magnetically linear system may not be as limiting as it might first appear. In many cases, most of the mmf is dropped across the air gap of the electromechanical device and therefore most of the field energy $W_f$ is stored in the air gap and air is a magnetically linear, conservative medium.

**Example 1D. Inductances of windings in relative motion.** Consider the windings in relative motion shown in Fig. 1D-1. The outer stator housing has been omitted from that shown in Fig. 1.5-1. We will often omit this housing for convenience. Winding 3 has been added to the rotor. Windings 2 and 3 are identical and displaced by $\frac{\pi}{2}$ relative to each other. Express $L_{11}$, $L_{22}$, $L_{33}$, $L_{12}$, $L_{13}$, $L_{23}$, and $v_1$.

We have $L_{11}$ and $L_{22}$ from (1.5-5) and (1.5-6), respectively. It follows that

$$L_{33} = \frac{N_3^2}{\mathcal{R}_{l3}} + \frac{N_3^2}{\mathcal{R}_m} \tag{1D-1}$$

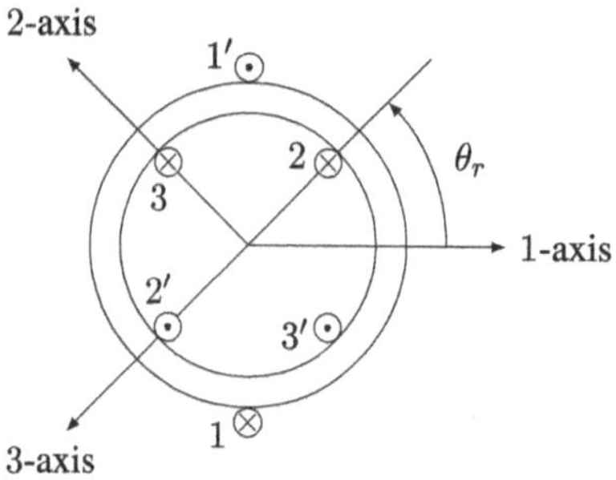


Figure 1D-1 Windings in relative motion (stator housing omitted).

Since windings 2 and 3 are identical $N_2 = N_3$ thus $L_{22} = L_{33}$. In the case of $L_{12}$ and $L_{13}$

$$L_{12} = -\frac{N_1 N_2}{\mathcal{R}_m} \sin\theta_r \tag{1D-2}$$

$$L_{13} = -\frac{N_1 N_3}{\mathcal{R}_m} \cos\theta_r \tag{1D-3}$$

Since windings 2 and 3 are orthogonal, $L_{23} = 0$. The voltage $v_1$ may be written by using (1.5-13) or (1.5-14) as a guide.

$$\begin{aligned} v_1 = r_1 i_1 + L_{11}\frac{di_1}{dt} - L_{sr}\sin\theta_r\frac{di_2}{dt} - L_{sr}\cos\theta_r\frac{di_3}{dt} \\ - i_2 L_{sr}\cos\theta_r\frac{d\theta_r}{dt} + i_3 L_{sr}\sin\theta_r\frac{d\theta_r}{dt} \end{aligned} \tag{1D-4}$$

where

$$L_{sr} = \frac{N_1 N_2}{\mathcal{R}_m} \tag{1D-5}$$

and

$$N_2 = N_3 \tag{1D-6}$$

---

**SP1.5-1.** Consider Fig. 1.5-1 $I_1 = 1\,\text{A}$, $L_{sr} = 0.1\,\text{H}$, $L_{11} = 10L_{sr}$, $\omega_r = 100\,\text{rad/s}$, $\theta_r(0) = 0$, and winding 2 is open circuited. Express $V_2$. [$V_2 = -10\sin 100t$]

**SP1.5-2.** Calculate the field energy for SP1.5-1. [$\frac{1}{2}$ J]

**SP1.5-3.** Repeat SP1.5-1 if $I_1 = 1\cos 100t$. [$V_2 = -10\sin 200t$]

**SP1.5-4.** Can we use phasors to solve steady-state versions of (1.5-13) or (1.5-14)? [Only if a single frequency is present.]

## 1.6 ELECTROMAGNETIC FORCE AND TORQUE

Our goal in this section is to derive an expression for the electromagnetic torque developed in rotational systems or force in translational systems. Although there are several approaches that could be used for this derivation, with the background that we have established, the "energy balance" approach is perhaps most convenient [2]. This method is quite direct and results in an easy to use expression for evaluating torque or force in magnetically linear electromagnetic systems.

Electromechanical devices consist of an electrical system, a coupling field, and a mechanical system. The electrical and the mechanical systems either supply or absorb energy by way of the coupling field. The coupling field can either be electromagnetic, which will be our focus, or electrostatic. The coupling field enables the electrical and mechanical systems to interact thereby providing a means of transferring energy between the two systems while at the same time providing a means of storing energy. A block diagram depicting this interaction and possible directions of energy flow is given in Fig. 1.6-1.

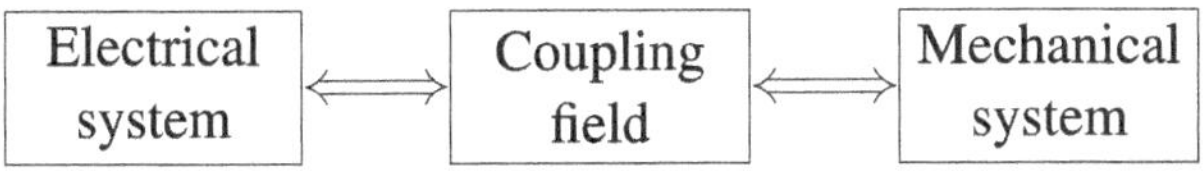


Figure 1.6-1 Block diagram of possible energy interchange in an elementary electromechanical system.

The energy balance is shown in Fig. 1.6-2. $W_E$ and $W_M$ are positive for energy being supplied to the electromechanical device from an external electrical system ($W_E$) and mechanical system ($W_M$). Likewise $W_e$ and $W_m$ are positive when supplying energy to the coupling field in the electrical form ($W_e$) and mechanical form ($W_m$).

In Fig. 1.6-2, the subscripts $E$ and $e$ pertain to the electrical system; $M$ and $m$ to the mechanical; and $f$ to the field. Also, $S$ and $L$ indicate energy stored and energy lost, respectively. The energy $W_{eL}$ is the energy lost due to $i^2r$; $W_{eS}$ is the energy stored in a field external to the coupling field. Some authors consider energy stored in the field of the leakage inductances as $W_{eS}$ while others will include the leakage flux in the coupling field. This is irrelevant since the energy stored in the field of the leakage inductance is generally not a function of the mechanical motion of the electromechanical device.

The energy $W_{mL}$ represents energy lost due to mechanical friction and windage losses. $W_{mS}$ is the energy stored as kinetic energy in the rotating mass (rotor) or as potential energy in a spring.

The energy $W_e$ is the energy coming into the coupling field from the electrical system. The energy $W_m$ is the energy coming into the coupling field from the

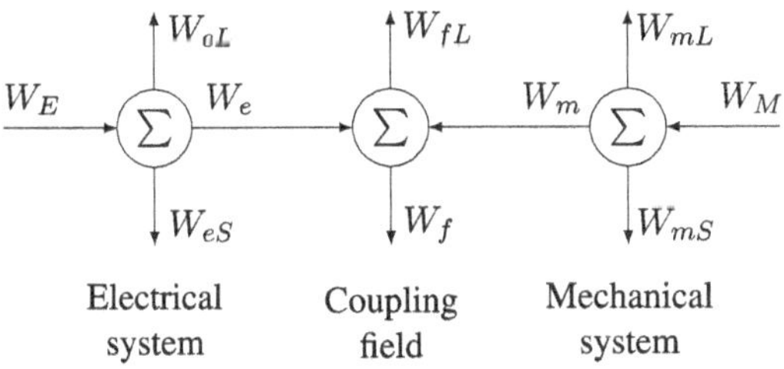


Figure 1.6-2 Energy balance.

mechanical system. The energy $W_{fL}$ is the energy lost due to hysteresis loss and circulating currents induced in the mechanical components of the coupling field. Electric machines are designed to minimize these losses making the $\lambda i$ relationship approach a single-valued function as shown in Fig. 1.3-3. We will neglect $W_{fL}$; whereupon, we can express the energy balance of the coupling field as

$$W_f = W_e + W_m \tag{1.6-1}$$

where $W_f$ is the energy stored in the coupling field [1].

Fig. 1.5-1a is shown again in Fig. 1.6-3 wherein the flux streamlines and the outside housing have been omitted from the cross-sectional view similarly in Fig. 1D-1. Also electromagnetic torque $T_e$, the load torque $T_L$, and the angular velocity of the rotor $\omega_r$ are indicated. Note that $T_e$ is positive in the direction of positive angular displacement of the rotor $\theta_r$ while the load torque $T_L$ is positive in the direction of negative $\theta_r$.

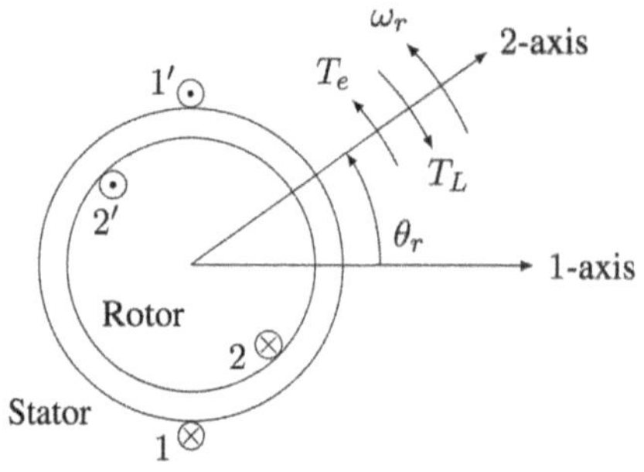


Figure 1.6-3 Elementary rotational electromechanical device.

The torque relationship between the electromechanical device and the mechanical system may be written

$$T_e = J\frac{d\omega_r}{dt} + B_m\omega_r + T_L \tag{1.6-2}$$

which is the rotational form of Newton's Second Law. Here, $J$ is the inertia of the rotor and any tightly mechanically coupled rotating mass, $B_m$ is the mechanical damping which is generally small and often neglected. Note that positive $T_e$ acts to increase $\omega_r$ whereas a positive load torque $T_L$ would act to retard $\omega_r$. Therefore, with $T_e$ positive for motor action, then positive $W_m$ entering the coupling field as shown in Fig. 1.6-2 and as expressed in (1.6-1), would be

$$W_m = -\int T_e d\theta_r \tag{1.6-3}$$

The energy entering the coupling field from the two electrical sources shown in Fig. 1.6-3 $W_e$ is

$$W_e = \int e_1 i_1 dt + e_2 i_2 dt \tag{1.6-4}$$

where

$$e_1 = \frac{d\lambda_1}{dt} \tag{1.6-5}$$

$$e_2 = \frac{d\lambda_2}{dt} \tag{1.6-6}$$

and $e_1$ and $e_2$ are $v_1 - r_1 i_1$ and $v_2 - r_2 i_2$, respectively. Substituting (1.6-5) and (1.6-6) into (1.6-4) yields

$$W_e = \int i_1 d\lambda_1 + i_2 d\lambda_2 \tag{1.6-7}$$

For $J$ electrical inputs (1.6-7) may be expressed

$$W_e = \int \sum_{j=1}^{J} i_j d\lambda_j \tag{1.6-8}$$

where $j$ is an index (1, 2, 3, … , $J$) not to be confused with the imaginary part of a complex number or with inertia or Joules. Thus, (1.6-1) may now be written

$$W_f = \int \sum_{j=1}^{J} i_j d\lambda_j - \int T_e d\theta_r \tag{1.6 9}$$

In differential form, which we will use extensively, (1.6-9) can be written

$$T_e d\theta_r = \sum_{j=1}^{J} i_j d\lambda_j - dW_f \tag{1.6-10}$$

Equation (1.3-25) may be written for multiple electrical inputs as

$$\sum_{j=1}^{J} \lambda_j i_j = W_c + W_f \tag{1.6-11}$$

where $W_c$ is the coenergy. If we take the derivative of (1.6 11) we have

$$\sum_{j=1}^{J} \lambda_j di_j + \sum_{j=1}^{J} i_j d\lambda_j = dW_c + dW_f \tag{1.6-12}$$

Solving (1.6-12) for $\sum_{j=1}^{J} i_j d\lambda_j$ and substituting the result into (1.6-10) yields

$$T_e d\theta_r = -\sum_{j=1}^{J} \lambda_j di_j + dW_c \tag{1.6-13}$$

Although (1.6-10) and (1.6-13) are valid for magnetically linear and nonlinear systems, we will consider only magnetically linear systems. Therefore, it is convenient to express the flux linkages in terms of currents and either $\theta_r$ for rotational systems or $x$ for translational systems. Hence, we will choose currents and $\theta_r$ or $x$

as independent variables. In this case, (1.6-13) is most convenient for obtaining an expression for $T_e$ which can be written

$$T_e(\mathbf{i}, \theta_r)d\theta_r = -\sum_{j=1}^{J} \lambda_j(\mathbf{i}, \theta_r)di_j + dW_c(\mathbf{i}, \theta_r) \tag{1.6-14}$$

for compactness,

$$\mathbf{i} = (i_1, i_2, i_3, \dots, i_J) \tag{1.6-15}$$

Equation (1.6-14) may now be written

$$T_e(\mathbf{i}, \theta_r)d\theta_r = -\sum_{j=1}^{J} \lambda_j(\mathbf{i}, \theta_r)di_j + \sum_{j=1}^{J} \frac{\partial W_c}{\partial i_j}(\mathbf{i}, \theta_r)di_j + \frac{\partial W_c(\mathbf{i}, \theta_r)}{\partial \theta_r}d\theta_r \tag{1.6-16}$$

Equating coefficients of $d\theta_r$

$$T_e(\mathbf{i}, \theta_r) = \frac{\partial W_c}{\partial \theta_r} \tag{1.6-17}$$

In case of a translational system

$$f_e(\mathbf{i}, x) = \frac{\partial W_c}{\partial x} \tag{1.6-18}$$

where $f_e$ is the electromagnetic force and $x$ is the displacement. Since we will consider only magnetically linear systems $W_c = W_f$.

**Example 1E. Force between pole faces of air gap.** Determine the force that exists between the pole faces of the elementary magnetic circuit given in Fig. 1.3-1 and the parameters given in Example 1B. In order to make use of (1.6-18) to calculate the electromagnetic force $f_e$ we must allow what is referred to as a virtual displacement. That is, we will assume that the air-gap length $x$ increases in the positive direction by $dx$; therefore, according to (1.6-18) $f_e$ will be positive if it acts to lengthen the air gap. If $f_e$ is negative, it acts to shorten the air gap.

From (1.3-11) the self-inductance is

$$L = L_l + L_m \tag{1E-1}$$

where $L_l$ is the leakage inductance and $L_m$ is the magnetizing inductance. In particular,

$$L = \frac{N^2}{\mathcal{R}_l} + \frac{N^2}{\mathcal{R}_i + \mathcal{R}_g} \tag{1E-2}$$

The coenergy is

$$W_c = \frac{1}{2}(L_l + L_m)i^2 \tag{1E-3}$$

In order to determine $f_e$, we must take the partial derivative of $W_c$ with respect to $x$. Since $L_l$ is not a function of x, we only need to concern ourselves with $L_m$. We can write $L_m$ from (1B-1) and (1B-2) as

$$L_m = \frac{N^2}{\mathcal{R}_i + \mathcal{R}_g} = \frac{N^2}{\dfrac{\ell_i}{\mu_{ri}\mu_0 A_i} + \dfrac{x}{\mu_0 A_g}} \tag{1E-4}$$

which can be written

$$L_m(x) = \frac{k_0}{k_1 + k_2 x} = \frac{\dfrac{k_0}{k_2}}{\dfrac{k_1}{k_2} + x} \tag{1E-5}$$

where

$$k_0 = N^2 = (1 \times 10^2)^2 = 1 \times 10^4 \text{ turns} \tag{1E-6}$$

$$k_1 = \frac{\ell_i}{\mu_{ri}\mu_0 A_i} = \frac{4 \times 10^{-1}}{(1 \times 10^3)(4\pi \times 10^{-7})(4 \times 10^{-4})} = 7.96 \times 10^5 \text{ H}^{-1} \tag{1E-7}$$

$$k_2 = \frac{1}{\mu_0 A_g} = \frac{1}{(4\pi \times 10^{-7})(4 \times 10^{-4})} = 1.99 \times 10^9 \text{ m}^{-1}\text{H}^{-1} \tag{1E-8}$$

Therefore

$$W_c = \frac{\frac{1}{2}\frac{k_0}{k_2} i^2}{\dfrac{k_1}{k_2} + x} \tag{1E-9}$$

Now

$$f_e(i,x) = \frac{\partial W_c}{\partial x} = -\frac{\frac{1}{2}\left(\frac{k_0}{k_2} i^2\right)}{\left(\frac{k_1}{k_2} + x\right)^2} = -\frac{\frac{1}{2}(5.025 \times 10^{-6} \times 10^2)}{(3.995 \times 10^{-4} + 3 \times 10^{-3})^2} = -21.74\,\text{N (an attractive force)} \tag{1E-10}$$

A force is established in a magnetic system to minimize the reluctance.

**SP1.6-1.** Neglect the reluctance of the ferromagnetic core in Example 1E and calculate $f_e$. [−27.92 N]

**SP1.6-2.** In a rotational system, the self-inductance of stator winding is $L_{11} = k_0 + k_1 \cos 2\theta_r$. There is only one stator winding and no rotor winding. Express $T_e$, which is referred to as the reluctance torque, if $I_1 = \cos \omega_e t$. $[-k_1 I_1^2 \sin 2\theta_r]$

**SP1.6-3.** Rearrange the magnetic circuit in Fig. 1.3-1 to produce a repelling force between pole faces. [The force will always be in the direction to minimize the reluctance of the magnetic system.]

## 1.7 ELEMENTARY ELECTROMECHANICAL DEVICE

It is instructive to derive the expression for the torque of the elementary electromagnetic device shown in Fig. 1.6-3. This will allow us to describe the steady-state torque-angle characteristics of this elementary device with constant currents and provides a means of determining the regions of stable operation of electric machines. Figure 1.6-3 is repeated here for convenience as Fig. 1.7-1. From (1.6-17),

$$T_e(\mathbf{i}, \theta_r) = \frac{\partial W_c(\mathbf{i}, \theta_r)}{\partial \theta_r} \tag{1.7-1}$$

where the coenergy $W_c$ is equal to the field energy $W_f$ since the system is assumed to be magnetically linear. Thus,

$$W_c(i_1, i_2, \theta_r) = \frac{1}{2} L_{11} i_1^2 + L_{12} i_1 i_2 + \frac{1}{2} L_{22} i_2^2 \tag{1.7-2}$$

where

$$L_{11} = L_{l1} + L_{m1} \tag{1.7-3}$$

$$L_{22} = L_{l2} + L_{m2} \tag{1.7-4}$$

$$L_{12} = L_{sr} \cos \theta_r \tag{1.7-5}$$

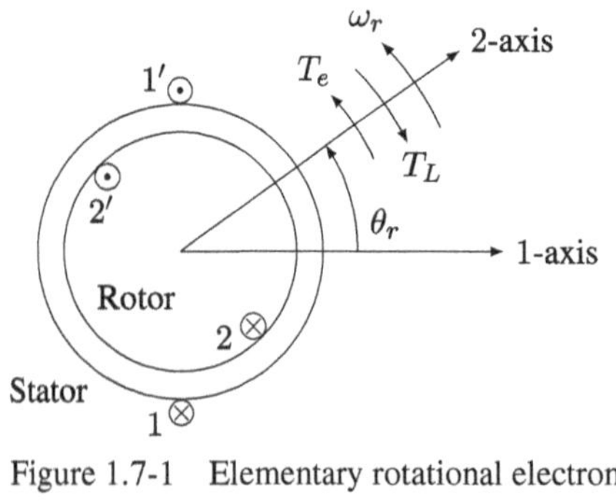


Figure 1.7-1 Elementary rotational electromechanical device.

The torque becomes

$$T_e(i_1, i_2, \theta_r) = -i_1 i_2 L_{sr} \sin\theta_r \tag{1.7-6}$$

Although (1.7-6) is valid regardless of the form of $i_1$ and $i_2$, let $i_1$ and $i_2$ be positive constants. For the positive direction of current shown, the torque would then be of the form

$$T_e = -K \sin\theta_r \tag{1.7-7}$$

where $K$ is a positive constant equal to $i_1 i_2 L_{sr}$. Positive $\theta_r$ in Fig. 1.7-1 is counterclockwise; we will let positive $\theta_r$ be from right to left in Fig. 1.7-2.

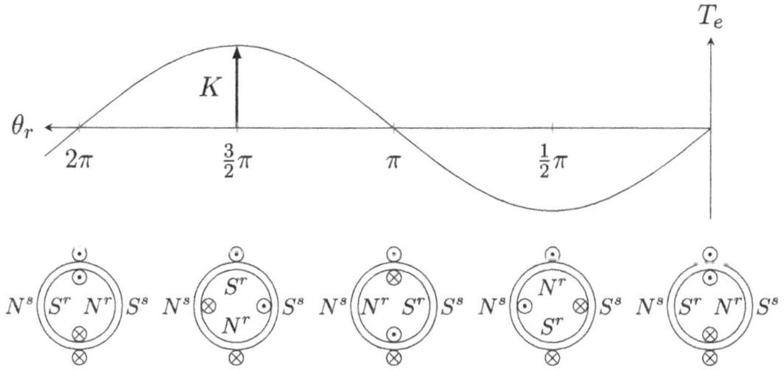


Figure 1.7-2 Electromagnetic torque versus angular displacement with constant winding currents.

Although the device being considered is not practical, Fig. 1.7-2, which depicts the torque and poles versus $\theta_r$, allows us to describe some important aspects of electromechanical devices. Depending on the type of electric machine the steady-state torque is often plotted versus displacement (angle) or speed. Stable operating regions can be determined from these plots which are similar to Fig. 1.7-2. Remember a positive torque $T_e$ will cause a positive movement in $\theta_r$, a negative torque will cause a decrease in $\theta_r$. If, for example, we hold the rotor at exactly $\theta_r = \pi$, the torque $T_e$ on the rotor would be zero; however, what happens when we let loose of the rotor? Well, any slight movement in $\theta_r$ due to a disturbance would cause a positive $T_e$ if $\theta_r$ increases slightly or a negative $T_e$ if $\theta_r$ decreases slightly. Therefore, regardless of the direction of the initial movement, a torque would occur to move the rotor away from $\theta_r = \pi$; thus at this position the rotor is in unstable equilibrium. If, however, we position the rotor at either $\theta_r = 0$ or $2\pi$ and release the rotor it will remain in that position. That is, if a slight disturbance should cause $\theta_r$ to increase, the resulting negative increase in $T_e$ would cause $\theta_r$ to decrease back to the zero $T_e$ position ($\theta_r = 0$ or $2\pi$). It follows that if the disturbance should cause a decrease in $\theta_r$, a positive increase in $T_e$ from zero would cause $\theta_r$ to increase back to its original position. With the positive directions of currents and $\theta_r$ that we have assumed and with the assumption of constant currents, stable operation occurs over the negative slope portion of the torque versus $\theta_r$ plot. Therefore, in Fig. 1.7-2 positive $\theta_r$ is from right to left, therefore stable operation occurs from $0 \le \theta_r < \frac{\pi}{2}$ and $\frac{3}{2}\pi < \theta_r \le 2\pi$. It is important to emphasize that if the currents change with a change of displacement or speed, then the region of stable operation may occur with the relative displacement between the stator and rotor poles greater than $\frac{\pi}{2}$. We will see this in the case of the electric machine considered in Chapters 2 and 4.

At $\theta_r = \frac{3}{2}\pi$ the torque is maximum positive and equal to

$$T_e = K = I_1 I_2 L_{sr} \tag{1.7-8}$$

which is motor action since positive torque is in the direction of counterclockwise rotation. This is the position of maximum torque per ampere.

**Example 1F. Inductance equations with clockwise rotation.** In Fig. 1F-1, $\theta_r$ is positive in the clockwise direction. The peak amplitude of the mutual inductance is $L_{sr}$. Express (*a*) the mutual inductance $L_{ab}$ and (*b*) the electromagnetic torque $T_e$.

(a) The maximum negative coupling occurs when $\theta_r = -\frac{\pi}{6}$. Therefore,

$$L_{ab} = -L_{sr} \cos\left(\theta_r + \frac{\pi}{6}\right) \tag{1F-1}$$

(b) The coenergy or field energy is

$$W_c = \frac{1}{2} L_{aa} i_a^2 + L_{ab} i_a i_b + \frac{1}{2} L_{bb} i_b^2 \tag{1F-2}$$

Since the self-inductances $L_{aa}$ and $L_{bb}$ are constant, the torque is

$$\begin{aligned} T_e &= \frac{\partial W_c(i_a, i_b, \theta_r)}{\partial \theta_r} \\ &= L_{sr} \sin\left(\theta_r + \frac{\pi}{6}\right) i_a i_b \end{aligned} \tag{1F-3}$$

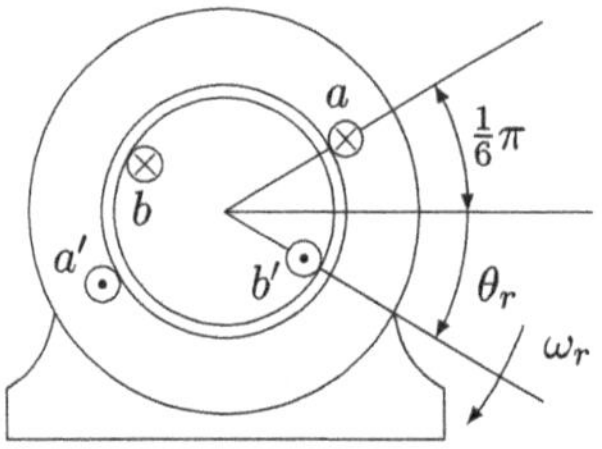


Figure 1F-1 A two-winding device with clockwise rotation.

---

**SP1.7-1.** In the system shown in Fig. 1.7-1, $L_{sr} = 0.1$ H, $i_1 = 2$ A, and $i_2 = 10$ A. (a) A torque of 1 N · m is applied in the clockwise direction. Calculate the steady-state value of $\theta_r$. (b) Repeat for an applied torque of 2 N · m. [(a) $\theta_r = -30°$; (b) unstable]

**SP1.7-2.** Reverse the direction of $i_1$ in Fig. 1.7-1 and determine the range of stable operation for constant, positive currents. [$\frac{1}{2}\pi < \theta_r < \frac{3}{2}\pi$]

**SP1.7-3.** In Fig. 1.7-1, winding 1 is moved so that ⊗ is at three o'clock and ⊙ at nine. Express $T_e$. [$T_e = i_1 i_2 L_{sr} \cos\theta_r$]

**SP1.7-4.** The currents $i_1$ and $i_2$ are positive constants in the device shown in Fig. 1.7-1. (a) An external torque is applied to increase $\theta_r$ to 45° and then released. Assume damping exists. What is the final position of the rotor? Repeat for $\theta_r$ increased to (b) 90°, (c) 120°, (d) 180°, and (e) 210°. [(a) $\theta_r = 0$; (b) 0; (c) 0; (d) 0 or $2\pi$; (e) $2\pi$]

## 1.8 TWO- AND THREE-PHASE SYSTEMS

High voltage transmission, most inverter-supplied electric drives, and the alternator of your car are examples of three-phase systems. Although two-phase systems are not common, a two-phase system is far less involved when it comes to machine analysis than its three-phase big sister. Fortunately, once the derivations have been set forth for a two-phase machine the extension to a three-phase machine is straightforward and easily achieved. This section is devoted to the introduction of two- and three-phase systems.

By definition, a two-phase set of variables is balanced if the variables are equal-amplitude sinusoidal quantities in time quadrature (90° out of time phase). A three-phase set of variables is balanced if the sinusoidal variables are equal-amplitude quantities that are 120° out of time phase with each other.

### Two-Phase Systems

In the broadest sense of the above definition, two-phase balanced sets may be expressed as

$$f_a(t) = \pm f(t)\cos\theta_{ef} \tag{1.8-1}$$

$$f_b(t) = \pm f(t)\sin\theta_{ef} \tag{1.8-2}$$

where

$$\theta_{ef}(t) = \int_0^t \omega_e(\xi)\,d\xi + \theta_{ef}(0) \tag{1.8-3}$$

In (1.8-1) and (1.8-2), $f(t)$ can represent voltage, current, or flux linkage. The amplitude may be any function of time; however, for steady-state balanced conditions $f(t)$ is a constant. In (1.8-3), $\omega_e$ is the electrical angular velocity and $\xi$ is a dummy variable of integration. Equations (1.8-1) and (1.8-2) express four balanced two-phase sets. Like signs of (1.8-1) and (1.8-2) define balanced sets where $f_a(t)$ leads $f_b(t)$ by 90°, an *ab* sequence; for unlike signs $f_a(t)$ lags $f_b(t)$ by 90°, a *ba* sequence.

For steady-state balanced conditions $\omega_e$ is constant and (1.8-3) becomes

$$\theta_{ef}(t) = \omega_e t + \theta_{ef}(0) \tag{1.8-4}$$

Whereupon (1.8-1) and (1.8-2) are written as

$$F_a(t) = \pm\sqrt{2}F\cos\left[\omega_e t + \theta_{ef}(0)\right] \tag{1.8-5}$$

$$F_b(t) = \pm\sqrt{2}F\sin\left[\omega_e t + \theta_{ef}(0)\right] \tag{1.8-6}$$

For like signs of (1.8-5) and (1.8-6) $F_a(t)$ leads $F_b(t)$ by 90°; thus $\tilde{F}_a = j\tilde{F}_b$. For unlike signs $F_a(t)$ lags $F_b(t)$; thus $\tilde{F}_a = -j\tilde{F}_b$.

Most large horsepower electric machines are three phase and smaller household machines are single phase. Therefore, it can be argued that there is no need to treat two-phase devices. This is an understandable position; however, we will find that it is convenient to first analyze a two-phase device and then extend this to a three-phase device. This turns out to be much less involved and more instructive than to start the analysis with the three-phase device. Although this is perhaps not evident at this juncture, it will become very evident as we get into the later chapters.

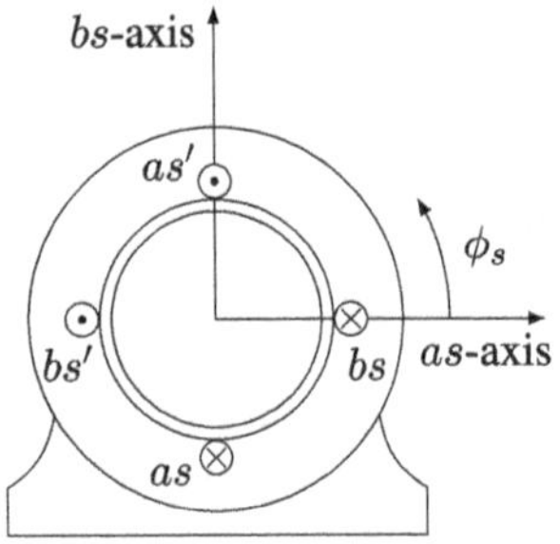


Figure 1.8-1 Elementary two-pole two-phase concentrated stator windings.

It is helpful to take a brief look at the stator winding arrangement of the two-phase machine, shown in Fig. 1.8-1. The windings are assumed to be concentrated as in Fig. 1.5-1. The displacement around the stator is denoted $\phi_s$. The two windings are displaced 90° degrees from each other. The voltage equations may be written (dropping the functional notation)

$$v_{as} = r_s i_{as} + \frac{d\lambda_{as}}{dt} \tag{1.8-7}$$

$$v_{bs} = r_s i_{bs} + \frac{d\lambda_{bs}}{dt} \tag{1.8-8}$$

In matrix form

$$\mathbf{v}_{abs} = \mathbf{r}_s \mathbf{i}_{abs} + p\boldsymbol{\lambda}_{abs} \tag{1.8-9}$$

The stator windings are identical, and the air gap is uniform. The magnetic axes are orthogonal, (thus $L_{asbs} = 0$) and the flux linkage equations may be written

$$\begin{aligned}\lambda_{as} &= L_{asas} i_{as} \\ &= (L_{ls} + L_{ms}) i_{as}\end{aligned} \tag{1.8-10}$$

$$\begin{aligned}\lambda_{bs} &= L_{bsbs} i_{bs} \\ &= (L_{ls} + L_{ms}) i_{bs}\end{aligned} \tag{1.8-11}$$

where $L_{ls}$ is the leakage inductance and $L_{ms}$ is the magnetizing inductance of the stator windings. In matrix form

$$\begin{bmatrix} \lambda_{as} \\ \lambda_{bs} \end{bmatrix} = \begin{bmatrix} L_{ss} & 0 \\ 0 & L_{ss} \end{bmatrix} \begin{bmatrix} i_{as} \\ i_{bs} \end{bmatrix} \tag{1.8-12}$$

or

$$\boldsymbol{\lambda}_{abs} = \mathbf{L}_s \mathbf{i}_{abs} \tag{1.8-13}$$

where

$$\mathbf{L}_s = \begin{bmatrix} L_{ss} & 0 \\ 0 & L_{ss} \end{bmatrix} \tag{1.8-14}$$

and

$$L_{ss} = L_{ls} + L_{ms} \tag{1.8-15}$$

An important feature of multiphase systems is that the instantaneous power is constant for balanced operation. You are asked to show this in SP1.8-2. Recall that in a single-phase system the instantaneous power has an average value and a double frequency component.

## Three-Phase Systems

A three-phase balanced set may be expressed as

$$f_a(t) = f(t)\cos\theta_{ef} \tag{1.8-16}$$

$$f_b(t) = f(t)\cos\left(\theta_{ef} - \frac{2}{3}\pi\right) \tag{1.8-17}$$

$$f_c(t) = f(t)\cos\left(\theta_{ef} + \frac{2}{3}\pi\right) \tag{1.8-18}$$

This set is referred to as an *abc* sequence, since $f_a(t)$ leads $f_b(t)$ by 120° and $f_b(t)$ leads $f_c(t)$ by 120°. An *acb* sequence is

$$f_a(t) = f(t)\cos\theta_{ef} \tag{1.8-19}$$

$$f_b(t) = f(t)\cos\left(\theta_{ef} + \frac{2}{3}\pi\right) \tag{1.8-20}$$

$$f_c(t) = f(t)\cos\left(\theta_{ef} - \frac{2}{3}\pi\right) \tag{1.8-21}$$

For steady-state balanced conditions, the *abc* sequence may be written

$$F_a(t) = \sqrt{2}F\cos[\omega_e t + \theta_{ef}(0)] \tag{1.8-22}$$

$$F_b(t) = \sqrt{2}F\cos\left[\omega_e t - \frac{2}{3}\pi + \theta_{ef}(0)\right] \tag{1.8-23}$$

$$F_c(t) = \sqrt{2}F\cos\left[\omega_e t + \frac{2}{3}\pi + \theta_{ef}(0)\right] \tag{1.8-24}$$

With $\tilde{F}_a = F\underline{/\theta_{ef}(0)}$, $\tilde{F}_b = F\underline{/-\frac{2}{3}\pi + \theta_{ef}(0)}$, and $\tilde{F}_c = F\underline{/+\frac{2}{3}\pi + \theta_{ef}(0)}$. For an *acb* sequence $\tilde{F}_b$ and $\tilde{F}_c$ are interchanged.

The windings of a three-phase stator are shown in Fig. 1.8-2. The magnetic axes of the windings are displaced 120° and the windings are often "wye-connected" as

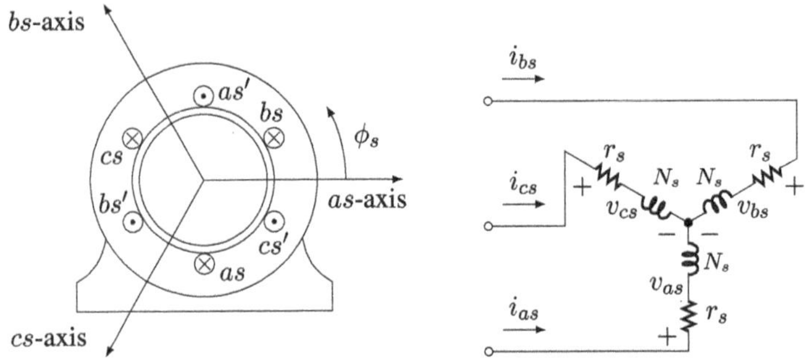


Figure 1.8-2 Elementary two-pole three-phase concentrated stator windings.

shown. We will talk more about three-phase connections later. The voltage equations may be written (again dropping the functional notation)

$$v_{as} = r_s i_{as} + \frac{d\lambda_{as}}{dt} \tag{1.8-25}$$

$$v_{bs} = r_s i_{bs} + \frac{d\lambda_{bs}}{dt} \tag{1.8-26}$$

$$v_{cs} = r_s i_{cs} + \frac{d\lambda_{cs}}{dt} \tag{1.8-27}$$

In matrix form

$$\mathbf{v}_{abcs} = \mathbf{r}_s \mathbf{i}_{abcs} + p\boldsymbol{\lambda}_{abcs} \tag{1.8-28}$$

Since the windings are displaced 120° from each other there is a mutual coupling between the stator windings. Recall that in Section 1.5 we approximated the mutual inductance between the two coils as $\cos\theta_r$. We can do something similar here. Let us assume that we can move to *bs*-winding clockwise through the iron until it is "on top" of the *as*-winding at $\phi_s = 0$. The coupling would be maximum positive. Now assume we can rotate the *bs*-winding counterclockwise back to $\phi_s = 120°$ where the mutual inductance between the *as*- and *bs*-windings would be

$$\begin{aligned} L_{asbs} &= L_{ms}\cos 120° \\ &= -\frac{1}{2}L_{ms} \end{aligned} \tag{1.8-29}$$

where $L_{ms}$ is the magnetizing inductance of the stator windings. Following this same approach we can express the flux-linkage matrix as

$$\begin{bmatrix} \lambda_{as} \\ \lambda_{bs} \\ \lambda_{cs} \end{bmatrix} = \begin{bmatrix} L_{ss} & -\frac{1}{2}L_{ms} & -\frac{1}{2}L_{ms} \\ -\frac{1}{2}L_{ms} & L_{ss} & -\frac{1}{2}L_{ms} \\ -\frac{1}{2}L_{ms} & -\frac{1}{2}L_{ms} & L_{ss} \end{bmatrix} \begin{bmatrix} i_{as} \\ i_{bs} \\ i_{cs} \end{bmatrix} \tag{1.8-30}$$

where

$$L_{ss} = L_{ls} + L_{ms} \tag{1.8-31}$$

Equation (1.8-30) may also be written as

$$\boldsymbol{\lambda}_{abcs} = \mathbf{L}_s \mathbf{i}_{abcs} \tag{1.8-32}$$

**Example 1G. Voltage equations for a three-wire system.** A three-phase stator similar to that given in Fig. 1.8-2 is connected to a three-phase source as shown in Fig. 1G-1. Assume the stator is symmetrical, that is, the windings have the same resistance and same number of turns and displaced 120°. The stator could be that of induction, synchronous, or brushless dc machines. The source voltages $e_{ga}$, $e_{gb}$, and $e_{gc}$ may be of any form. Express $v_{as}$, $v_{bs}$, and $v_{cs}$ in terms of $e_{ga}$, $e_{gb}$, and $e_{gc}$.

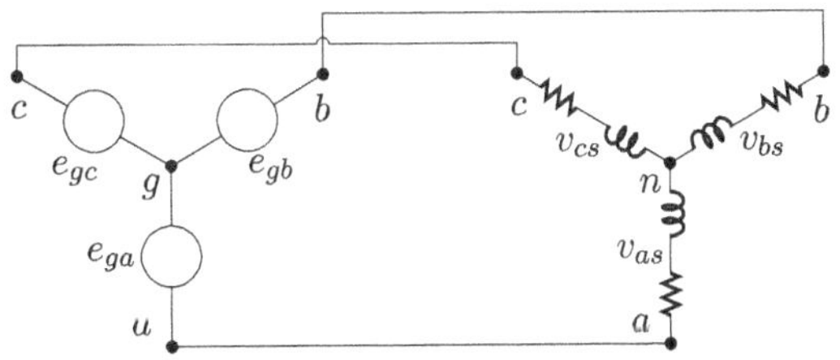


Figure 1G-1 Three-phase source connected to symmetrical stator windings.

From Fig. 1G-1 we can write

$$e_{ga} = v_{as} + v_{ng} \tag{1G-1}$$

$$e_{gb} = v_{bs} + v_{ng} \tag{1G-2}$$

$$e_{gc} = v_{cs} + v_{ng} \tag{1G-3}$$

where $e_{ga}$ is a voltage rise and $v_{as}$ and $v_{ng}$ are voltage drops, etc. Adding (1G-1) through (1G-3) yields

$$e_{ga} + e_{gb} + e_{gc} = v_{as} + v_{bs} + v_{cs} + 3v_{ng} \tag{1G-4}$$

Let us look at $v_{as} + v_{bs} + v_{cs}$. From (1.8-25) through (1.8-27)

$$v_{as} + v_{bs} + v_{cs} = r_s(i_{as} + i_{bs} + i_{cs}) + p(\lambda_{as} + \lambda_{bs} + \lambda_{cs}) \tag{1G-5}$$

In a three-wire, wye-connected stator the sum of $i_{as} + i_{bs} + i_{cs}$ must be zero regardless of the form of the currents. Now from (1.8-30)

$$\lambda_{as} + \lambda_{bs} + \lambda_{cs} = L_{ss}(i_{as} + i_{bs} + i_{cs}) - L_{ms}(i_{as} + i_{bs} + i_{cs}) \tag{1G-6}$$

Thus,

$$v_{as} + v_{bs} + v_{cs} = 0 \tag{1G-7}$$

The question we have is will this be the case when we bring the rotor into play? We will find that for the electromechanical devices we will consider, it will be true. Substituting (1G-7) into (1G-4) yields

$$v_{ng} = \frac{1}{3}(e_{ga} + e_{gb} + e_{gc}) \tag{1G-8}$$

Going back to (1G-1) through (1G-3) we can write

$$
\begin{aligned}
v_{as} &= e_{ga} - v_{ng} \\
&= \frac{2}{3}e_{ga} - \frac{1}{3}(e_{gb} + e_{gc}) \qquad (1G\text{-}9)\\
v_{bs} &= e_{gb} - v_{ng} \\
&= \frac{2}{3}e_{gb} - \frac{1}{3}(e_{gc} + e_{ga}) \qquad (1G\text{-}10)\\
v_{cs} &= e_{gc} - v_{ng} \\
&= \frac{2}{3}e_{gc} - \frac{1}{3}(e_{ga} + e_{gb}) \qquad (1G\text{-}11)
\end{aligned}
$$

We will make use of these equations in Chapter 4 on electric drives.

---

**SP1.8-1.** In Fig. 1G-1, let $e_{ga} = 1$, $e_{gb} = 0$, $e_{gc} = \cos\omega_e t$. Determine $v_{as}$, $v_{bs}$, and $v_{cs}$. $[\frac{2}{3} - \frac{1}{3}\cos\omega_e t; -\frac{1}{3} - \frac{1}{3}\cos\omega_e t; \frac{2}{3}\cos\omega_e t - \frac{1}{3}]$. Note that $v_{as} + v_{bs} + v_{cs} = 0$.

**SP1.8-2.** In a two-phase system, let $V_a = \sqrt{2}V\cos[\omega_e t + \theta_{ev}(0)]$, $I_a = \sqrt{2}I\cos[\omega_e t + \theta_{ei}(0)]$, $V_b = \sqrt{2}V\sin[\omega_e t + \theta_{ev}(0)]$, and $I_b = \sqrt{2}I\sin[\omega_e t + \theta_{ei}(0)]$. Show that the total instantaneous power is $P = 2VI\cos[\theta_{ev}(0) - \theta_{ei}(0)]$.

## REFERENCES


[1] P. C. Krause, O. Wasynczuk, and S. D. Pekarek, *Electromechanical Motion Devices*, 2nd edition, Wiley-IEEE Press, New York, 2012.

[2] D. C. White and H. H. Woodson, *Electromechanical Energy Conversion*, John Wiley and Sons, New York, 1959.


## PROBLEMS

1. Using the same circuit parameters as given in Example 1A; $R = 6\,\Omega$, $L = 20$ mH, and $C = 1 \times 10^3\,\mu$F, consider the circuit shown in Fig. 1.10-1. Let the 60 Hz source voltage be $\tilde{V}_a = 110\underline{/0°}$. Determine all phasors and draw the phasor diagram.

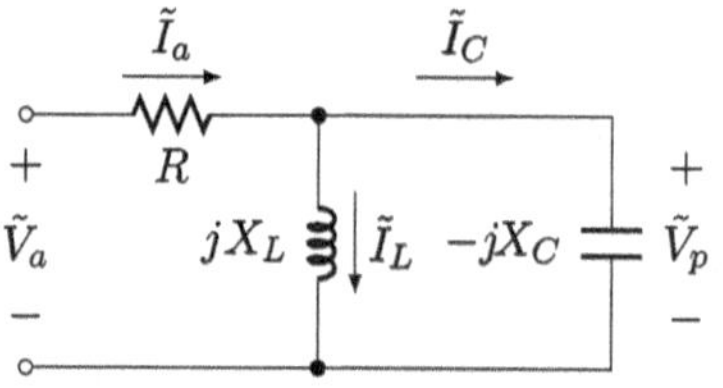


Figure 1.10-1 Series-parallel circuit.

2. For the magnetic system in Example 1B, let $V(t) = \sqrt{2}10\cos 377t$. Establish $I(t)$ and express $W_f(t)$.

**3.** During the short-circuit test of the transformer in Example 1C, which is a step-down distribution transformer, it was noticed that $I_2$ was 6 A. In re-doing the open-circuit test, the open-circuit current $|\tilde{I}_1|$ was not 1.05 A but 0.6 A. Upon checking the wiring, it was discovered that the lower terminal of $N_1$ was mistakenly connected to the top terminal of $N_2$ (See Fig. 1.4-1) and the 110 V source was connected between the top terminal of $N_1$ and the bottom terminal of $N_2$. Justify that $|\tilde{I}_1| = 0.6$ A. The equivalent "T" circuit will not yield the correct answer.

**4.** During the open-circuit test performed in Example 1C the rms voltage across the open-circuit 2-winding was 34.8 V. Determine $X_{m2}$ ($\omega_e L_{m2}$).

**5.** Express the field energy $W_f$ stored in the fields of the coils shown in Fig. 1.4-1 in terms of $L_{l1}$, $L'_{l2}$, $L_{m1}$, $i_1$, and $i'_2$. Identify the energy stored in the coupling field.

**6.** Winding 4 is added to the stator of the device given in Fig. 1D-1 as shown in Fig. 1.10-2. It is identical to winding 1 and displaced $\frac{\pi}{2}$ radians from it. Express $L_{44}$, $L_{14}$, $L_{24}$, $L_{34}$, and $v_2$. Let us mention in passing that the device in Fig. 1.10-2 has the winding configuration of an induction motor. There are many induction motors serving you in your home; furnace fan, garbage disposal, air conditioner, etc.

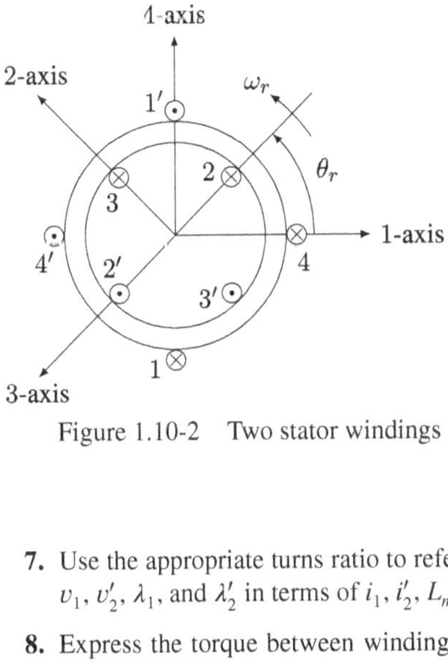


Figure 1.10-2 Two stator windings and two rotor windings.

**7.** Use the appropriate turns ratio to refer (1.5-1) through (1.5-4) to the 1-winding. Express $v_1$, $v'_2$, $\lambda_1$, and $\lambda'_2$ in terms of $i_1$, $i'_2$, $L_{m1}$, $L'_{\ell 2}$, $L_{\ell 1}$, $r'_2$, turns ratios, and $\theta_r$.

**8.** Express the torque between windings 2 and 3 of Fig. 1D-1. Use the concept of virtual rotation with $\beta$, the angle between the positive magnetic axes of the windings.

**9.** Use the device depicted in Fig. 1.6-3 to show that the first two terms on the right-hand side of (1.6-16) sum to zero. That is, show that the coefficients of the *di* terms are equal.

**10.** Consider the device shown in Fig. 1F-1. (a) Identify the stable operating region for $i_a$ and $i_b$ positive constants. (b) Identify the stable region if $i_b$ is a negative constant. (c) Reverse the direction of positive current in the rotor winding and identify the region of stable operation for constant positive currents.

**11.** Consider the device shown in Fig. 1.10-3. The 3-winding which is identical to the 1-winding has been added to the stator of Fig. 1.7-1. The device now has the winding configuration of an elementary synchronous machine. Express $T_e(i_1, i_2, i_3, \theta_r)$.

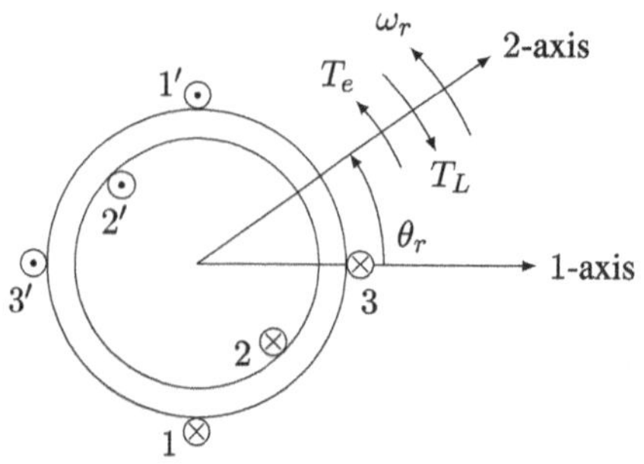


Figure 1.10-3 The stator windings and one rotor winding.

**12.** Consider Example 1G and Fig. 1G-1. The load is symmetrical and the source voltages are given in Fig. 1.10-4. (a) Plot $v_{ng}$, $v_{as}$, $v_{bs}$, and $v_{cs}$. (b) Connect $n$ to $g$ in Fig. 1G-1 and repeat part (a).

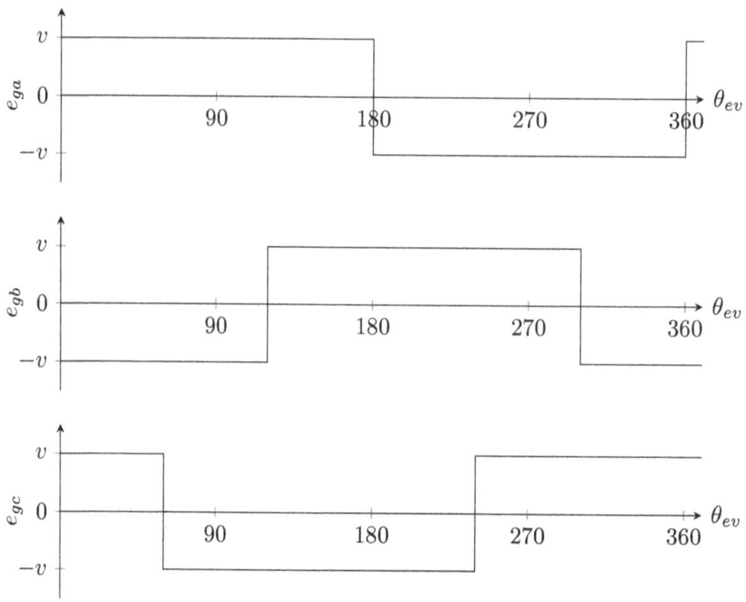


Figure 1.10-4 Waveforms of the source voltages of Fig. 1G-1.

CHAPTER 2

# ELECTRIC MACHINES

## 2.1 INTRODUCTION

It is told that as a young student in Graz, Austria, Nikola Tesla tried to convince his professor that there must be a better electric machine design than the dc motor being demonstrated to the class. Tesla argued that the arcing of the commutator was very undesirable and inefficient. The commutator, along with a carbon-brush arrangement, mechanically switched the direction of the currents flowing in the rotor circuits in order to create a stationary magnet field, orthogonal with the magnetic field created by direct current flowing in the stator winding. The professor chided young Tesla and told the class there was no other way to design an electric motor. It took Tesla several years to develop the concept of the rotating magnetic field which is also referred to as the rotating air-gap mmf and/or rotating poles [1]. This led to his invention of alternating-current machines and multiphase power transmission and he and Westinghouse used this concept to light the 1893 World Columbian Exposition in Chicago. This marked the birth of the ac power system as we know it today and the beginning of the demise of the dc power system being promoted by Edison.

In 1929, Robert H. Park published a change of variables [2] that revolutionized the analysis of ac machines and ac power systems. It is interesting that this change of variables or transformation is readily obtained from the expression of Tesla's rotating magnetic field. Although this transformation was originally used in power system analysis, it also set the stage for the analysis of electric drive systems which began to evolve with the advent of the controlled rectifier in the 1950s. Although Tesla and Park are deceased, their work will live as long as there are ac power systems and electric drives.

In this chapter, Tesla's rotating magnet field and Park's transformation are introduced. This is achieved by starting with a two-phase system, which allows these concepts to be set forth without the complication of tedious trigonometry. Once this is done, the extension to a three-phase machine/system is straightforward and direct. This is a logical approach since for balanced conditions the expression for Tesla's rotating magnetic field (air-gap mmf) differs only by a constant when going from a two-phase to a three-phase machine.

*Introduction to Electric Power and Drive Systems*, First Edition.
Paul Krause, Oleg Wasynczuk, Timothy O'Connell and Maher Hasan.
© 2017 by The Institute of Electrical and Electronics Engineers, Inc. Published 2017 by John Wiley & Sons, Inc.
Companion website: www.wiley.com/go/krause/electricpower

These basic concepts establish a foundation for the analysis of electric machines, power systems, and electric drives that are now being used in ground vehicles, aircraft, spacecraft, and industrial applications. Once this foundation is established, a permanent-magnet ac machine is analyzed. This choice of machine allows us to set the stage for the analysis of the brushless dc electric drive (Chapter 4) and the synchronous generator which is the backbone of the ac power system (Chapter 5).

Although the transformation or change of variables is straightforward and easy to implement, it can be de-emphasized or skipped without sacrificing the salient steady-state characteristics during balanced operation of the machines. A procedure for such a study focus is suggested in Sections 2.4 and 2.5.

## 2.2 FUNDAMENTALS OF ELECTRIC MACHINE ANALYSIS

### Concentrated Winding

The analysis of ac electromechanical motion devices is based upon the concept of the rotating air-gap mmf (rotating magnetic field) which was invented by Nikola Tesla in the late nineteenth century. Most ac electric machines, particularly synchronous generators, are designed so that the stator windings produce a relatively good approximation of a sinusoidally distributed air-gap mmf, thereby minimizing the voltage and current harmonics. To establish a sinusoidal air-gap mmf waveform (often referred to as a *space sinusoid*), the physical distribution of the winding must approximate a sinusoidal space distribution. We have not heard of many of these terms before; hopefully, this chapter will help us.

Back in Section 1.5, we considered two concentrated windings in relative motion for a "first look" at the mutual inductance between stator and rotor windings. Our purpose now is to look a little further into the magnetic characteristics of the windings and how we are going to treat the windings in our analysis. Let us start by considering just a concentrated stator winding of $N$ turns as shown in the cutaway view of the machine in Fig. 2.2-1a and a cross-sectional view of the machine taken at some point along the length of the rotor shown in Fig. 2.2-1b. In Fig. 2.2-1a, $r$ is

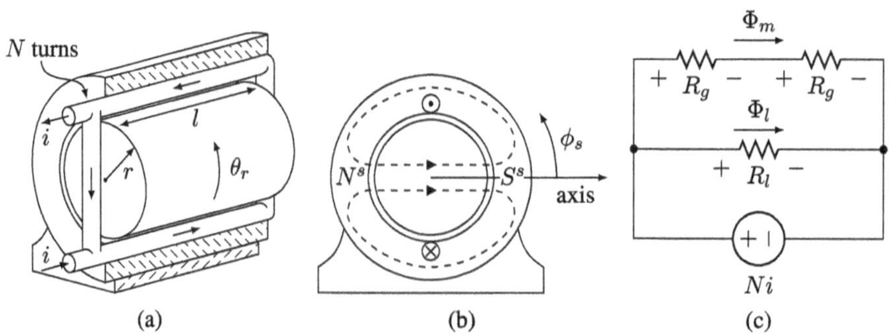


Figure 2.2-1 Concentrated winding: (a) cutaway view; (b) cross-sectional view; (c) equivalent magnetic circuit.

the radius and $\ell$ the axial length of the rotor. The stator winding consists of a coil of insulated wire of $N$ turns in one slot. Current entering the bottom part of the coil is assumed to be positive, which is shown as ⊗ in Fig. 2.2-1b and exiting the top part of the coil, shown as ⊙ in Fig. 2.2-1b. It is important to realize that the length of the air gap is exaggerated relative to the rest of the machine; in fact, the radius of the inside circumference of the stator is essentially the radius of the rotor.

The streamlines of the flux established by current flowing in the stator winding are shown in Fig. 2.2-1b. We see that flux leaves the stator where the winding exists and enters the air gap between $\frac{\pi}{2} < \phi_s < \frac{3}{2}\pi$, where $\phi_s$ is the displacement from the magnetic axis of the winding. This is the north pole of the stator, $N^s$, and it follows that $S^s$ exists from $-\frac{\pi}{2} < \phi_s < \frac{\pi}{2}$.

The equivalent magnetic circuit is shown in Fig. 2.2-1c, where the reluctance of the iron has been neglected. The leakage flux, $\Phi_l$, is the flux that fails to cross the air gap; instead occurring inside the stator slot itself and at the end turns of the coil. A detailed calculation of leakage flux is beyond the scope of this text. The reader may wish to refer to [3, 4] for a comprehensive analysis. Since this flux path is mostly in air, $\Phi_l$ is small compared to $\Phi_m$. Now, what about $\mathcal{R}_g$? Well, in many respects the configurations of Fig. 2.2-1b and Fig. 1.3-3 are similar. Each has two air gaps and since the two gaps in each figure are identical, one-half of the mmf is dropped across each air gap. That is, the air-gap mmf of each air gap is $\frac{1}{2}(Ni)$. That is logical; however, we have yet to determine $\mathcal{R}_g$ for Fig. 2.2-1b. Now, an acceptable approximation of the cross-sectional area of the flux path across each of the air gaps is

$$A_{ag} = \frac{1}{2}(2\pi r\ell) \tag{2.2-1}$$

where $r$ is essentially the radius of the rotor and $\ell$ is the axial length of the rotor. The reluctance is

$$\mathcal{R}_g = \frac{g}{\mu_{ri}\mu_0\pi r\ell} \tag{2.2-2}$$

The flux $\Phi_m$ is

$$\begin{aligned}\Phi_m &= \frac{Ni}{2\mathcal{R}_g}\\ &= \frac{Ni\mu_{ri}\mu_0\pi r\ell}{2g}\end{aligned} \tag{2.2-3}$$

In (2.2-3) we have calculated the flux, $\Phi_m$, as the total mmf drop across both air gaps, which are in series, divided by the total reluctance. Clearly, we would get the same answer if we calculated $\Phi_m$ flowing across one air gap. In this case, the air-gap mmf is $\frac{1}{2}Ni$ and the reluctance is $\mathcal{R}_g$.

Before becoming too involved with winding distribution, let us talk a little more about air-gap mmf of an electric machine. For this purpose, we will use Fig. 2.2-2 which is a repeat of Fig. 2.2-1b. The meter with an arrow indicator in Fig. 2.2-2a is our newly invented mmfmeter, which we patterned after an old voltmeter. As in Fig. 2.2-1b, the flux streamlines are shown and a positive (+) and a negative (−) meter deflection indicates the mmf drop across the air gap (air-gap mmf). It has a red

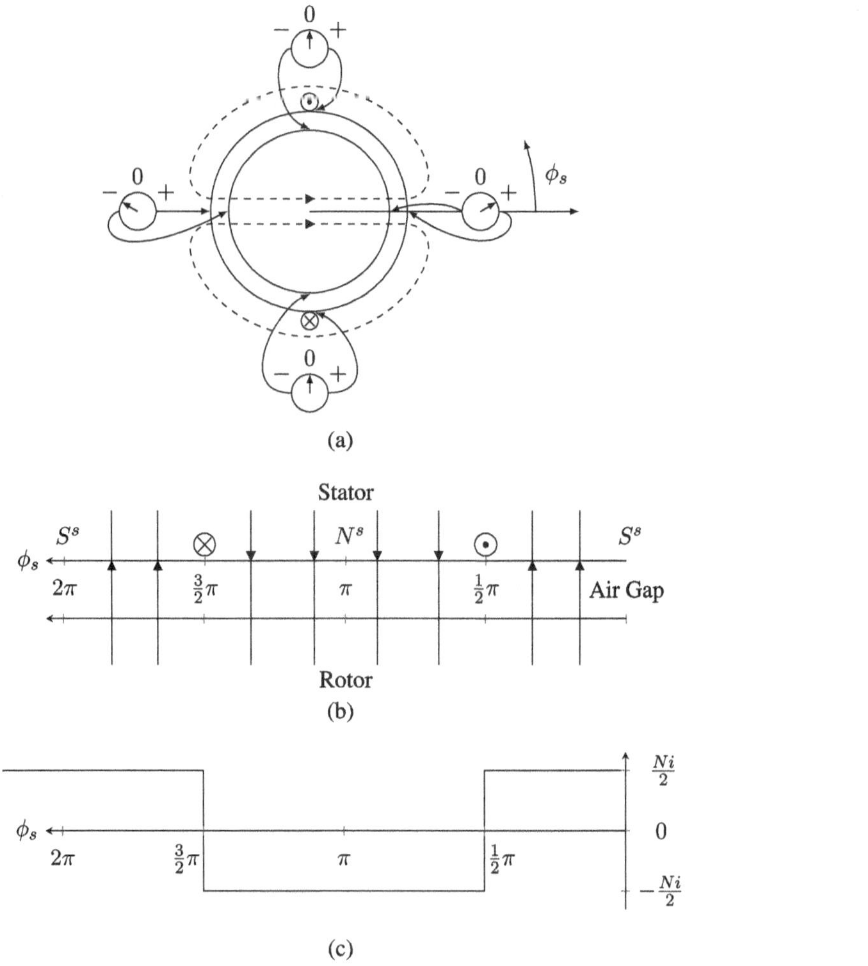


Figure 2.2-2 Air-gap mmf of concentrated stator winding with constant positive current: (a) "measuring" the air-gap mmf; (b) developed-winding view; (c) plot of air-gap mmf versus $\phi_s$.

probe (R) or (−) and a black probe (B) or (+). If we place the R-probe on the rotor side of the air gap and the B-probe on the stator side and if the constant current is positive ⊗ to ⊙, we would observe a positive deflection at $\phi_s = 0$. This is an mmf drop across the air gap from rotor to stator. If now we move the probes counterclockwise in unison around the air gap the meter reading remains constant until we reach $\phi_s = 90°$. At $\phi_s = 90°$, it will decrease to zero since there is no air-gap mmf at $\phi_s = 90°$ and therefore flux does not cross the air gap. Moving the probes beyond $\phi_s = 90°$, the deflection will be negative and equal in magnitude to the positive deflection we saw at $\phi_s = 0$. The deflection will remain constant until we reach $\phi_s = 270°$. The deflection will then increase back to zero. Thus, for a positive coil current, a positive air-gap mmf exists from $-\frac{\pi}{2} \leq \phi_s \leq \frac{\pi}{2}$ and negative from $\frac{\pi}{2} \leq \phi_s \leq \frac{3}{2}\pi$. Recall that this is only if the coil current is positive. We are going to let the current be sinusoidal later in this section. As we have mentioned, the flux, like current, must have a closed path; there is no point flux source.

The so-called developed winding is shown in Fig. 2.2-2c. Here, the machine is "cut" along the radius at $\phi_s = 0$ and the machine is "stretched out" with counterclockwise displacement about the stator viewed from right to left for positive $\phi_s$ displacement. This orientation allows us to place the stator above the rotor. The linear approximation of the circular air gap is allowed since the length of air gap is much less than the radius, $r$, of the rotor; that is $g \ll r$. Note the location of the stator poles in Fig. 2.2-2b.

The plot of the air-gap mmf versus $\phi_s$ is shown in Fig. 2.2-2c. Here, in accordance with established convention, positive air-gap mmf occurs when the mmf drop is from rotor to stator (as in Fig. 2.2-2a) with the positive assumed current (flux) direction. We will make extensive use of the developed winding and the air-gap mmf versus $\phi_s$ plot.

## Distributed Windings

Thus far, we have assumed that the stator windings are concentrated in one stator slot; however, the phase windings would normally be distributed over a 30° – 60° arc. Let us redraw Fig. 2.2-1b with the following changes. We will now call the winding *as*-winding and assume that it is distributed in slots over the inner circumference of the stator, as shown in Fig. 2.2-3.

The winding is depicted as a series of individual coils. Each coil is placed in a slot in the stator steel. If we follow the path of assumed positive current $i_{as}$ flowing in the *as*-winding, we see that current enters $as_1$, depicted by ⊗, to indicate that the assumed direction of positive current is down the length of the stator in an axial direction (into the paper). Current flows down the length of the stator, loops at the

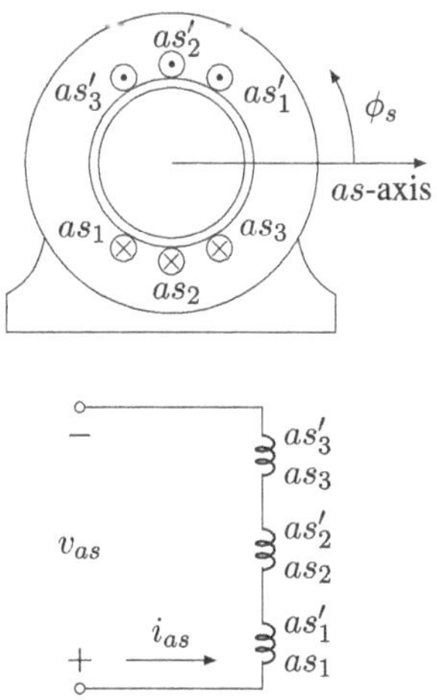


Figure 2.2-3 Elementary two-pole single-phase stator winding.

end, and flows back down the length of the stator and out at $as_1'$, depicted by ⊙. Note that $as_1$ and $as_1'$ are placed in stator slots that span $\pi$ radians. This is referred to as the *winding pitch* of $\pi$ radians which is characteristic of a two-pole machine. We will talk more about poles later. Now, $as_1$ around to $as_1'$ is referred to as a *coil* and $as_1$ or $as_1'$ is a *coil side*. In practice, a coil will contain more than one conductor. Current flows into $as_1$ in a conductor and out $as_1'$ via the same conductor. The conductor, which is insulated, may then be looped back to $as_1$ and the winding of the conductor around the $as_1 - as_1'$ path repeated, thereby forming a coil with numerous turns. The number of conductors in a coil side tells us the number of turns in the coil, which is denoted $nc_s$.

Once we have wound $nc_s$ turns in the $as_1 - as_1'$ coil, we will take the same conductor and repeat this winding process to form the $as_2 - as_2'$ coil. We will assume that the same number of turns ($nc_s$) make up the $as_2 - as_2'$ coil as the $as_1 - as_1'$ coil and, similarly, for $as_3 - as_3'$. We could have wound a different number of turns in each coil but we will assume that this was not done. With the same number of turns ($nc_s$) in each of the coils, the winding is said to be uniformly distributed over an arc from $as_1$ to $as_3$, or 60°. Once the winding is wound, we can use the right-hand rule to give a meaning to the *as*-axis shown in Fig. 2.2-3. It is, by definition, the principal direction of the magnetic flux established by the assumed positive current flowing in the *as*-winding. It is said to indicate the assumed positive direction of the magnetic axis of the *as*-winding.

Let us add coils to the *as*-winding shown in Fig. 2.2-3, whereupon we have Fig. 2.2-4a. Although we may be going a bit far by adding coils that cover an arc of $\frac{2}{3}\pi$, it helps to illustrate our point. For the purpose of establishing an expression for the air-gap mmf, it is convenient to employ the developed diagram of the cross-sectional view of the device. Again, this developed diagram is obtained by flattening out the rotor and stator as previously depicted in Fig. 2.2-2.

From our earlier discussion, each circle represents $nc_s$ conductors or coils. Figure 2.2-4c is the sinusoidal approximation of the developed winding distribution shown in Fig. 2.2-4b. From this figure, we see that the physical distribution or turn density of the winding from $0 < \phi_s < \pi$ may be approximated by the expression

$$N_{as} = N_p \sin \phi_s \quad \text{for } 0 < \phi_s < \pi \tag{2.2-4}$$

and from $\pi < \phi_s < 2\pi$ as

$$N_{as} = -N_p \sin \phi_s \quad \text{for } \pi < \phi_s < 2\pi \tag{2.2-5}$$

where $N_p$ is the peak turns density in turns/radian that could be determined from a Fourier analysis of the actual winding distribution. If $N_s$ represents the number of turns of the equivalent sinusoidally distributed winding, then, using $\xi$ as a dummy variable of integration

$$N_s = \int_0^{\pi} N_p \sin \xi d\xi = 2N_p \tag{2.2-6}$$

It is important to note that the number of turns is obtained by integrating the turns density $N_{as}$ from 0 to $\pi$ rather than from 0 to $2\pi$. Also, $N_s$ represents not the total turns of the winding, but those of the equivalent sinusoidally distributed winding

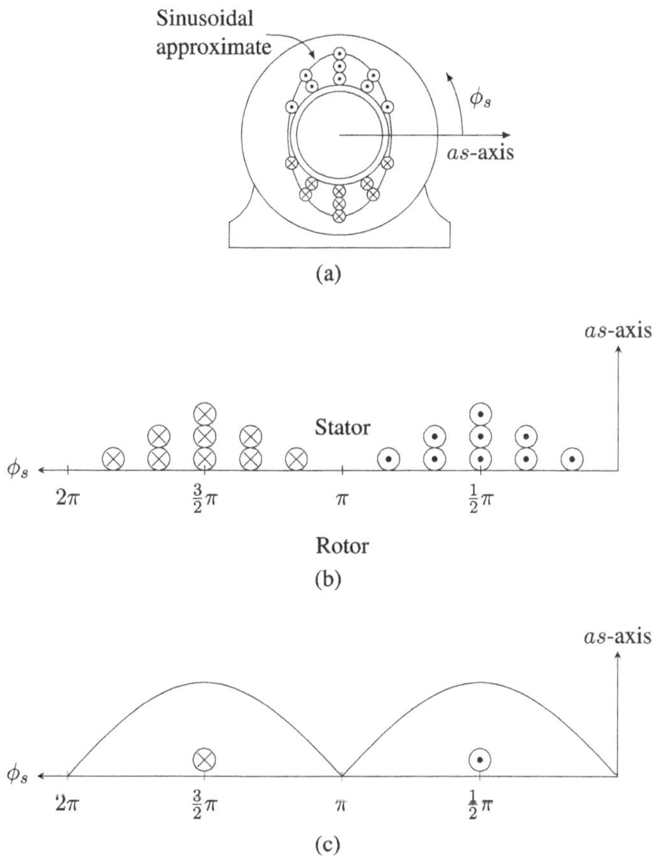


Figure 2.2-4 Sinusoidally distributed stator winding: (a) approximate sinusoidal distribution; (b) developed diagram of (a); (c) sinusoidal approximation of (b).

that corresponds to the fundamental component of the actual winding distribution, which, depending on the winding distribution, could be different than the total number of turns.

Our goal is to develop an expression for the air-gap mmf as a function of displacement $\phi_s$ about the air gap due to the spatial distribution of the winding. The winding distribution introduced in Fig. 2.2-4a is shown again in Figs. 2.2-5a through 2.2-5d wherein the mmf drop across the air gap is depicted at $\phi_s = 0°$, $30°$, $60°$, and $90°$, respectively. Recall from Section 1.3 that according to Ampere's Law the ampere-turns enclosed within the closed line integral is the mmf dropped around that path. In Figs. 2.2-5a through 2.2-5d, the path of Ampere's closed line integration is shown. Note that the path spans one pole pitch; that is $\pi$ radians and the path is rotated as shown to account for all coils.

In Fig. 2.2-5a where $\phi_s = 0$, the total mmf around the closed path is $N_s i_{as}$ and this mmf is dropped across the air gap at $\phi_s = 0$ and $\phi_s = 180°$. Assuming the reluctance of the iron is negligible, the mmf across each air gap (air-gap mmf) is equal

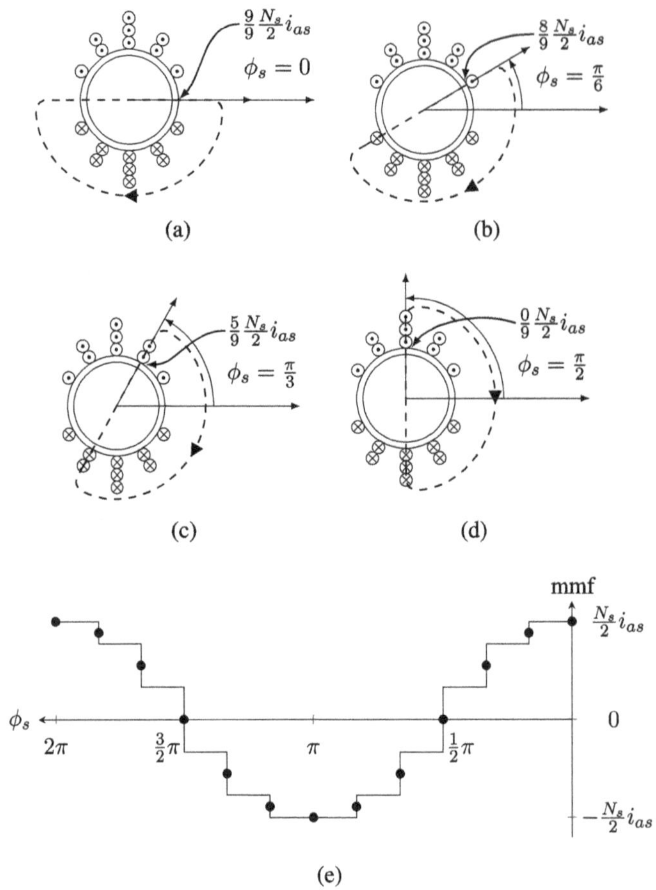


Figure 2.2-5 Air-gap mmf of *as*-winding. Air-gap mmf at (*a*) $\phi_s = 0$, (*b*) $\phi_s = 30°$, (*c*) $\phi_s = 60°$, and (*d*) $\phi_s = 90°$. (*e*) Developed diagram of air-gap mmf.

to $\frac{N_s}{2}i_{as}$. Now, by our convention, an mmf drop from rotor to stator is assumed to be positive. This occurs at $\phi_s = 0$; thus the air-gap mmf at $\phi_s = 180°$ is negative since the mmf drop is from stator to rotor. Ok, but what does the $\frac{9}{9}$ in Fig. 2.2-4a represent? Well, this is the fraction of the total number of turns that is contributing to the air-gap mmf at $\phi_s = 0$ and 180°; that is, all 9 coils. Therefore, $\frac{9}{9}\frac{N_s}{2}i_{as}$ is plotted at $\phi_s = 0$ and $-\frac{9}{9}\frac{N_s}{2}i_{as}$ at $\phi_s = 180°$. We have established two points in Fig. 2.2-5e.

At $\phi_s = 30°$ (Fig. 2.2-5b) we have one less coil in our Ampere's Law path. The air-gap mmf at $\phi_s = 30°$ and 210° is $\frac{8}{9}\frac{N_s}{2}i_{as}$. Thus, we have two more points for Fig. 2.2-5e. We are inclined to connect the points; however, let us look at the air-gap mmf just prior to $\phi_s = 30°$. There we still have $\frac{9}{9}\frac{N_s}{2}i_{as}$; however, at $\phi_s = 30°$ the coil at that position does not contribute to the drop across the air gap at that position;

therefore, we have $\frac{8}{9}\frac{N_s}{2}i_{as}$. This is the point plotted at $\phi_s = 30°$ in Fig. 2.2-5e. Now, if we look at $\phi_s$ slightly larger than 30° we see 8 − 1 or 7 effective coils; thus $\frac{7}{9}\frac{N_s}{2}i_{as}$. Therefore, as $\phi_s$ increases from zero to slightly less than 30° we have $\frac{9}{9}\frac{N_s}{2}i_{as}$ and when $\phi_s$ is slightly larger than 30° we have $\frac{7}{9}\frac{N_s}{2}i_{as}$; however, at 30° we have $\frac{8}{9}\frac{N_s}{2}i_{as}$. At $\phi_s = 30°$, we get the mid-point between the air-gap mmf at $\phi_s < 30°$ and $\phi_s > 30°$. We realize that, by portraying the diameter of the coils greatly exaggerated relative to the diameter of the rotor, we would have a negatively sloping air-gap mmf at 30° from $\frac{9}{9}\frac{N_s}{2}i_{as}$ to $\frac{7}{9}\frac{N_s}{2}i_{as}$; however, in a practical machine this transition would be abrupt, nearly as shown by the solid lines in Fig. 2.2-5e.

At $\phi_s = 60°$, we have 6 coils with ⊗ and one with ⊙, thus, only $\frac{5}{9}\frac{N_s}{2}i_{as}$ is the "good" air-gap mmf at $\phi_s = 60°$ and 240°. Since at $\phi_s = 90°$ and 270° there are as many ⊗ as ⊙ involved, the mmf is $\frac{0}{9}\frac{N_s}{2}i_{as}$. We have established the air-gap mmf for $0 \le \phi_s \le \frac{\pi}{2}$ and $\pi \le \phi_s \le \frac{3}{2}\pi$. We could continue increasing $\phi_s$ to obtain the air-gap mmf for $\frac{\pi}{2} \le \phi_s \le \pi$ and $\frac{3}{2}\pi \le \phi_s \le 2\pi$ and obtain Fig. 2.2-5e; however, we will not since it is straightforward. The air-gap mmf in Fig. 2.2-5e is a stepped waveform; therefore, we can approximate the air-gap mmf due to the *as*-winding as

$$\text{mmf}_{as} = \frac{N_s}{2}i_{as}\cos\phi_s \tag{2.2-7}$$

In fact, if we connect the "mid-point" dots in Fig. 2.2-5e we would closely approximate (2.2-7).

Although we have established the location of the poles for a concentrated winding, let us take a minute to locate the magnetic poles that would be established by the distributed *as*-winding shown in Fig. 2.2-5a. The *as*-axis is located at the maximum positive air-gap mmf. For the assumed direction of positive current, this is at $\phi_s = 0$ where positive flux is crossing the air gap from the rotor to the stator and positive flux leaves the stator and returns to the rotor at $\phi_s = \pi$. Using our "locate the poles rule" from Section 1.3, we put ourselves on the member where the winding exists and where the assumed positive flux, by the right-hand rule, leaves this member and enters the air. This is a north pole. Hence, for positive $i_{as}$, the north pole $N^s$ (north stator pole) is located between $\pi/2 < \phi_s < 3\pi/2$ and the south pole $S^s$ is between $-\frac{\pi}{2} < \phi_s < \pi/2$. Obviously, the device shown in Fig. 2.2-5 is a two-pole device.

**Example 2A. Air-gap mmf for a uniformly distributed winding.** Consider the uniformly distributed winding arrangement shown in Fig. 2.2-3. Each coil has *N* turns and the current in each turn is *I* with the positive direction as shown. Follow the procedure used in Fig. 2.2-5 to establish the air-gap mmf. Using the "locate the poles rule," verify that a stator south pole $S^s$ exists from $\frac{3}{2}\pi \le \phi_s \le \frac{\pi}{2}$ and a stator north pole $N^s$ from $\frac{\pi}{2} \le \phi_s \le \frac{3}{2}\pi$. The pertinent Ampere's law paths and resultant air-gap mmf are depicted in Fig. 2A-1.

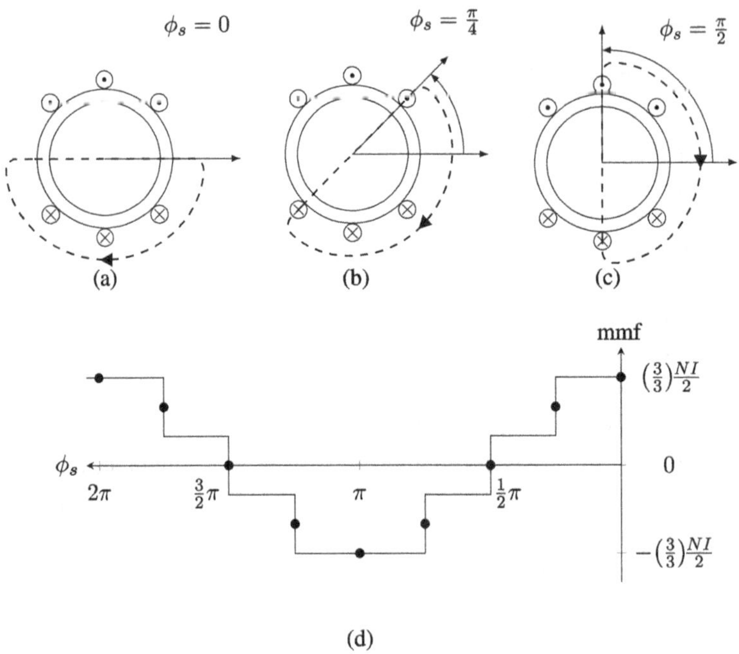


Figure 2A-1 Developed diagram of air-gap mmf.

## Rotating Air-Gap MMF – Tesla's Rotating Magnetic Field

In the case of a single-phase stator winding, the sinusoidal approximation of the air-gap mmf is given by (2.2-7). Let us think about this for a minute; (2.2-7) is a product of $i_{as}$ and $\cos \phi_s$; whereupon, the air-gap mmf will be nonexistent whenever $i_{as} = 0$. This is not too desirable since we will find that in order to produce a constant steady-state torque we need a constant amplitude rotating air-gap mmf (north and south poles) produced by the stator. Is it possible to place another winding on the stator and arrange its position and the time sequence of the current flowing in it so that its mmf would complement the mmf of the as-winding resulting in a rotating sinusoidally distributed air-gap mmf whose amplitude is constant? Yes, fortunately, Tesla showed us how, and the symmetrical two-pole two-phase stator is the most elementary example.

## Two-Pole Two-Phase Stator

In the case of a symmetrical two-pole, two-phase stator, the *bs*-winding is located $\frac{\pi}{2}$ radians from the *as*-winding and it is identical in distribution and parameters to the *as*-winding. Let us assume it is positioned as illustrated in Fig. 2.2-6. Therein, a circle placed at the position of maximum turn density is now used to represent a sinusoidally distributed winding. The air-gap mmf created by $i_{bs}$ is approximated as

$$\text{mmf}_{bs} = \frac{N_s}{2} i_{bs} \sin \phi_s \tag{2.2-8}$$

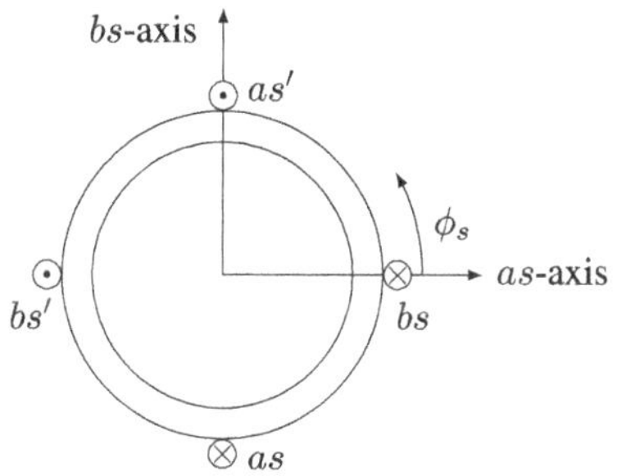


Figure 2.2-6 Elementary two-pole two-phase sinusoidally distributed stator windings.

Thus, the total air-gap mmf due to the two-pole stator windings would be

$$\begin{aligned} \text{mmf}_s &= \text{mmf}_{as} + \text{mmf}_{bs} \\ &= \frac{N_s}{2}(i_{as}\cos\phi_s + i_{bs}\sin\phi_s) \end{aligned} \tag{2.2-9}$$

This relationship is perhaps the most important equation we will run into in this text, since it is key to the operation and analysis of ac machines.

From Section 1.8, a two-phase steady-state balanced set of stator currents (*ab* sequence) may be expressed

$$I_{as} = \sqrt{2}I_s\cos[\omega_e t + \theta_{esi}(0)] \tag{2.2-10}$$

$$I_{bs} = \sqrt{2}I_s\sin[\omega_e t + \theta_{esi}(0)] \tag{2.2-11}$$

where $I_s$ is the rms value of the phase current, $\omega_e$ is the electrical angular velocity in rad/s, and $\theta_{esi}(0)$ is the phase angle of the currents. For this balanced set $\tilde{I}_{bs} = -j\tilde{I}_{as}$.

It appears that the concept of Tesla's rotating field can be explained by a very familiar trigonometric identity. If (2.2-10) and (2.2-11) are substituted into (2.2-9) and if $\phi_s$ is set equal to $\theta_{esi}$ then, since $\cos^2 x + \sin^2 x = 1$, (2.2-9) becomes a constant. Although this concept is indeed logical, one wonders what this all entails. We will try to sort some of this out.

Since Tesla's magnetic field rotates relative to the stator, perhaps, looking at it from a rotating axis will help us visualize what is going on. In Fig. 2.2-7 we have added a third axis, the $q$-axis, to Fig. 2.2-6. We will explain why we use "$q$" later. As shown in Fig. 2.2-7, the concept of a $q$-axis is no different from the *as*- or *bs*-axes, except that it can rotate at any angular velocity $\omega$, or it can be stationary ($\omega = 0$). We are free to specify it. The angle from the *as*-axis to the $q$-axis is $\theta$ and for a constant $\omega$

$$\theta = \omega t + \theta(0) \tag{2.2-12}$$

The displacement $\phi$ shown in Fig. 2.2-7 is from the $q$-axis just as $\phi_s$ is the displacement from the *as*-axis. Let us relate a rotating position to an adjacent position on the stator. From Fig. 2.2-7

$$\phi_s = \theta + \phi \tag{2.2-13}$$

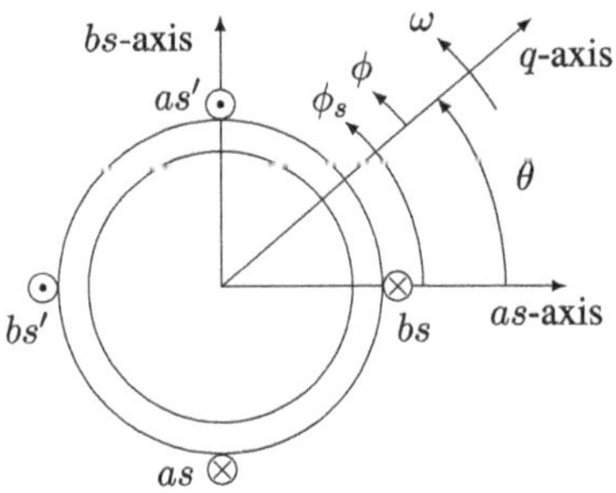


Figure 2.2-7 Elementary two-pole two-phase sinusoidally distributed stator windings with a third magnetic axis ($q$-axis).

If, for example, $\omega$ and $\theta(0)$ are zero, the $q$-axis is stationary and coincides with the $as$-axis; thus $\phi_s = \phi$. If instead, we select $\theta$ to be $\omega_e t$, where $\omega_e$ is the angular velocity (frequency) of the electrical variables, an adjacent stator position would be $\phi_s = \omega_e t + \phi_e$. We have added an $e$ subscript to $\phi$ in order to indicate that $\phi_e$ is a displacement from the $q$-axis that is now rotating at $\omega_e$.

If we substitute (2.2-13) into (2.2-9) we obtain

$$\text{mmf}_s = \frac{N_s}{2}\left[i_{as}\cos(\theta + \phi) + i_{bs}\sin(\theta + \phi)\right] \tag{2.2-14}$$

Substituting (2.2-10) and (2.2-11) into (2.2-14) for $i_{as}$ and $i_{bs}$, respectively and (2.2-12) for $\theta$ and after some work

$$\text{mmf}_s = \frac{N_s}{2}\sqrt{2}I_s \cos\left[(\omega_e - \omega)t + \theta_{esi}(0) - \theta(0) - \phi\right] \tag{2.2-15}$$

Recall that $\omega_e$ is the angular frequency of the electrical system and $\omega$ is the angular velocity of the $q$-axis, $\theta(0)$ is the time-zero position of the $q$-axis which will be considered to be zero unless otherwise specified, and $\phi$ is the displacement from the $q$-axis.

If we let the angular velocity of the $q$-axis be zero ($\omega = 0$) and $\theta(0) = 0$, then $\phi = \phi_s$ and we would be viewing Tesla's rotating magnetic field as a stationary observer. In this case, (2.2-15) becomes

$$\text{mmf}_s^s = \frac{N_s}{2}\sqrt{2}I_s \cos\left[\omega_e t + \theta_{esi}(0) - \phi_s\right] \tag{2.2-16}$$

where we have added a superscript $s$ to emphasize that we are observing Tesla's rotating magnetic field from a stationary frame of reference. Since both $\theta_{esi}(0)$ and $\phi_s$ are constants, $\text{mmf}_s^s$ is a sinusoidal variation of frequency $\omega_e$. If, for example, $\phi_s$ is zero we would be positioned (fixed) at the $as$-axis ($\phi_s = 0$ in Fig. 2.2-7) and the $\text{mmf}_s^s$ would be pulsating at $\omega_e$.

If we now let $\omega = \omega_e$, (2.2-15) becomes

$$\text{mmf}_s^e = \frac{N_s}{2}\sqrt{2}I_s \cos[\theta_{esi}(0) - \phi_e] \tag{2.2-17}$$

where $\phi_e$ is the displacement from the $q$-axis which is rotating at $\omega_e$. Since $\theta_{esi}(0)$ is constant for balanced steady-state conditions, (2.2-17) is a sinusoidal function of $\phi_e$ (the displacement from the $q$-axis) with its maximum value occurring at $\phi_e = \theta_{esi}(0)$ which is the phase angle of the balanced steady-state stator currents. In other words, we are observing Tesla's rotating magnetic field as we run at $\omega_e$ in the counterclockwise direction around the air gap. Since $\omega_e$ is the angular frequency of $I_{as}$ and $I_{bs}$ and the $q$-axis is also rotating at $\omega_e$ or in synchronism with the electrical system, it is referred to as the synchronously rotating frame of reference. Therein, since we have 360° vision, Tesla's rotating magnetic field appears to us as a sinusoidal function of $\phi_e$ with the maximum value at $\phi_e = \theta_{esi}(0)$ from the $q$-axis and stationary relative to us. This is depicted in Fig. 2.2-8 where $\theta_{esi}(0)$ is negative for inductive circuits. A positive $\text{mmf}_s^e$ is a south pole due to the stator currents ($S^s$) with the maximum intensity at $\phi_e = \theta_{esi}(0)$. A negative $\text{mmf}_s^e$ represents a north pole ($N^s$).

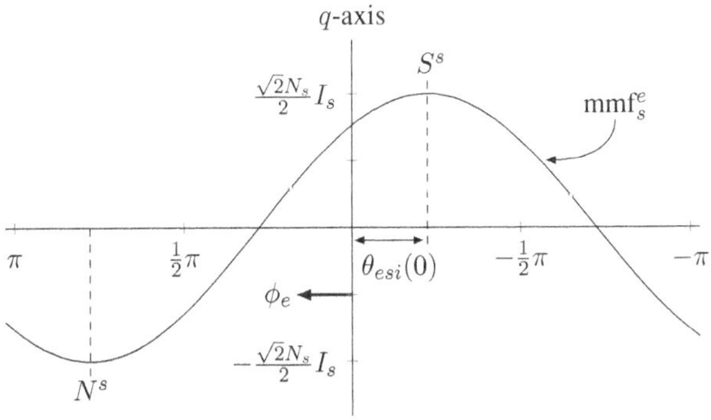


Figure 2.2-8 Tesla's rotating magnetic field ($\text{mmf}_s^e$) viewed from $-\pi$ to $\pi$ by an observer rotating counterclockwise about the air gap at $\omega_e$ with $\theta(0) = 0$ and $\theta_{esi}(0)$ negative.

We have just opened the door to a concept of viewing Tesla's rotating magnetic field from a frame of reference other than that of a stationary observer; in this case, a frame of reference rotating at $\omega_e$ in the counterclockwise direction. The importance of this cannot be over emphasized since we will find that this concept not only enables a clear visualization of the operation of electric machines but it provides the basis for modern analysis of electric machines. This is not the first time we have viewed a sinusoidal variation from a reference frame rotating at $\omega_e$. Recall in Section 1.2, Example 1A, and specifically in Fig. 1A-2, it was illustrated that a phasor rotates at $\omega_e$ in the counterclockwise direction and when we consider the phasor as a constant we are running counterclockwise at $\omega_e$. We will talk more of this in later sections; however, before proceeding there are a couple things that should be mentioned.

It is important not to confuse the magnetic axes of the stator windings (*as*- and *bs*- axes) which are stationary, with the phasors representing the stator currents ($\tilde{I}_{as}$ and $\tilde{I}_{bs}$). However, it is interesting to note the relative position of the *as*-axis and *bs*-axis versus the relative position of $\tilde{I}_{as}$ and $\tilde{I}_{bs}$ in a phasor diagram for a constant, counterclockwise $\text{mmf}_s$ to occur. The *bs*-axis is displaced $\frac{1}{2}\pi$ ahead of the *as*-axis, as illustrated in Fig. 2.2-6; however, from (2.2-10) and (2.2-11) we see that $\tilde{I}_{bs}$ *lags* $\tilde{I}_{as}$ by $\frac{1}{2}\pi$. The direction of rotation of the air-gap mmf ($\text{mmf}_s$) may readily be determined by forgetting about $\theta_{esi}(0)$ and locating the position of $\frac{N_s}{2}\sqrt{2}I_s$ when time has progressed to where $I_{bs}$ is maximum and $I_{as}$ is zero ($\omega_e t = \frac{1}{2}\pi$). That is, the maximum mmf$_s$ has rotated counterclockwise from the *as*-axis to the *bs*-axis. Note, that the negative of both (2.2-10) and (2.2-11) would also produce a counterclockwise mmf$_s$. In this case, $\tilde{I}_{as}$, would be at 180° and $\tilde{I}_{bs}$ at 90°. In other words, maximum $I_{as}$ followed $\frac{\pi}{2}$ electrical degrees by maximum $I_{bs}$ produces counterclockwise rotation of mmf$_s$ in the device shown in Fig. 2.2-6.

**Example 2B. Components of mmf$_s^e$.** Express mmf$_{as}^e$ and mmf$_{bs}^e$ with $\theta(0) = 0$. Add the resulting relationships to obtain mmf$_s^e$ which is (2.2-17). The currents may be expressed

$$I_{as} = \sqrt{2}I_s \cos[\omega_e t + \theta_{esi}(0)] \tag{2B-1}$$

$$I_{bs} = \sqrt{2}I_s \sin[\omega_e t + \theta_{esi}(0)] \tag{2B-2}$$

To view the mmf as an observer running at $\omega_e$ counterclockwise with the *q*-axis (Fig. 2.2-7)

$$\phi_s = \omega_e t + \phi_e \tag{2B-3}$$

where $\phi_e$ is the displacement from the *q*-axis when it is rotating at $\omega_e$. Thus, from (2.2-7)

$$\begin{aligned} \text{mmf}_{as}^e &= \frac{N_s}{2}\sqrt{2}I_s\{\cos[\omega_e t + \theta_{esi}(0)]\cos(\omega_e t + \phi_e)\} \\ &= \frac{N_s}{2}\sqrt{2}I_s\left(\frac{1}{2}\right)\{\cos[2\omega_e t + \theta_{esi}(0) + \phi_e] + \cos[\theta_{esi}(0) - \phi_e]\} \end{aligned} \tag{2B-4}$$

From (2.2-8) and (2B-2) we can express mmf$_{bs}^e$ as

$$\text{mmf}_{bs}^e = \frac{N_s}{2}\sqrt{2}I_s\left(\frac{1}{2}\right)\{\cos[\theta_{esi}(0) - \phi_e] - \cos[2\omega_e t + \theta_{esi}(0) + \phi_e]\} \tag{2B-5}$$

Now, from (2.2-9)

$$\begin{aligned} \text{mmf}_s^e &= \text{mmf}_{as}^e + \text{mmf}_{bs}^e \\ &= \frac{N_s}{2}\sqrt{2}I_s\cos[\theta_{esi}(0) - \phi_e] \end{aligned} \tag{2B-6}$$

The double frequency terms cancel and we obtain (2.2-17).

## Three-Phase Stator

The arrangements of the stator windings of a two-pole three-phase device are shown in Fig. 2.2-9. The windings are connected in wye and they are identical, sinusoidally distributed with $N_s$ equivalent turns and with their magnetic axes displaced by $\frac{2}{3}\pi$. The positive direction of the magnetic axes is selected so as to achieve counterclockwise rotation of the rotating air-gap mmf with balanced stator currents of the *abc*

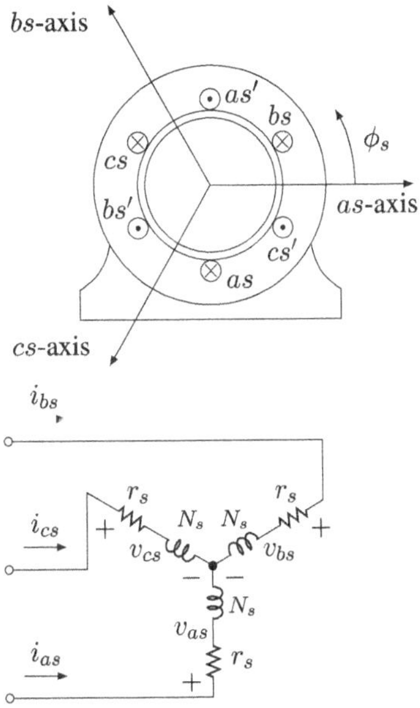


Figure 2.2-9 Elementary two-pole three-phase sinusoidally distributed stator windings.

sequence. We shall see this in just a moment. The air-gap mmfs established by the stator phases may be expressed by inspection of Fig. 2.2-9. In particular,

$$\text{mmf}_{as} = \frac{N_s}{2} i_{as} \cos \phi_s \tag{2.2-18}$$

$$\text{mmf}_{bs} = \frac{N_s}{2} i_{bs} \cos \left( \phi_s - \frac{2}{3}\pi \right) \tag{2.2-19}$$

$$\text{mmf}_{cs} = \frac{N_s}{2} i_{cs} \cos \left( \phi_s + \frac{2}{3}\pi \right) \tag{2.2-20}$$

As before, $N_s$ is the number of turns of the equivalent sinusoidally distributed stator windings and $\phi_s$ is the angular displacement about the stator measured from the *as*-axis. The total air-gap $\text{mmf}_s$ is

$$\text{mmf}_s = \frac{N_s}{2} \left[ i_{as} \cos \phi_s + i_{bs} \cos \left( \phi_s - \frac{2}{3}\pi \right) + i_{cs} \cos \left( \phi_s + \frac{2}{3}\pi \right) \right] \tag{2.2-21}$$

For balanced steady-state conditions, the stator currents for an *abc* sequence may be expressed as

$$I_{as} = \sqrt{2} I_s \cos[\omega_e t + \theta_{esi}(0)] \tag{2.2-22}$$

$$I_{bs} = \sqrt{2} I_s \cos \left[ \omega_e t - \frac{2}{3}\pi + \theta_{esi}(0) \right] \tag{2.2-23}$$

$$I_{cs} = \sqrt{2} I_s \cos \left[ \omega_e t + \frac{2}{3}\pi + \theta_{esi}(0) \right] \tag{2.2-24}$$

Substituting (2.2-22) through (2.2-24) into (2.2-21) and after some trigonometric manipulations, we can obtain an expression for the rotating air-gap mmf (Tesla's rotating field) established by balanced steady-state stator currents as viewed by a stationary observer

$$\text{mmf}_s^s = \frac{N_s}{2}\sqrt{2}I_s\frac{3}{2}\cos[\omega_e t + \theta_{esi}(0) - \phi_s] \tag{2.2-25}$$

If $\text{mmf}_s^s$ for the three-phase device given by (2.2-25) is compared with $\text{mmf}_s^s$ for a two-phase device given by (2.2-16), we see that they are identical except that the amplitude for the three-phase device is $\frac{3}{2}$ times that of a two-phase device. Actually, it can be shown that the amplitude for multiphase devices changes from that of a two-phase device in proportion to the number of phases divided by two. It has been pointed out that we will first analyze the two-phase machine since the trigonometry is less involved and the extension to the three-phase machine, once we have considered its "little sister," is straightforward. It follows that we need not repeat the work we did regarding introducing the $q$-axis and the derivation leading up to (2.2-15). The only difference would be that (2.2-15) would be multiplied by $\frac{3}{2}$. Similarly, Fig. 2.2-8 would apply to a three-phase stator if the amplitude of the field is multiplied by $\frac{3}{2}$.

It is important to note that, with the assumed positive directions of the magnetic axes, a counterclockwise rotating air-gap mmf is obtained with a three-phase set of balanced stator currents of the *abc* sequence. As in the two-phase case, it is also important to note the relative positions of the magnetic axes versus the relative positions of the phasors representing the currents in order to establish a constant amplitude counterclockwise rotating air-gap mmf. From (2.2-18) through (2.2-20) or Fig. 2.2-9 we see that the *bs*-axis is stationary at 120°, whereas the *cs*-axis is stationary at −120°. From (2.2-22) through (2.2-24), $\tilde{I}_{as}$, $\tilde{I}_{bs}$, and $\tilde{I}_{cs}$ are 120° out of phase; however, in order for the constant-amplitude air-gap mmf to rotate in the counterclockwise direction, $\tilde{I}_{bs}$ lags $\tilde{I}_{as}$ by 120° and $\tilde{I}_{cs}$ lags $\tilde{I}_{as}$ by 240°. In other words, maximum positive $I_{as}$ is followed by maximum $I_{bs}$ and then maximum $I_{cs}$ for an *abc* sequence. With the magnetic axis positioned as shown in Fig. 2.2-9, this sequence of maximum currents produces a counterclockwise rotation of the air-gap mmf.

## P-Pole Machines

Thus far, we have considered only two-pole electromechanical motion devices; however, electromechanical devices may have any even number of poles (2, 4, 6, 8, ...) up to more than 50 in the case of large hydroturbine generators. We may at first be a little hesitant to tackle the analysis of machines with more than two poles, since it appears that this may complicate matters before we even get started. Fortunately, this is not the case. We will find that with a simple change of variables we can analyze all machines as if they were two-pole devices. We need only to modify the expression for evaluating torque and show that, physically, the actual rotor speed of a machine with more than two poles will be a multiple less than that which we determine from our two-pole analysis.

The rotating air-gap mmf of a machine with more than two poles can be determined by considering the four-pole device shown in Fig. 2.2-10. In this figure, each phase winding consists of two series-connected windings, each of which is assumed to be sinusoidally distributed. For example, $as1'$ represents a group of conductors sinusoidally distributed over $0 < \phi_s < \frac{1}{2}\pi$, and so on. The phase windings consist of $N_s$ turns, with $N_s/\frac{P}{2}$ turns in each of the two series-connected windings. There may be some confusion in regard to notation. When considering the coils of a winding, notation $as_1, bs_1, \ldots$ was used. In Fig. 2.2-10, the notation $as1$, $bs1$, … is used to denote sinusoidally distributed windings. Thus, a subscripted number denotes a coil, whereas a number that is not a subscript denotes a sinusoidally distributed winding.

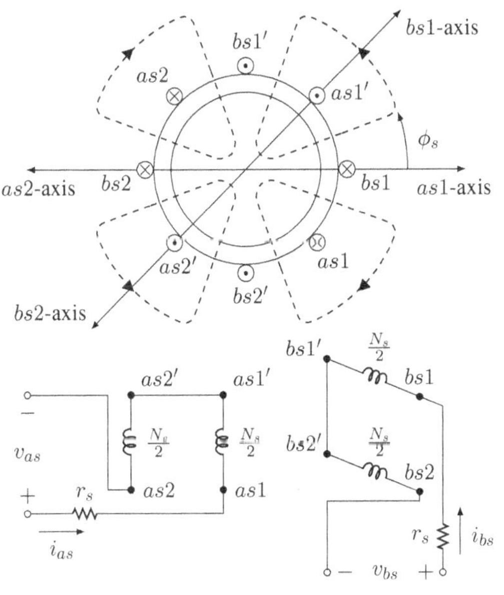


Figure 2.2-10 Stator winding arrangement of a four-pole two-phase symmetrical electromechanical device. Flux streamlines shown for $I_{as}$ at maximum positive value and $I_{bs}$ at zero.

Note that each of the two series windings per phase spans $\frac{1}{2}\pi$ radians and each phase winding establishes two magnetic systems. As shown in Fig. 2.2-10, at the instant when $I_{bs} = 0$ and $I_{as}$ is a positive maximum, the $as1 - as1'$ part of the $as$ winding produces positive flux in the $as1$-axis direction whereas the $as2 - as2'$ part of the $as$ winding produces positive flux in the $as2$-axis direction. South poles occur from $-\frac{1}{4}\pi < \phi_s < \frac{1}{4}\pi$ and $\frac{3}{4}\pi < \phi_s < \frac{5}{4}\pi$. Now, as shown in Fig 2.2-10 by the streamlines, half of the flux that enters the stator steel from $-\frac{1}{4}\pi < \phi_s < \frac{1}{4}\pi$ reenters the air gap between $\frac{5}{4}\pi < \phi_s < \frac{7}{4}\pi$ and the other half between $\frac{1}{4}\pi < \phi_s < \frac{3}{4}\pi$. The flux that enters the stator from $\frac{3}{4}\pi < \phi_s < \frac{5}{4}\pi$ divides similarly. Hence, for $I_{as} = \sqrt{2}I_s$ and $I_{bs} = 0$, two north poles occur; one from $\frac{1}{4}\pi < \phi_s < \frac{3}{4}\pi$ and from $\frac{5}{4}\pi < \phi_s < \frac{7}{4}\pi$.

The air-gap mmf established by each phase is a sinusoidal function of $2\phi_s$ for a four-pole machine, or, in general, $(P/2)\phi_s$, where $P$ is the number of poles. In particular,

$$\text{mmf}_{as} = \frac{2}{P}\frac{N_s}{2} i_{as} \cos \frac{P}{2}\phi_s \tag{2.2-26}$$

$$\text{mmf}_{bs} = \frac{2}{P}\frac{N_s}{2} i_{bs} \sin \frac{P}{2}\phi_s \tag{2.2-27}$$

where $N_s$ is the total equivalent turns per stator phase. For balanced steady-state operation, the stator currents may be expressed as

$$I_{as} = \sqrt{2}I_s \cos[\omega_e t + \theta_{esi}(0)] \tag{2.2-28}$$

$$I_{bs} = \sqrt{2}I_s \sin[\omega_e t + \theta_{esi}(0)] \tag{2.2-29}$$

Equations (2.2-28) and (2.2-29) are the same expressions for the stator currents as given for the two-pole case. Also, (2.2-16) would become

$$\text{mmf}_s^s = \left(\frac{2}{P}\right)\frac{N_s}{2}\sqrt{2}I_s \cos\left[\omega_e t + \theta_{esi}(0) - \frac{P}{2}\phi_s\right] \tag{2.2-30}$$

With the stator arranged as in Fig. 2.2-10, balanced steady-state stator currents of frequency $\omega_e$ produce a four-pole ($P$-pole) magnetic system that rotates about the air gap at $(2/4)\omega_e$ or $(2/P)\omega_e$ relative to a stationary observer. How do we know that? Well in order for (2.2-30) to be a constant, the argument of the cosine must be a constant and this occurs when $\frac{P}{2}\phi_s$ is equal to $\omega_e t$. In other words, when $\phi_s = \frac{2}{P}\omega_e t$. Therefore, the speed of the rotating mmf (rotating poles) is now $(2/P)\omega_e$ which is the synchronous speed of the machine; however, the stator variables are unaware of this. To the electric system, $\omega_e$ is synchronous speed; however, the magnetic system of the rotor must "lock in" with the magnetic system established by the stator.

A four-pole stator winding and a four-pole rotor winding are shown in Fig. 2.2-11. The mutual inductance between the *as*- and *ar*-windings may be expressed

$$L_{asar} = L_{sr} \cos 2\theta_{rm} \tag{2.2-31}$$

where $\theta_{rm}$ is the actual rotor displacement. In the two-pole case (Fig. 1.5-1) we used $\theta_r$. We are going to give a new meaning to $\theta_r$ in just a minute. Let us first generalize the argument of the cosine in (2.2-31). Thus,

$$L_{asar} = L_{sr} \cos \frac{P}{2}\theta_{rm} \tag{2.2-32}$$

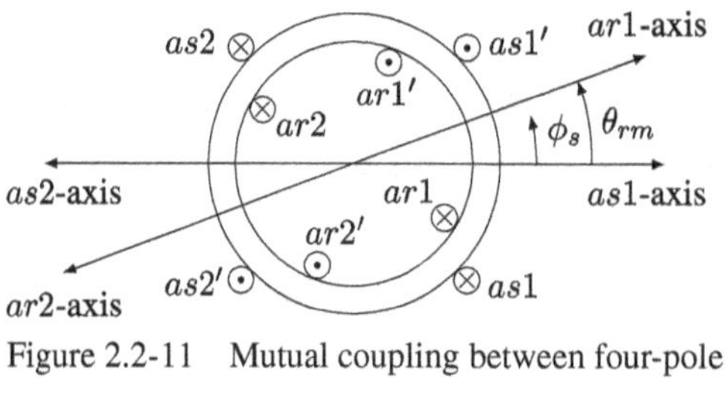


Figure 2.2-11 Mutual coupling between four-pole stator and rotor windings.

If we let

$$\frac{P}{2}\theta_{rm} = \theta_r \tag{2.2-33}$$

then (2.2-32) becomes what we had for the two-pole case. Now, $\theta_{rm}$ is the actual rotor displacement and $\theta_r$ is what is referred to as the electrical angular displacement of the rotor. What in the world does that mean? Well, let us be stator variables; we do not know or care how many pole pairs the machine has, so we will assume that both the stator and rotor have two poles. Thus, the electrical system is rotating at $\omega_e$ and it thinks the rotor speed is $\omega_r$ since we assumed it to be a two-pole stator and rotor. Therefore, $\omega_r$ is now the rotor speed "referred" to the angular velocity of the electrical system and (2.2-33) applies regardless of how many even-number poles are on the stator and rotor.

Although we see that if we substitute (2.2-33) into (2.2-32), the result is the mutual inductance for a two-pole system as given by (1.5-9); however, what about the mutual inductance between stator windings as shown in Fig. 2.2-10? In the case of the two-pole machine, $L_{asbs} = 0$, since the magnetic axes are orthogonal. Note, however, that in the four-pole machine in Fig. 2.2-10 the *bs*-winding is displaced 45° counterclockwise from the *as*-winding. If we assume, for a minute, that in Fig. 2.2-10, we rotated the *bs* winding clockwise 45°, it would be atop the *as*-winding and $L_{asbs}$ would be similar to (2.2-32) except $L_{asar}$ would be replaced by $L_{asbs}$, $L_{sr}$ would be replaced by $L_{ms}$ and $\theta_{rm}$ by the angle between *as*- and *bs*- windings. If now we rotate the *bs*-winding counterclockwise by an angle of 45° then $L_{asbs}$ would be zero. From here on out, we will treat all electric machines as two-pole devices, knowing full well that (2.2-33) is the relationship between $\theta_r$ and $\theta_{rm}$ and

$$\frac{P}{2}\omega_{rm} = \omega_r \tag{2.2-34}$$

where $\omega_r$ is referred to as the electrical angular velocity of the rotor.

Now, since we are going to treat all machines as two-pole devices and use $\theta_r$ in all the voltage equations we have to do one last thing in order to make things work out. Back in Section 1.6 we expressed the change in energy entering the coupling field from the mechanical system as

$$dW_m = -T_e d\theta_r \tag{2.2-35}$$

and the torque $T_e$ as

$$T_e(\mathbf{i}, \theta_r) = \frac{\partial W_c(\mathbf{i}, \theta_r)}{\partial \theta_r} \tag{2.2-36}$$

Equation (2.2-35) must now be written in terms of $\theta_{rm}$

$$dW_m = -T_e d\theta_{rm} \tag{2.2-37}$$

and since $\theta_{rm} = \frac{2}{P}\theta_r$, (2.2-36) becomes

$$T_e(\mathbf{i}, \theta_r) = \frac{P}{2}\frac{\partial W_c(\mathbf{i}, \theta_r)}{\partial \theta_r} \tag{2.2-38}$$

Analyzing a $P$-pole machine has turned out to be not as troublesome as we might have first thought. We are fortunate.

## Machine Inductances

Although we have introduced the sinusoidal distributed winding, we have not derived the mutual and self-inductances of sinusoidal distributed windings. In Chapter 1, we considered only concentrated windings and assumed that the mutual inductance between windings in relative motion varied sinusoidally. This, of course, was a first-order approximation; however, now that we have sinusoidally distributed windings, we will accept sinusoidal variations in the mutual inductances without proof. In Appendix B, we show this to be the case. The interested reader may want to take a few minutes to jump ahead and go through the derivation given there.

**Example 2C. Torque calculation for four-pole machine.** Consider the four-pole device shown in Fig. 2.2-11. $L_{sr} = 0.1$ H, $N_s = 2N_r$, $L_{ls} = 2L_{lr}$ where $L_{ls} = 0.1L_{ms}$, $i_{as} = 2i_{ar} = 2$ A. Determine (a) the coenergy $W_c$ and (b) the torque $T_e$.

From Section 1.5 with $\mathcal{R}_g$ the total reluctance of the two air gaps,

$$L_{sr} = \frac{N_s N_r}{\mathcal{R}_g} = 0.1 \text{ H} \tag{2C-1}$$

$$L_{ms} = \frac{N_s^2}{\mathcal{R}_g} \tag{2C-2}$$

$$L_{mr} = \frac{N_r^2}{\mathcal{R}_g} \tag{2C-3}$$

Therefore,

$$L_{ms} = \frac{N_s}{N_r} L_{sr} = 2L_{sr} = 0.2 \text{ H} \tag{2C-4}$$

and

$$L_{mr} = L_{ms}\left(\frac{N_r}{N_s}\right)^2 = L_{ms}\left(\frac{1}{2}\right)^2 = 0.05 \text{ H} \tag{2C-5}$$

Also

$$L_{ls} = 0.1L_{ms} = 0.02 \text{ H} \tag{2C-6}$$

$$L_{lr} = 0.5L_{ls} = 0.01 \text{ H} \tag{2C-7}$$

Now,

$$W_c = \frac{1}{2}L_{ss}i_{as}^2 + L_{asar}i_{as}i_{ar} + \frac{1}{2}L_{rr}i_{ar}^2 \tag{2C-8}$$

where

$$L_{ss} = L_{ls} + L_{ms} = 0.02 + 0.2 = 0.22 \text{ H} \tag{2C-9}$$

$$L_{asar} = 0.1 \cos \theta_r \tag{2C-10}$$

$$L_{rr} = L_{lr} + L_{mr} = 0.01 + 0.05 = 0.06 \text{ H} \tag{2C-11}$$

$$\begin{aligned} W_c &= \left(\frac{1}{2}\right)(0.22)(2)^2 + (0.1)(2)(1)\cos\theta_r + \left(\frac{1}{2}\right)(0.06)(1)^2 \\ &= 0.44 + 0.2\cos\theta_r + 0.03 = 0.47 + 0.2\cos\theta_r \text{ J} \end{aligned} \tag{2C-12}$$

The torque may be calculated using (2.2-38), that is

$$\begin{aligned} T_e(\mathbf{i}, \theta_r) &= \frac{P}{2}\frac{\partial W_c(\mathbf{i}, \theta_r)}{\partial \theta_r} \\ &= \left(\frac{4}{2}\right)\frac{\partial(0.2\cos\theta_r)}{\partial\theta_r} \\ &= -0.4\sin\theta_r \text{ N}\cdot\text{m} \end{aligned} \tag{2C-13}$$

---

**SP2.2-1.** Determine the direction of rotation of mmf$_s$ for the winding arrangement shown in Fig. 2.2-6 for (a) $\tilde{I}_{as} = j\tilde{I}_{bs}$, (b) $\tilde{I}_{as} = -j\tilde{I}_{bs}$, (c) $\tilde{I}_{bs} = j\tilde{I}_{as}$, (d) $\tilde{I}_{bs} = -j\tilde{I}_{as}$. Equations not needed. [(a) ccw, (b) cw, (c) cw, (d) ccw]

**SP2.2-2.** Determine the direction of rotation of mmf$_s$ for the winding arrangement shown in Fig. 2.2-9 for (a) an *abc* sequence, (b) an *acb* sequence. Equations not needed. [(a) ccw, (b) cw]

**SP2.2-3.** Repeat Example 2A with the ⊗ and ⊙ reversed. [mmf plot inverted and $N^s$ and $S^s$ reversed]

**SP2.2-4.** Consider the four-pole device shown in Fig. 2.2-10. $I_{as} = \cos\omega_e t$ and $I_{bs} = \sin\omega_e t$. At $t = 0$ you are at $\phi_s = \frac{\pi}{4}$. You start running at $\omega_e/2$ ccw at $t > 0$. (a) Determine the value of mmf$_s$ you will see. (b) Repeat for starting at $\phi_s = 0$. $\left[(a)\text{ zero, }(b)\ \frac{N_s}{2\sqrt{2}}I_s\right]$

**SP2.2-5.** A device has ten stator poles. How many poles must the rotor have? Why? [10, each stator pole pair must have a rotor pole pair with which to interact]

**SP2.2-6.** Consider Fig. 2.2-10. If $N_s$ is the total number of turns per phase and $\mathcal{R}_g$ the reluctance of the total air gap, show that $L_{ms} = 2N_s^2/\mathcal{R}_g$.

## 2.3 TWO-PHASE PERMANENT-MAGNET AC MACHINE

Before starting the analysis of the permanent-magnet ac machine, it is appropriate to take a few minutes to justify our choice of machines to be considered. Rotational electromechanical devices fall into three general categories: direct-current, synchronous, and induction. As discussed in Section 1.7, we will not treat the dc machine other than to point out that the magnetic axes of the dc machine are stationary and orthogonal thus producing the maximum torque per ampere. Although the magnetic systems are rotating in ac machines, the goal of some modern controls is to emulate the orthogonal positioning of the rotating magnetic axes.

Synchronous machines are so called because they develop a constant torque only when the rotor is rotating in synchronism (at synchronous speed) with the rotating air-gap mmf (rotating magnet field) that is established by balanced currents flowing in the stator windings. Examples are the reluctance machine, permanent-magnet ac machine, and the ac generator. Even though these machines are of the synchronous type, the machine commonly referred to as the synchronous machine or ac generator is the device used to generate electric power. Large electric power generators are synchronous generators and are called so. We will talk more about the synchronous machine (generator) in Chapter 5 on power systems.

The induction machine, which was invented by Tesla, is a means of converting energy from electrical to mechanical motion. It does not develop torque at synchronous speed. The windings on the rotor are short-circuited and, in order for current to be induced in these windings and thereby to produce torque by interacting with the air-gap mmf established by the stator windings, the rotor must rotate at a speed other than synchronous speed. Unlike the synchronous machine, the induction machine family does not include a large number of relatives.

Since it is not practical to consider all machines in the synchronous and induction categories, it is our goal to treat the machines that are characteristic of the drives and power systems areas. In the case of power systems, the choice must be the synchronous generator. In the drives area the choice is between the induction machine and the permanent-magnet ac machine. In this text, the permanent-magnet ac machine is selected, since in many low-power drive applications, it has emerged as the electric machine of choice. Moreover, it provides an excellent example of power electronic control when operated as a brushless dc motor; moreover, the analysis of the permanent-magnet ac machine is similar in many respects to that of the synchronous generator.

A two-pole two-phase permanent-magnet ac machine is shown in Fig. 2.3-1. The stator windings are identical, sinusoidally distributed windings each with $N_s$ equivalent turns and resistance, $r_s$. The magnetic axes of the stator windings are denoted as the *as*- and the *bs*-axes. The *d*-axis (direct axis) is fixed at the center of the magnetic flux that enters the air gap from the permanent-magnet rotor and the *q*-axis (quadrature axis) is displaced $\frac{1}{2}\pi$ counterclockwise from the *d*-axis. This is the same *q*-axis we introduced in Section 2.2. The angular displacement about the stator is denoted $\phi_s$, referenced to the *as*-axis. The angular displacement about the rotor is $\phi_r$, referenced to the *q*-axis. The angular velocity of the rotor is $\omega_r$ and $\theta_r$ is the angular displacement of the rotor measured from the *as*-axis to the *q*-axis. Thus, a given point on the rotor surface at the angular position $\phi_r$ may be related to an adjacent point on the inside stator surface with angular position $\phi_s$ as

$$\phi_s = \theta_r + \phi_r \tag{2.3-1}$$

which is (2.2-13) with $\theta$ and $\phi$ set equal $\theta_r$ and $\phi_r$, respectively.

The electromechanical torque $T_e$ and the load torque $T_L$ are also indicated in Fig. 2.3-1. The torque $T_e$ is assumed positive in the direction of increasing $\theta_r$; $T_L$ is

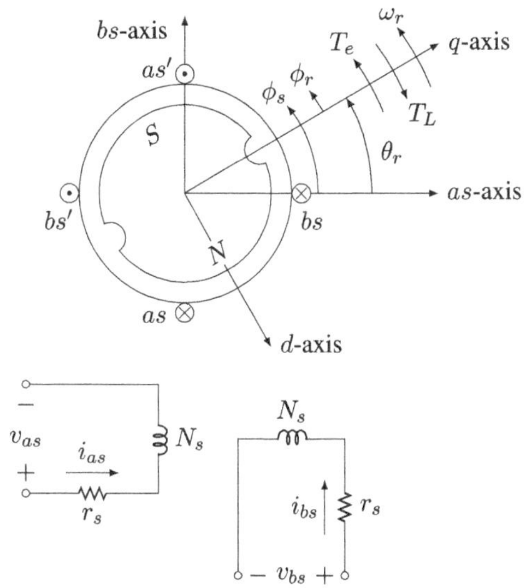


Figure 2.3-1 Two-pole two-phase permanent-magnet ac machine.

positive in the opposite direction. Electromagnetic torque is produced by the interaction of the poles of the permanent-magnet rotor and the poles resulting from the rotating air-gap mmf established by currents flowing in the stator windings. The rotating mmf ($\text{mmf}_s$) established by symmetrical two-phase stator windings carrying balanced two-phase currents is given by (2.2-16).

A four-pole, three-phase 28-V, $\frac{1}{3}$–hp permanent-magnet ac machine is shown in Fig. 2.3-2. The disassembled motor is shown in Fig. 2.3-2a, wherein the stator windings are visible. The opposite end of the stator housing is shown in Fig. 2.3-2b. Housed therein are Hall-effect sensors, which are used to determine the position of the permanent magnet, and the drive inverter, filter capacitor, and logic circuitry. We will talk more about the sensors and inverter in Chapter 4

The voltage equations for the two-pole two-phase permanent-magnet ac machine shown in Fig. 2.3-1 may be expressed as

$$v_{as} = r_s i_{as} + \frac{d\lambda_{as}}{dt} \tag{2.3-2}$$

$$v_{bs} = r_s i_{bs} + \frac{d\lambda_{bs}}{dt} \tag{2.3-3}$$

In matrix form,

$$\mathbf{v}_{abs} = \mathbf{r}_s \mathbf{i}_{abs} + p\boldsymbol{\lambda}_{abs} \tag{2.3-4}$$

where $p$ is the operator $d/dt$, and for voltages, currents, and flux linkages

$$(\mathbf{f}_{abs})^T = [f_{as} \;\; f_{bs}] \tag{2.3-5}$$

(a)

(b)

Figure 2.3-2 Four-pole three-phase 28-V $\frac{1}{3}$–hp permanent-magnet ac machine: (a) disassembled motor; (b) stator housing showing Hall-effect sensors. Reproduced with permission of McGraw-Hill.

with

$$\mathbf{r}_s = \begin{bmatrix} r_s & 0 \\ 0 & r_s \end{bmatrix} \tag{2.3-6}$$

The flux linkage equations may be expressed as

$$\lambda_{as} = L_{asas} i_{as} + L_{asbs} i_{bs} + \lambda_{asm} \tag{2.3-7}$$

$$\lambda_{bs} = L_{bsas} i_{as} + L_{bsbs} i_{bs} + \lambda_{bsm} \tag{2.3-8}$$

In matrix form,

$$\boldsymbol{\lambda}_{abs} = \mathbf{L}_s \mathbf{i}_{abs} + \boldsymbol{\lambda}'_m \tag{2.3-9}$$

where $\boldsymbol{\lambda}'_m$ is the column vector

$$\boldsymbol{\lambda}'_m = \begin{bmatrix} \lambda_{asm} \\ \lambda_{bsm} \end{bmatrix} = \lambda'_m \begin{bmatrix} \sin\theta_r \\ -\cos\theta_r \end{bmatrix} \tag{2.3-10}$$

In (2.3-10), $\lambda'_m$ is the amplitude of the flux linkages established by the permanent magnet as viewed from the stator phase windings. In other words, the magnitude of $\lambda'_m$ is proportional to the magnitude of the open-circuit sinusoidal voltage induced in each stator phase winding due to the rotation of the rotor. The rotor displacement $\theta_r$ is expressed as

$$\frac{d\theta_r}{dt} = \omega_r \tag{2.3-11}$$

We will assume that the reluctance of the rotor of the permanent-magnet ac machine is the same in the $q$- and $d$-axes. Actually, this assumption may be an over-simplification in some cases; however, it markedly reduces our work and allows us to establish the operating principles of the permanent-magnet ac machine without significant error.

With the assumption of a uniform air gap, the mutual inductance between the $as$- and $bs$- windings is zero. Since the windings are identical, the self-inductances $L_{asas}$ and $L_{bsbs}$ are equal and denoted as $L_{ss}$. As in the case of a transformer, the self-inductance is made up of a leakage $L_{ls}$ and a magnetizing inductance $L_{ms}$. Thus,

$$L_{ss} = L_{ls} + L_{ms} \tag{2.3-12}$$

The machine is designed to minimize the leakage inductance, which generally makes up approximately 10 percent of $L_{ss}$. Since the air gap is assumed to be uniform, the magnetizing inductances of the $as$- and $bs$-windings are constant and equal; therefore, the self-inductance matrix $\mathbf{L}_s$ may be expressed as

$$\mathbf{L}_s = \begin{bmatrix} L_{ss} & 0 \\ 0 & L_{ss} \end{bmatrix} \tag{2.3-13}$$

We can derive an expression for torque $T_e$ by first expressing the coenergy for a magnetically linear system.

$$\begin{aligned} W_c &= W_f \\ &= \frac{1}{2}L_{asas}i_{as}^2 + \frac{1}{2}L_{bsbs}i_{bs}^2 + W_{pm} + \lambda_{asm}i_{as} + \lambda_{bsm}i_{bs} \end{aligned} \tag{2.3-14}$$

where $W_{pm}$ is the energy associated with the permanent magnet which is assumed to be constant and $\lambda_{asm}$ and $\lambda_{bsm}$ are given by (2.3-10). These flux linkages may be thought of as a mutual inductance times a current corresponding to the strength of the permanent magnet. Taking the partial derivative with respect to $\theta_r$ and substituting the result into (2.2-38) yields

$$\begin{aligned} T_e &= \frac{P}{2}\frac{\partial W_c}{\partial \theta_r} \\ &= \frac{P}{2}\lambda'_m(i_{as}\cos\theta_r + i_{bs}\sin\theta_r) \end{aligned} \tag{2.3-15}$$

The above expression is positive for motor action. The torque and rotor speed may be related as

$$T_e = J\left(\frac{2}{P}\right)\frac{d\omega_r}{dt} + B_m\left(\frac{2}{P}\right)\omega_r + T_L \tag{2.3-16}$$

where $J$ is in kg · m$^2$; it is the inertia of the rotor and rigidly connected mechanical load. Since we will be concerned primarily with motor action, the load torque $T_L$ is assumed positive, as indicated in Fig. 2.3-1. The constant $B_m$ is a damping coefficient associated with the rotational system of the machine and mechanical load. It has the units N · m · s/rad and it is generally small and often neglected in the case of the machine but may be considerable for the mechanical load.

**Example 2D. Measuring $\lambda'_m$ of a permanent-magnet machine.** The parameters of a four-pole two-phase permanent-magnet ac machine are $r_s = 3.4\,\Omega$, $L_{ls} = 1.1$ mH, and $L_{ms} = 11$ mH. When the device is driven at 1000 r/min, the open-circuit phase voltage is sinusoidal with a peak-to-peak value of 34.6 V. Determine $\lambda'_m$. The actual rotor speed at which the measurement was taken is

$$\omega_{rm} = \frac{(r/\min)(\text{rad}/r)}{s/\min}$$

$$= \frac{(1000)(2\pi)}{60} = \frac{100}{3}\pi \text{ rad/s} \tag{2D-1}$$

From (2.2-34), the electrical angular velocity is

$$\omega_r = \frac{P}{2}\omega_{rm}$$

$$= \frac{4}{2}\frac{100\pi}{3} = \frac{200}{3}\pi \text{ rad/s} \tag{2D-2}$$

With the phases open-circuited, $i_{as} = i_{bs} = 0$. Thus, from (2.3-2) and (2.3-10),

$$v_{as} = \frac{d(\lambda'_m \sin\theta_r)}{dt} = \lambda'_m \omega_r \cos\theta_r \tag{2D-3}$$

Now the peak-to-peak voltage is 34.6 V; hence, from (2D-3), with the peak-to-peak voltage divided by 2, we have

$$\frac{34.6}{2} = \lambda'_m\left(\frac{200}{3}\pi\right) \tag{2D-4}$$

Solving for $\lambda'_m$ yields

$$\lambda'_m = \frac{(34.6)(3)}{(2)(200\pi)} = 0.0826 \text{ V} \cdot \text{s/rad} \tag{2D-5}$$

---

**SP2.3-1.** The stator windings of the permanent-magnet ac machine shown in Fig. 2.3-1 are open-circuited. The rotor is driven clockwise and $V_{as} = -10\sin 100t$. Determine $V_{bs}$. [$V_{bs} = -10\cos 100t$]

**SP2.3-2.** Determine $\lambda'_m$ for SP2.3-1. [$\lambda'_m = 0.1$ V · s/rad]

## 2.4 ANALYSIS OF A TWO-PHASE PERMANENT-MAGNET AC MACHINE

We are unaware of the thought process that led Park to the change of variables that eliminated the time varying coefficients in the flux linkage equations [2]. It is instructive, however, to build on work we have done in an attempt to give some insight into this change of variables rather than setting down the transformation as if it suddenly appeared from out of the blue. In Section 2.2, we introduced Tesla's rotating magnetic field. In this section, we will show how the change of variables and Tesla rotating magnetic field are related and how torque is produced. If, however, the focus is the analysis of the steady-state performance, a study procedure is outlined following (2.4-10).

Repeating (2.2-9) for convenience

$$\text{mmf}_s = \frac{N_s}{2}(i_{as}\cos\phi_s + i_{bs}\sin\phi_s) \tag{2.4-1}$$

where $\phi_s$ is the displacement about the stator as shown in Fig. 2.3-1 and defined by (2.3-1) as

$$\phi_s = \theta_r + \phi_r \tag{2.4-2}$$

In Fig. 2.3-1, the direct-axis ($d$-axis) is defined as the magnetic axis of the permanent magnet and the quadrature-axis ($q$-axis) is displaced $\frac{1}{2}\pi$ counterclockwise from the $d$-axis.

If we select $\phi_r = 0$, then from (2.4-2) $\phi_s = \theta_r$ and (2.4-1) becomes the value of $\text{mmf}_s$ at $\theta_r$, or the $\text{mmf}_s$ that rotates in unison with the $q$-axis

$$\text{mmf}_s|_{\phi_r=0} = \frac{N_s}{2}(i_{as}\cos\theta_r + i_{bs}\sin\theta_r) \tag{2.4-3}$$

Recall from (2.2-15) that during balanced steady-state operation, $\omega_e$ is equal to $\omega_r$ and (2.4-3) is a constant; the value depends on the time zero positions of $\theta_r$ and the currents. The units of (2.4-3) are ampere-turns. Please note that (2.4-3) is valid for transient and steady-state operation.

The mmf or current is directed along the $q$-axis at $\phi_r = 0$; we will call the current part of (2.4-3) $i_{qs}^r$; that is

$$i_{qs}^r = i_{as}\cos\theta_r + i_{bs}\sin\theta_r \tag{2.4-4}$$

Since (2.4-3) is constant during balanced steady-state conditions, so must $i_{qs}^r$. The subscript $q$ indicates that the current acts in the $q$-axis, the $s$ indicates that it is due to stator currents, and the raised $r$ tells us it is flowing in a winding that is rotating at $\omega_r$. It has the same connotation as $i_{as}(i_{bs})$ flowing in stationary windings and acting along the $as$-axis ($bs$-axis); except during balanced steady-state conditions $i_{as}$ and $i_{bs}$ are sinusoidal, whereas $i_{qs}^r$ is constant.

Since we are replacing two currents, $i_{as}$ and $i_{bs}$, we need a second substitute variable. To do this we will again use the expression for Tesla's rotating magnetic field given by (2.4-1). We can obtain a value of $\text{mmf}_s$ rotating with the $d$-axis shown in Fig. 2.3-1 by evaluating mmf$_s$ (2.4-1) at $\phi_r = -\frac{\pi}{2}$. From (2.4-2), with $\phi_r = -\frac{\pi}{2}$, $\phi_s = \theta_r - \frac{\pi}{2}$, whereupon (2.4-1) becomes

$$\text{mmf}_s|_{(\phi_r=-\frac{\pi}{2})} = \frac{N_s}{2}(i_{as}\sin\theta_r - i_{bs}\cos\theta_r) \tag{2.4-5}$$

thus, the current in the $d$-axis windings which must also be constant during balanced, steady-state conditions, is

$$i_{ds}^r = i_{as}\sin\theta_r - i_{bs}\cos\theta_r \tag{2.4-6}$$

From (2.4-4) and (2.4-6) we can express the substitute currents in matrix form as

$$\begin{bmatrix} i_{qs}^r \\ i_{ds}^r \end{bmatrix} = \begin{bmatrix} \cos\theta_r & \sin\theta_r \\ \sin\theta_r & -\cos\theta_r \end{bmatrix} \begin{bmatrix} i_{as} \\ i_{bs} \end{bmatrix} \tag{2.4-7}$$

What do (2.4-4) and (2.4-6) tell us? First, these two equations define $i_{qs}^r$ and $i_{ds}^r$, respectively, and thus the change of variables. Moreover, each equation establishes two points (one positive and one negative) in each axis ($q$- and $d$-axes) of the rotating magnetic field and viewed as a space sinusoid by an observer with 360° vision, traveling with the rotor ($\omega_r = \omega_e$). This, of course, is similar to Tesla's rotating magnetic field shown in Fig. 2.2-8. This space sinusoid is completely defined by these four points for sinusoidally distributed $as$- and $bs$-windings. It is a function of $\phi_r$ with the mmf in the $q$-axis equal to + or $-\frac{N_s}{2}i_{qs}^r$ and + or $-\frac{N_s}{2}i_{ds}^r$ in the $d$-axis for transient or steady-state operation.

## Transformation

Now, since the $qs^r$ and $ds^r$ circuits/currents must have associated voltages and flux linkages the change of variables given by (2.4-7) will also apply to voltages and flux linkages; that is, $v_{as}$ and $v_{bs}$ transform to $v_{qs}^r$ and $v_{ds}^r$ and so on. Thus, for any electrical quantity $f$ (voltage, current, or flux-linkage), the transformation between substitute and physical variables can be written in matrix form as

$$\begin{bmatrix} f_{qs}^r \\ f_{ds}^r \end{bmatrix} = \begin{bmatrix} \cos\theta_r & \sin\theta_r \\ \sin\theta_r & -\cos\theta_r \end{bmatrix} \begin{bmatrix} f_{as} \\ f_{bs} \end{bmatrix} \tag{2.4-8}$$

or

$$\mathbf{f}_{qds}^r = \mathbf{K}_s^r \mathbf{f}_{abs} \tag{2.4-9}$$

where $f$ can represent either voltage, current, or flux linkage and $\theta_r$ is defined by (2.3-11). It follows that

$$\mathbf{f}_{abs} = (\mathbf{K}_s^r)^{-1}\mathbf{f}_{qds}^r \tag{2.4-10}$$

where, for (2.4-8), $(\mathbf{K}_s^r)^{-1} = \mathbf{K}_s^r$. In the case of $\mathbf{f}_{abs}$, the $s$ subscript denotes stator variables; however, in the case of $\mathbf{f}_{qds}^r$ the $s$ subscript denotes substitute variables related to the physical stator variables. The $r$ superscript indicates that the transformation is to a frame of reference that rotates at the electrical angular velocity of the rotor $\omega_r$.

If the focus is steady-state operation, most of the material from here to (2.4-47) and (2.4-48) can be de-emphasized or skipped; however, it is necessary to consider some of the interim material. In particular, the expression for torque which is valid for transient and steady-state calculations is given by (2.4-26). Also, of importance are the expressions from (2.4-41) through (2.4-46). Although these expressions are written in lower case, they are valid for steady-state operation where in (2.4-46) $\theta_{esv} = \theta_{esv}(0)$ and $\theta_r = \theta_r(0)$ which is zero, thus $\phi_v = \theta_{esv}(0)$ which is the phase angle of $\tilde{V}_{as}$ relative to the *as*-axis. For steady-state calculations, (2.4-49) for torque and (2.4-61) for $\tilde{V}_{as}$ along with (2.4-60) for $\tilde{E}_a$ are all key. Also, Example 2E should be studied in detail.

Before incorporating the change of variables in the analysis of a permanent-magnet ac machine, let us take a little closer look at the transformation and what we might expect from it. From our work in this section, one might conclude that the transformation (change of variables) is valid only if the windings are sinusoidally distributed in space and if the voltages, currents, and flux linkages are sinusoidal time varying quantities. This is not the case; the transformation given by (2.4-8) is nothing more than a change of variables and can be implemented regardless of the form of the variables, the winding distribution, or even if a machine is not involved. It is, however, especially useful in the analysis of machines since we will find that, in the case of sinusoidally distributed windings, the time varying coefficients which are due to position dependent mutual inductances, are not present in the equations that describe machine operation. We are not too surprised by this since during balanced steady-state conditions, we found that $i_{qs}^r$ and $i_{ds}^r$ are constants and it seems reasonable that the associated voltages, $v_{qs}^r$ and $v_{ds}^r$, as well as the flux linkages, $\lambda_{qs}^r$ and $\lambda_{ds}^r$, would also be constants during balanced steady-state operation. Tesla's rotating magnetic field given by (2.4-1) along with (2.4-2) led us to this transformation.

Some prefer to portray the first row of (2.4-8) as depicted in Fig. 2.4-1 where the stator currents are projected onto the $q$-axis. The second row of (2.4-8) is likewise portrayed as shown in Fig. 2.4-2. Perhaps Figs. 2.4-1 and 2.4-2 are not necessary since we already know from our discussion surrounding (2.4-4) for $i_{qs}^r$ and (2.4-6) for $i_{ds}^r$ that they are constant during balanced, steady-state operation. Unfortunately, Figs. 2.4-1 and 2.4-2 do not help us to visualize this; nevertheless, these figures are given to satisfy convention.

It is time to derive the $qs^r$- and $ds^r$-equations for a permanent-magnet ac machine. In particular, we will incorporate the transformation, given by (2.4-8), into

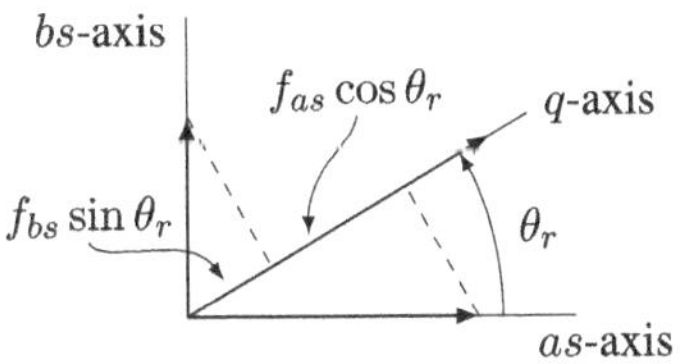


Figure 2.4-1 Stator phase currents referred to the $q$-axis portraying the first row of (2.4-8).

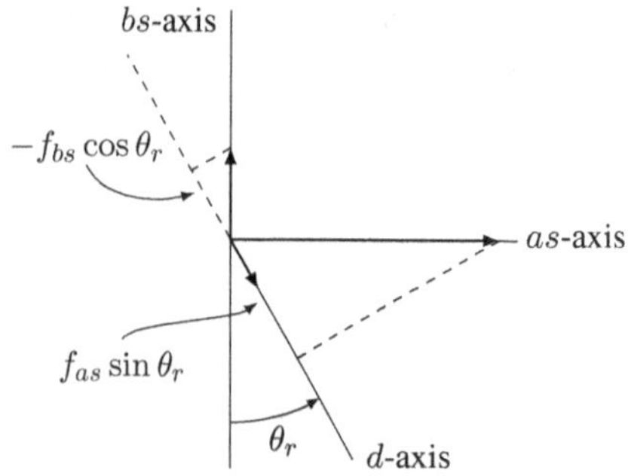


Figure 2.4-2 Stator phase currents referred to the $d$-axis portraying the second row of (2.4-8).

the analysis of the two-phase permanent-magnet ac machine by substituting (2.4-10) into (2.3-4). Thus,

$$(\mathbf{K}_s^r)^{-1}\mathbf{v}_{qds}^r = \mathbf{r}_s(\mathbf{K}_s^r)^{-1}\mathbf{i}_{qds}^r + p[(\mathbf{K}_s^r)^{-1}\boldsymbol{\lambda}_{qds}^r)] \tag{2.4-11}$$

Multiplying (2.4-11) by $\mathbf{K}_s^r$ yields

$$\mathbf{v}_{qds}^r = \mathbf{r}_s\mathbf{i}_{qds}^r + \omega_r\boldsymbol{\lambda}_{dqs}^r + p\boldsymbol{\lambda}_{qds}^r \tag{2.4-12}$$

where, it is left to the reader to show that $\mathbf{K}_s^r\mathbf{r}_s(\mathbf{K}_s^r)^{-1}$ is $\mathbf{r}_s$, and

$$(\boldsymbol{\lambda}_{dqs}^r)^T = \begin{bmatrix}\lambda_{ds}^r & -\lambda_{qs}^r\end{bmatrix} \tag{2.4-13}$$

The last two terms of (2.4-12) came from $\mathbf{K}_s^r$ times the last term of (2.4-11). That is, by the chain rule

$$\mathbf{K}_s^r p[(\mathbf{K}_s^r)^{-1}\boldsymbol{\lambda}_{qds}^r] = \mathbf{K}_s^r[p(\mathbf{K}_s^r)^{-1}]\boldsymbol{\lambda}_{qds}^r + \mathbf{K}_s^r(\mathbf{K}_s^r)^{-1}p\boldsymbol{\lambda}_{qds}^r \tag{2.4-14}$$

It is left to the reader to show that the right-hand side of (2.4-14) reduces to the last two terms of (2.4-12).

For a magnetically linear system, the stator flux linkages are expressed by (2.3-9). Substituting the change of variables into (2.3-9) yields

$$(\mathbf{K}_s^r)^{-1}\boldsymbol{\lambda}_{qds}^r = \mathbf{L}_s(\mathbf{K}_s^r)^{-1}\mathbf{i}_{qds}^r + \boldsymbol{\lambda}_m' \tag{2.4-15}$$

Pre-multiplying by $\mathbf{K}_s^r$ and substituting (2.3-12) and (2.3-13) for $\mathbf{L}_s$ yields

$$\boldsymbol{\lambda}_{qds}^r = \begin{bmatrix} L_{ls}+L_{ms} & 0 \\ 0 & L_{ls}+L_{ms}\end{bmatrix}\begin{bmatrix} i_{qs}^r \\ i_{ds}^r\end{bmatrix} + \lambda_m'^r\begin{bmatrix}0\\1\end{bmatrix} \tag{2.4-16}$$

Make sure you agree that $\mathbf{K}_s^r\mathbf{L}_s(\mathbf{K}_s^r)^{-1}$ is $\mathbf{L}_s$. To be consistent with our previous notation, we have added the superscript $r$ to $\lambda_m'$. We see from (2.4-16) that, in our new system of variables, the flux linkage created by the rotating permanent magnet appears constant. Hence, the $qs^r$- and $ds^r$- circuits are fixed relative to the permanent magnet and, therefore fixed with the rotor speed $\omega_r$. We have accomplished the goal of eliminating flux linkages that vary with $\theta_r$. One more thing before we leave (2.4-16), please note that the stator leakage inductance is contained in $\lambda^r_{qs}$ and $\lambda^r_{ds}$. This tells us that we must be careful where we conceptualize the location of these windings. If we consider them to be fixed on the rotor there would not be a stator leakage inductance. Therefore, we have little choice but to think of these windings as "rotating through iron" or as fixed in the stator with the stator "rotating at $\omega_r$"; however, it is common to refer to the $qs^r$- and $ds^r$-windings as "fixed in the rotor."

In expanded form, the voltage equations are

$$v^r_{qs} = r_s i^r_{qs} + \omega_r \lambda^r_{ds} + p\lambda^r_{qs} \tag{2.4-17}$$

$$v^r_{ds} = r_s i^r_{ds} - \omega_r \lambda^r_{qs} + p\lambda^r_{ds} \tag{2.4-18}$$

where

$$\lambda^r_{qs} = L_{ss} i^r_{qs} \tag{2.4-19}$$

$$\lambda^r_{ds} = L_{ss} i^r_{ds} + \lambda'^r_m \tag{2.4-20}$$

and, as before,

$$L_{ss} = L_{ls} + L_{ms} \tag{2.4-21}$$

Equations (2.4-17) through (2.4-21) suggest the $qs^r$- and $ds^r$- equivalent machines and the equivalent circuits shown in Fig. 2.4-3.

Substituting (2.4-16) into (2.4-17) and (2.4-18) and, since $\lambda'^r_m$ is constant, $p\lambda'^r_m = 0$.

$$v^r_{qs} = (r_s + pL_{ss}) i^r_{qs} + \omega_r L_{ss} i^r_{ds} + \omega_r \lambda'^r_m \tag{2.4-22}$$

$$v^r_{ds} = (r_s + pL_{ss}) i^r_{ds} - \omega_r L_{ss} i^r_{qs} \tag{2.4-23}$$

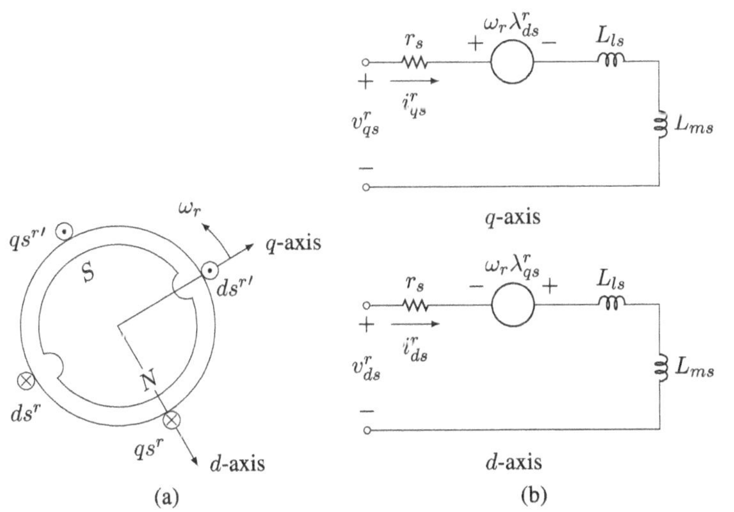


Figure 2.4-3 (a) The $qs^r$- and $ds^r$- equivalent permanent-magnet ac machine; (b) quadrature- and direct-axis equivalent circuits.

The above equations may be rewritten in matrix form as

$$\begin{bmatrix} v^r_{qs} \\ v^r_{ds} \end{bmatrix} = \begin{bmatrix} r_s + pL_{ss} & \omega_r L_{ss} \\ -\omega_r L_{ss} & r_s + pL_{ss} \end{bmatrix} \begin{bmatrix} i^r_{qs} \\ i^r_{ds} \end{bmatrix} + \begin{bmatrix} \omega_r \lambda'^r_m \\ 0 \end{bmatrix} \tag{2.4-24}$$

It should be mentioned that the forms of the transformation and the voltage equations are not unique. They depend on the selected positions of the positive *as*- and *bs*-axes, the position of the orthogonal $q$- and $d$-axes, and the order in which the variables appear in the matrix equations. Although we are not going to get involved in a discussion of this, it is touched upon in the problems at the end of the chapter. It is, however, necessary that the transformation yields constants in the synchronously rotating reference frame during balanced steady-state conditions. It is important to note that regardless of where the orthogonal $q$- and $d$-axes are placed, the transformation between the stator variables and the $qs^r$- and $ds^r$-variables can be readily determined from the expression of Tesla's rotating magnet field and the relationship between the stator and rotor displacements, $\phi_s$ and $\phi_r$, and the rotor position, $\theta_r$. This is true regardless of the number of stator phases involved.

The expression for electromagnetic torque positive for motor action is given by (2.3-15) as

$$T_e = \frac{P}{2} \lambda'^r_m (i_{as} \cos \theta_r + i_{bs} \sin \theta_r) \tag{2.4-25}$$

which can now be expressed as

$$T_e = \frac{P}{2} \lambda'^r_m i^r_{qs} \tag{2.4-26}$$

In Section 1.6, we derived an expression for torque wherein the voltage equations were of the form

$$v = ri + p\lambda \tag{2.4-27}$$

We used the differential form of the energy balanced equation

$$dW_f = dW_e + dW_m \tag{2.4-28}$$

where, with positive torque for motor action

$$dW_m = -\frac{2}{P} T_e d\theta_r \tag{2.4-29}$$

Now that we are working with Park's equations, the voltage equations are (2.4-17) and (2.4-18). Thus, as a result of the change of variables $p\lambda_{as}$ and $p\lambda_{bs}$ terms of (2.3-2) and (2.3-3) are replaced by the last two terms of (2.4-17) and (2.4-18). Therefore $W_e$, the energy entering the coupling field from the electrical system, becomes

$$W_e = \int e^r_{qs} i^r_{qs} dt + \int e^r_{ds} i^r_{ds} dt \tag{2.4-30}$$

where we have arbitrarily taken $r_s$ outside of the coupling field as we did in Section 1.6 and

$$e^r_{qs} = \omega_r \lambda^r_{ds} + p\lambda^r_{qs} \tag{2.4-31}$$

$$e^r_{ds} = -\omega_r \lambda^r_{qs} + p\lambda^r_{ds} \tag{2.4-32}$$

Now, $dW_e$ may be expressed

$$\begin{aligned} dW_e &= \lambda^r_{ds} i^r_{qs} d\theta_r + i^r_{qs} d\lambda^r_{qs} - \lambda^r_{qs} i^r_{ds} d\theta_r + i^r_{ds} d\lambda^r_{ds} \\ &= (\lambda^r_{ds} i^r_{qs} - \lambda^r_{qs} i^r_{qs}) d\theta_r + i^r_{qs} d\lambda^r_{qs} + i^r_{ds} d\lambda^r_{ds} \end{aligned} \quad (2.4\text{-}33)$$

The field energy may be expressed

$$W_f = \int i^r_{qs} d\lambda^r_{qs} + \int i^r_{ds} d\lambda^r_{ds} + W_{pm} \quad (2.4\text{-}34)$$

where, as before, $W_{pm}$ is the energy associated with the permanent magnet which is assumed to be constant and since the $qs^r$- and $ds^r$-windings are orthogonal, there is no mutual coupling. Therefore,

$$dW_f = i^r_{qs} d\lambda^r_{qs} + i^r_{ds} d\lambda^r_{ds} \quad (2.4\text{-}35)$$

Now, substituting (2.4-35) for $dW_f$, (2.4-33) for $dW_e$, and (2.4-29) for $dW_m$ into (2.4-28) we obtain, for a $P$-pole machine

$$i^r_{qs} d\lambda^r_{qs} + i^r_{ds} d\lambda^r_{ds} = (\lambda^r_{ds} i^r_{qs} - \lambda^r_{qs} i^r_{ds}) d\theta_r + i^r_{qs} d\lambda^r_{qs} + i^r_{ds} d\lambda^r_{ds} - \frac{2}{P} T_e d\theta_r \quad (2.4\text{-}36)$$

Solving for $T_e$

$$T_e = \frac{P}{2} (\lambda^r_{ds} i^r_{qs} - \lambda^r_{qs} i^r_{ds}) \quad (2.4\text{-}37)$$

The expression for torque given by (2.4-37) is valid in general; unbalanced, magnetically nonlinear, and transient conditions.

Let us take a minute to consider the power balance

$$P_e = P_{\text{loss}} + P_m \quad (2.4\text{-}38)$$

where $P_e$ is the power from the electrical system, $P_{\text{loss}}$ is the power loss, in this case the $i^2 r$ loss, and $P_m$ is the power to the mechanical system, which is $\frac{2}{P} T_e \omega_r$ for a rotational system. With the positive direction of currents we have assumed $P_e$ and $P_m$ are positive for motor action.

The power input is

$$\begin{aligned} P_e &= v^r_{qs} i^r_{qs} + v^r_{ds} i^r_{ds} \\ &= r_s i^{r2}_{qs} + \omega_r \lambda^r_{ds} i^r_{qs} + i^r_{qs} p \lambda^r_{qs} \\ &\quad + r_s i^{r2}_{ds} - \omega_r \lambda_{qs} i^r_{ds} + i^r_{ds} p \lambda^r_{ds} \\ &= r_s (i^{r2}_{qs} + i^{r2}_{ds}) + (\lambda^r_{ds} i^r_{qs} - \lambda^r_{qs} i^r_{ds}) \omega_r \\ &\quad + i^r_{qs} p \lambda^r_{qs} + i^r_{ds} p \lambda^r_{ds} \end{aligned} \quad (2.4\text{-}39)$$

Equation (2.4-39) may be written

$$P_e = P_{\text{loss}} + \frac{2}{P} T_e \omega_r + i^r_{qs} p \lambda^r_{qs} + i^r_{ds} p \lambda^r_{ds} \quad (2.4\text{-}40)$$

Now we know for balanced steady-state conditions $P_e$ is constant, the losses are constant, $\omega_r$ is constant, and $\lambda^r_{qs}$ and $\lambda^r_{ds}$ are constants, so if $P_e$ is constant $P_m$ is constant and thus $T_e$ is constant.

Let us now assume that the applied stator voltages are

$$v_{as} = v_s \cos \theta_{esv} \tag{2.4-41}$$

$$v_{bs} = v_s \sin \theta_{esv} \tag{2.4-42}$$

where $v_s$ is the amplitude of the stator applied voltages and

$$\omega_e = \frac{d\theta_{esv}}{dt} \tag{2.4-43}$$

Transforming these voltages to the rotor reference frame yields

$$v^r_{qs} = v_s \cos \phi_v \tag{2.4-44}$$

$$v^r_{ds} = -v_s \sin \phi_v \tag{2.4-45}$$

where

$$\phi_v = \theta_{esv} - \theta_r \tag{2.4-46}$$

We will find that if $\phi_v$ is zero, the device has the operating characteristics ($T_e$ vs. $\omega_r$) similar to that of a permanent-magnet dc machine; this is the reason a permanent-magnet ac machine, which has no brushes or commutator, is called a "brushless dc machine." We will talk more about this in Chapter 4.

Before becoming involved with steady-state analysis of the permanent-magnet ac machine, it is appropriate to mention a few more things. First, in Park's original paper [2] he analyzed a synchronous generator since his focus was the power system industry. Therefore, he assumed positive currents out of the machine in the direction of a voltage rise; whereupon, the torque $T_e$ was positive for generator action. He also positioned the $q$- and $d$-axes differently; however, the concept is the same.

It may first appear that the transformation is for mathematical convenience when dealing with the synchronous machine. Fortunately, the concept of a change of variables has far-reaching fundamental philosophical importance for symmetrical systems. Although we are not going to become involved in reference frame theory, the transformation for a generalized change of variables can be obtained by expressing $\phi_s$ by (2.2-13) rather than (2.4-2). Whereupon, $\theta_r$ in (2.4-7) becomes $\theta$. This transformation is also referred as the transformation to an arbitrary reference frame [3]. For symmetrical stator circuits the $qs$- and $ds$-voltage equations are of the same form in all reference frames regardless of the angular velocity $\omega$ of the arbitrary reference frame; however, the variables will differ in waveform depending upon the choice of $\omega$. If $\omega$ is selected equal to $\omega_r$ the variables are constant in the steady-state whereupon $\omega_r = \omega_e$; for all other constant values of $\omega$ the steady-state variables will be sinusoidal and balanced with the frequency corresponding to $(\omega_e - \omega)$. It is also interesting that the expression for Tesla's rotating magnetic field (2.4-1) along with (2.2-13) encompass all known transformations used in machine analysis including Park's and the arbitrary reference frame.

In advanced courses, we will find that this generalized theory of a change of variables simplifies the analysis and has become useful in the dynamic analysis of ac machines, power systems, and modern machine control theory [3]. We will leave all this to advanced courses. Our present concern is the permanent-magnet ac machine and we will assume that the stator windings are sinusoidally distributed and the stator variables are balanced. It is, however, important to realize that the change of variables used in machine analysis can be thought of as a means of observing the machine variables and associated circuits from different frames of reference. Tesla's rotating magnetic field was perhaps the forerunner of this concept when he realized that a balanced set of ac variables associated with symmetrical stator circuits gave rise to a constant amplitude rotating air-gap mmf and it follows that associated with this constant rotating mmf would be constant voltages and currents. One can only imagine as to the depth and extent of Tesla's understanding and visualization of the analysis potential of his concept.

## Steady-State Analysis

For balanced, steady-state conditions, $v^r_{qs}$ and $v^r_{ds}$ are constants and the voltage equations given by (2.4-22) and (2.4-23) become

$$V^r_{qs} = r_s I^r_{qs} + \omega_r L_{ss} I^r_{ds} + \omega_r \lambda'^r_m \tag{2.4-47}$$

$$V^r_{ds} = r_s I^r_{ds} - \omega_r L_{ss} I^r_{qs} \tag{2.4-48}$$

where $pi^r_{qs}$ and $pi^r_{ds}$ are zero and capital letters denote steady-state (constant) quantities, except for $\lambda'^r_m$ which is assumed to always be constant. If (2.4-47) and (2.4-48) are solved for $I^r_{qs}$ and the result substituted into the expression for electromagnetic torque given by (2.4-26), we obtain

$$T_e = \frac{P}{2}\frac{\lambda'^r_m r_s}{r_s^2 + \omega_r^2 L_{ss}^2}\left(V^r_{qs} - \frac{\omega_r L_{ss} V^r_{ds}}{r_s} - \omega_r \lambda'^r_m\right) \tag{2.4-49}$$

It is instructive to take a moment to derive a steady-state voltage equation in phasor form. To accomplish this, consider the stator variables during steady-state balanced operation. In particular, during steady-state operation $\omega_e = \omega_r$, thus

$$F_{as} = \sqrt{2}F_s \cos[\omega_r t + \theta_{esf}(0)] \tag{2.4-50}$$

$$F_{bs} = \sqrt{2}F_s \sin[\omega_r t + \theta_{esf}(0)] \tag{2.4-51}$$

Substituting $F_{as}$ and $F_{bs}$ into the equation of transformation, (2.4-8), yields

$$F^r_{qs} = \sqrt{2}F_s \cos[\theta_{esf}(0) - \theta_r(0)] \tag{2.4-52}$$

$$F^r_{ds} = -\sqrt{2}F_s \sin[\theta_{esf}(0) - \theta_r(0)] \tag{2.4-53}$$

The phasor that represents $F_{as}$ given by (2.4-50) is

$$\tilde{F}_{as} = F_s e^{j\theta_{esf}(0)} \tag{2.4-54}$$

which may be written as

$$\tilde{F}_{as} = F_s \cos\theta_{esf}(0) + jF_s \sin\theta_{esf}(0) \tag{2.4-55}$$

Since we can choose our time zero as we wish, let us select it so that $\theta_r(0) = 0$. In other words, our time zero will always be selected at the time the $q$-axis in Fig. 2.3-1 is horizontal to the right; whereupon $F^r_{qs}$ and $-F^r_{ds}$ are, respectively, the same as $\sqrt{2}$ times the real and imaginary parts of (2.4-55). Thus,

$$\sqrt{2}\tilde{F}_{as} = F^r_{qs} - jF^r_{ds} \tag{2.4-56}$$

Here, we have equated a phasor, which represents a sinusoidal quantity, to $F^r_{qs}$ and $F^r_{ds}$ which are constants. However, without the $e^{j\omega_r t}$ $(\omega_r = \omega_e)$ which gives rotation, the phasor is simply a complex number.

Substituting (2.4-47) and (2.4-48) into (2.4-56) yields

$$\begin{aligned}\sqrt{2}\tilde{V}_{as} &= V^r_{qs} - jV^r_{ds} \\ &= r_s I^r_{qs} + \omega_r L_{ss} I^r_{ds} + \omega_r \lambda'^r_m - j(r_s I^r_{ds} - \omega_r L_{ss} I^r_{qs})\end{aligned} \tag{2.4-57}$$

From (2.4-56)

$$j\sqrt{2}\tilde{F}_{as} = F^r_{ds} + jF^r_{qs} \tag{2.4-58}$$

Hence, (2.4-57) may be written as

$$\tilde{V}_{as} = (r_s + j\omega_r L_{ss})\tilde{I}_{as} + \frac{1}{\sqrt{2}}\omega_r \lambda'^r_m e^{j0} \tag{2.4-59}$$

We can define

$$\tilde{E}_a = \frac{1}{\sqrt{2}}\omega_r \lambda'^r_m e^{j0} \tag{2.4-60}$$

and write (2.4-59) as

$$\tilde{V}_{as} = (r_s + j\omega_r L_{ss})\tilde{I}_{as} + \tilde{E}_a \tag{2.4-61}$$

We can use the phasor diagram to advantage in portraying the steady-state operation of the permanent-magnet ac machine. During balanced steady-state operation, the rotor speed is synchronous; thus $\omega_r = \omega_e$ and since the rotor and phasor diagram are rotating in unison at $\omega_e$ they can be superimposed. Moreover, since the rotating magnetic field is also rotating at $\omega_e$ it can also be superimposed with the rotor and the phasor diagram. Now, since all are rotating at $\omega_e$ and the voltage and current phasors are constant we can stop the rotation at any time or we can run at $\omega_e$ and observe the operation of the machine. We will do as we did when dealing with a phasor diagram for an electric circuit in Section 1.2. In particular, we will select $t = 0$ (stop rotation) when the $q$-axis is horizontal to the right. That is $\theta_r = 0$ at $t = 0$. Now, the magnitude and phase of the stator voltages will be controlled by the drive inverter, Chapter 4, and the phasor $\tilde{V}_{as}$ would be located at $\theta_{esv}(0)$ which is $\phi_v$ from (2.4-46) since $\theta_r = 0$ at our time zero and $\tilde{I}_{as}$ would be located at $\theta_{esi}(0)$. Recall that $\theta_{esi}(0)$ also locates the south pole of the "rotating" magnetic field, which is (2.2-17) with the $e$ superscript replaced by $r$ and $\phi_e$ with $\phi_r$.

In Fig. 2.4-4 we have located the phasors associated with (2.4-61) and the stator poles relative to the stator current. We have assumed that fundamental component of the north pole of the rotor is fixed in the $d$-axis. Does Fig. 2.4-4 tell us anything about torque? Yes. We have motor action and we can see this since the poles established by the stator, which are rotating counterclockwise at $\omega_e(\omega_r)$, are "trying" to pull the rotor poles in to alignment with them. This "pulling" is being resisted by the load torque. When this "pulling torque" is equal to the load torque on the rotor shaft, steady-state operation occurs with torque in the counterclockwise direction; motor action. Is it clear that complete alignment occurs only if there is no load torque (see Fig. 1.7-2)?

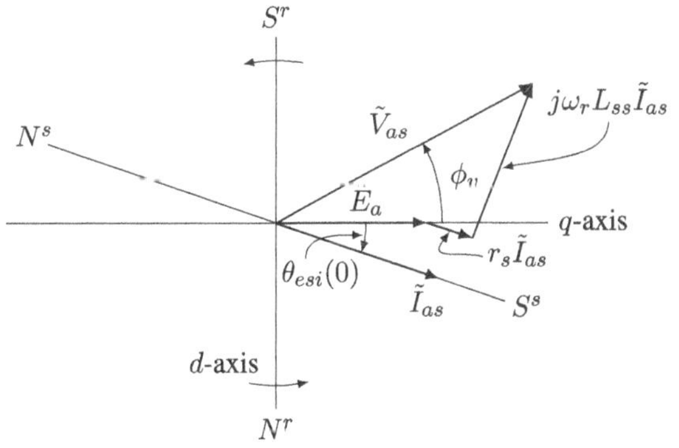


Figure 2.4-4 Phasor diagram showing voltages, currents, and rotor stator poles.

**Example 2E. Steady-state operation of a two-phase permanent-magnet ac machine.** Consider a four-pole two-phase permanent-magnet ac machine. The rotor electrical angular velocity $\omega_r = 50$ rad/s. $V_s = 40$ V, $I_s = 1$ A, $r_s = 4\ \Omega$, $L_{ss} = 20$ mH, and $\theta_r(0)$ is selected equal to zero. If $V_{ds}^r$ is controlled (Chapter 4) equal to zero, ($a$) calculate $\tilde{E}_a$ and draw the phasor diagram including the location of the stator and rotor poles, and ($b$) plot the magnetic fields as viewed running counterclockwise at $\omega_e$ which is $\omega_r$.

Since $V_{ds}^r$ and $\theta_r(0)$ are zero, then from (2.4-45)

$$0 = -\sqrt{2}V_s \sin[\theta_{esv}(0) - 0] \tag{2E-1}$$

Thus, $\theta_{esv}(0) = 0$ and $\tilde{V}_{as}$ and $\tilde{E}_a$ (2.4-60) are in phase and from (2.4-46) $\phi_v = 0$. Now, since $\tilde{V}_{as}$ is controlled to be in phase with $\tilde{E}_a$, (we will see how this is done by inverter control in Chapter 4), the current lags $\tilde{V}_{as}$ ($q$-axis) by the impedance phase angle. That is

$$\theta_{esi}(0) = -\tan^{-1}\frac{\left(\frac{4}{2}\right)(50)(20\times 10^{-3})}{4} = -26.6° \tag{2E-2}$$

From (2.4-59) and (2.4-60)

$$\tilde{E}_a = \tilde{V}_{as} - (r_s + j\omega_r L_{ss})\tilde{I}_{as}$$

$$= 40\underline{/0^\circ} - \left[4 + j\left(\frac{4}{2}\right)(50)(20 \times 10^{-3})\right] 1\underline{/-26.6^\circ}$$

$$= 40 - (4 + j2)\underline{/-26.6^\circ} = 35.5\underline{/0^\circ}\ \text{V} \tag{2E-3}$$

The phasor diagram is shown in Fig. 2E-1.

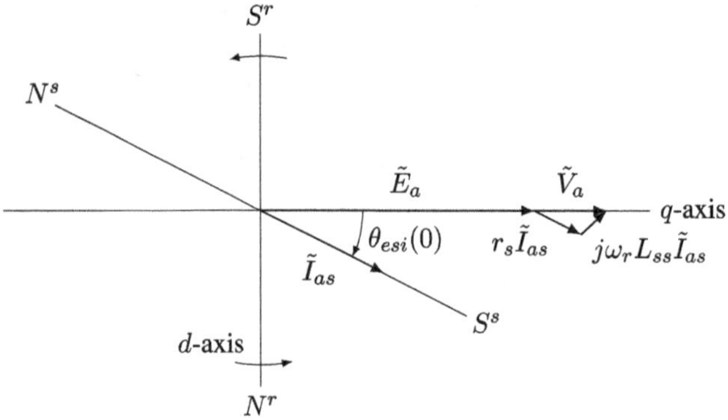


Figure 2E-1 Phasor diagram for steady-state operation of two-phase permanent-magnet ac machine.

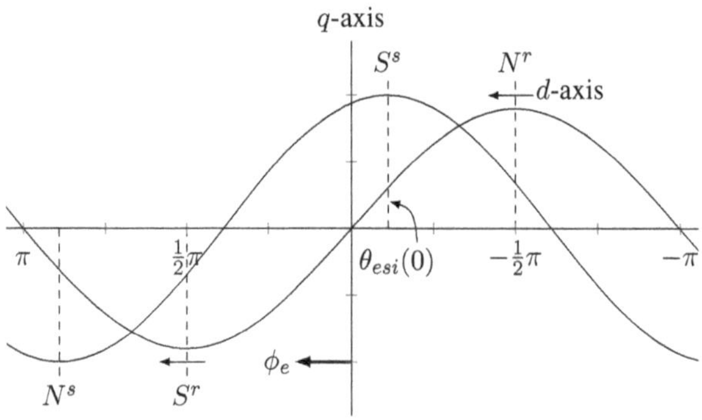


Figure 2E-2 Observing the stator and rotor rotating magnetic field while running at $\omega_r(\omega_e)$ counterclockwise with 360° vision.

We see from Figs. 2E-1 and 2E-2, that the rotor poles are being "pulled" by the stator poles and the direction of the torque is counterclockwise; motor action.

---

**SP2.4-1.** In the drive application of the permanent-magnet ac machine the frequency of $v_{as}$ and $v_{bs}$ corresponds to the electrical angular velocity of the rotor. Therefore, assume that $I_{as} = \sqrt{2}I_s \cos[\omega_r t + \theta_{esi}(0)]$, $I_{bs} = \sqrt{2}I_s \sin[\omega_r t + \theta_{esi}(0)]$, and $\theta_r = \omega_r t + \theta_r(0)$. Express $T_e$ using (2.4-26). $\left\{\frac{P}{2}\lambda_m' \sqrt{2}I_s \cos[\theta_{esi}(0) - \theta_r(0)]\right\}$

**SP2.4-2.** Let $f_{as} = -\cos\theta_r$ and $f_{bs} = -\sin\theta_r$. Determine $f_{qs}^r$ and $f_{ds}^r$. [$f_{qs}^r = -1$; $f_{ds}^r = 0$]

**SP2.4-3.** Let $f_{as} = \cos\theta_{esf}$, $f_{bs} = \sin\theta_{esf}$, and $\theta_r = \omega_r t$. Express $\theta_{esf}$ for $f_{qs}^r = 0$ and $f_{ds}^r = 1$. $\left[\theta_{esf} = \omega_r t - \frac{1}{2}\pi\right]$

**SP2.4-4.** If the speed $\omega_r$ is constant and if $I_{as} = \cos\theta_r$ and $I_{bs} = \sin\theta_r$, what will be the waveform of $I_{qs}^r$ and $I_{ds}^r$ (steady-state currents)? [dc]

**SP2.4-5.** Express $\phi_v$ if $\theta_{esv} = \omega_r t + \theta_{esv}(0)$ and $\theta_r = \omega_r t$. [$\phi_v = \theta_{esv}(0)$]

**SP2.4-6.** Assuming a two-phase permanent magnet ac machine, start with (2.4-37) and obtain (2.4-26).

## 2.5 THREE-PHASE PERMANENT-MAGNET AC MACHINE

We have mentioned several times that we will analyze the two-phase machine and then extend our work to the three-phase device. We pointed out that the trigonometry involved for the two-phase machine was much less when we incorporated Park's transformation. However, this alone does not justify treating the two-phase device before the three-phase device. The question arises why even consider the two-phase device since it is not used widely? Well, when we derived Tesla's rotating magnetic field we noted that during balanced conditions the two- and three-phase symmetrical stator each produced rotating magnetic fields differing only in magnitude. Although this similarity was interesting, we really did not pay much attention to all this until we realized that Park's transformation for a two-phase stator is contained in the expression of Tesla's rotating magnetic field and that the $qs^r$ and $ds^r$ two-phase currents completely define this rotating magnetic field. Now, let us think three-phase; the rotating magnetic field differs from the two-phase only by a constant. Does not it seem logical that in this section we will find that the rotating magnetic field for a three-phase device can also be completely defined by two-phase $qs^r$- and $ds^r$-variables? In other words, as far as Tesla's rotating magnetic field is concerned, the three-phase system is a two-phase system.

A two-pole three-phase permanent-magnet ac machine is shown in Fig. 2.5-1. The stator windings are identical, displaced relative to each other by $\frac{2}{3}\pi$. The windings are sinusoidally distributed, each with $N_s$ equivalent turns and resistance $r_s$. Electromagnetic torque is produced by the interaction of the poles of the permanent-magnet rotor and the poles resulting from the rotating air-gap mmf established by currents flowing in the stator windings. The rotating air-gap mmf ($\text{mmf}_s$), Tesla's rotating magnetic field, established by symmetrical three-phase stator windings carrying balanced three-phase currents is given by (2.2-25).

In Chapter 4, we will drive the three-phase permanent-magnet ac machine from the forced commutated inverter that is set forth in Chapter 3. The firing of the inverter is such that the permanent-magnet ac machine has the torque-speed characteristics of a dc machine and is referred to as a brushless dc machine. It is important to emphasis that in Chapter 4 we will refer to this section to set the stage for the analysis and simulation of the brushless dc drive.

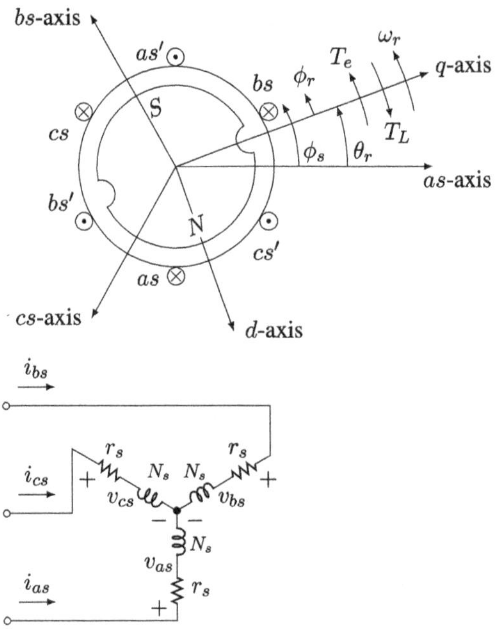


Figure 2.5-1 Two-pole three-phase permanent-magnet ac machine.

## Voltage Equations and Winding Inductances

The voltage equations for the two-pole, three-phase, permanent-magnet ac machine shown in Fig. 2.5-1 may be expressed as

$$v_{as} = r_s i_{as} + \frac{d\lambda_{as}}{dt} \tag{2.5-1}$$

$$v_{bs} = r_s i_{bs} + \frac{d\lambda_{bs}}{dt} \tag{2.5-2}$$

$$v_{cs} = r_s i_{cs} + \frac{d\lambda_{cs}}{dt} \tag{2.5-3}$$

In matrix form,

$$\mathbf{v}_{abcs} = \mathbf{r}_s \mathbf{i}_{abcs} + p\boldsymbol{\lambda}_{abcs} \tag{2.5-4}$$

For voltages, currents, and flux linkages

$$(\mathbf{f}_{abcs})^T = [f_{as}, f_{bs}, f_{cs}] \tag{2.5-5}$$

and

$$\mathbf{r}_s = \begin{bmatrix} r_s & 0 & 0 \\ 0 & r_s & 0 \\ 0 & 0 & r_s \end{bmatrix} \tag{2.5-6}$$

The flux linkage equations may be expressed as

$$\lambda_{as} = L_{asas}i_{as} + L_{asbs}i_{bs} + L_{ascs}i_{cs} + \lambda_{asm} \tag{2.5-7}$$

$$\lambda_{bs} = L_{bsas}i_{as} + L_{bsbs}i_{bs} + L_{bscs}i_{cs} + \lambda_{bsm} \tag{2.5-8}$$

$$\lambda_{cs} = L_{csas}i_{as} + L_{csbs}i_{bs} + L_{cscs}i_{cs} + \lambda_{csm} \tag{2.5-9}$$

In matrix form,

$$\boldsymbol{\lambda}_{abcs} = \mathbf{L}_s\mathbf{i}_{abcs} + \boldsymbol{\lambda}'_m \tag{2.5-10}$$

where $\boldsymbol{\lambda}'_m$ is the column vector

$$\boldsymbol{\lambda}'_m = \begin{bmatrix} \lambda_{asm} \\ \lambda_{bsm} \\ \lambda_{csm} \end{bmatrix} = \lambda'_m \begin{bmatrix} \sin\theta_r \\ \sin\left(\theta_r - \frac{2}{3}\pi\right) \\ \sin\left(\theta_r + \frac{2}{3}\pi\right) \end{bmatrix} \tag{2.5-11}$$

In (2.5-11), $\lambda'_m$ is the amplitude of the flux linkages established by the permanent magnet as viewed from the stator phase windings. In other words, the magnitude of $p\lambda_{asm}$, $p\lambda_{bsm}$, and $p\lambda_{csm}$ would be the magnitude of the open-circuit voltage induced in the respective stator phase windings if the rotor were driven. The rotor displacement $\theta_r$ is defined by (2.3-11).

For our purposes, we will assume that the reluctance is the same in the $q$- and $d$-axes; whereupon, $\mathbf{L}_s$ may be written as (1.8-30)

$$\mathbf{L}_s = \begin{bmatrix} L_{ls} + L_{ms} & -\frac{1}{2}L_{ms} & -\frac{1}{2}L_{ms} \\ -\frac{1}{2}L_{ms} & L_{ls} + L_{ms} & -\frac{1}{2}L_{ms} \\ -\frac{1}{2}L_{ms} & -\frac{1}{2}L_{ms} & L_{ls} + L_{ms} \end{bmatrix} \tag{2.5-12}$$

where $L_{ls}$ represents the leakage inductance and $L_{ms}$ the magnetizing inductance per phase. The off-diagonal terms are $-\frac{1}{2}L_{ms}$ since the mutual inductance between two stator windings displaced by $\frac{2}{3}\pi$ is $\cos\frac{2}{3}\pi$ times the mutual inductance when their magnetic axis coincide, that is, $L_{ms}\cos\frac{2}{3}\pi$.

## Torque

Since the system is assumed to be magnetically linear, the field and coenergy are equal, thus

$$W_c = \frac{1}{2}L_{ss}(i_{as}^2 + i_{bs}^2 + i_{cs}^2) - \frac{1}{2}L_{ms}(i_{as}i_{bs} + i_{bs}i_{cs} + i_{cs}i_{as}) + \lambda'_m i_{as}\sin\theta_r$$
$$+ \lambda'_m i_{bs}\sin\left(\theta_r - \frac{2}{3}\pi\right) + \lambda'_m i_{cs}\sin\left(\theta_r + \frac{2}{3}\pi\right) + W_{pm} \tag{2.5-13}$$

where $W_{pm}$ is the energy associated with the permanent magnet, which is assumed constant. Taking the partial derivative with respect to $\theta_r$ and multiplying by $P/2$ yields

$$T_e = \frac{P}{2}\lambda'_m \left[ i_{as}\cos\theta_r + i_{bs}\cos\left(\theta_r - \frac{2}{3}\pi\right) + i_{cs}\cos\left(\theta_r + \frac{2}{3}\pi\right)\right] \quad (2.5\text{-}14)$$

The above expression is positive for motor action. The torque and speed may be related as given by (2.3-16).

## The $qs^r$-, $ds^r$-, and 0s- Equations

Since there are three stator variables ($f_{as}$, $f_{bs}$, and $f_{cs}$), we must use three substitute variables in the transformation of the stator variables to the rotor reference frame. In particular,

$$\mathbf{f}^r_{qd0s} = \mathbf{K}^r_s \mathbf{f}_{abcs} \quad (2.5\text{-}15)$$

where

$$(\mathbf{f}^r_{qd0s})^T = [f^r_{qs}, f^r_{ds}, f_{0s}] \quad (2.5\text{-}16)$$

$$\mathbf{K}^r_s = \frac{2}{3}\begin{bmatrix} \cos\theta_r & \cos\left(\theta_r - \frac{2}{3}\pi\right) & \cos\left(\theta_r + \frac{2}{3}\pi\right) \\ \sin\theta_r & \sin\left(\theta_r - \frac{2}{3}\pi\right) & \sin\left(\theta_r + \frac{2}{3}\pi\right) \\ \frac{1}{2} & \frac{1}{2} & \frac{1}{2} \end{bmatrix} \quad (2.5\text{-}17)$$

where the rotor displacement $\theta_r$ is defined by (2.3-11). The inverse of $\mathbf{K}^r_s$ is

$$(\mathbf{K}^r_s)^{-1} = \begin{bmatrix} \cos\theta_r & \sin\theta_r & 1 \\ \cos\left(\theta_r - \frac{2}{3}\pi\right) & \sin\left(\theta_r - \frac{2}{3}\pi\right) & 1 \\ \cos\left(\theta_r + \frac{2}{3}\pi\right) & \sin\left(\theta_r + \frac{2}{3}\pi\right) & 1 \end{bmatrix} \quad (2.5\text{-}18)$$

It is important to note that the same notation ($\mathbf{K}^r_s$) used for the two-phase transformation is also used for the three-phase transformation. Equation (2.5-17) is essentially Park's original transformation for the synchronous generator [2]. The zero variable $f_{0s}$ is the third substitute variable and it is zero for balanced and even some unbalanced conditions. Also, $f_{0s}$ is not a function of $\theta_r$; therefore, the 0s-variables are associated with stationary circuits and a raised index is not incorporated with the zero variables.

It is also important to point out that when $\text{mmf}_{as}$, $\text{mmf}_{bs}$, and $\text{mmf}_{cs}$ given by (2.2-18) through (2.2-20), respectively, are added we obtain (2.2-21). If we now follow the same procedure as in the case of the two-phase machine and substitute $\phi_s = \theta_r + \phi_r$ and set $\phi_r = 0$ we obtain the first row of (2.5-15) which is $i^r_{qs}$. When $\phi_r = -\frac{\pi}{2}$ we obtain $i^r_{ds}$ which is the second row of (2.5-15). Clearly, the discussion given at the beginning of Section 2.4 regarding the two-phase transformation applies also to the three-phase transformation. Note quite so fast. What about the $\frac{2}{3}$ factor in (2.5-17). Tesla does not have that in (2.2-21). Park, however, choose to add the $\frac{2}{3}$, which essentially makes the three-phase machine a two-phase. Ok, but it is then necessary to multiple the input power and torque by $\frac{3}{2}$ for the three-phase machine. One can argue both ways on this; but, perhaps Park was looking ahead knowing that all machines can be analyzed as two-phase devices so he made the input power and torque that of a two-phase machine. Regardless, our only concern is to remember to multiply by $\frac{3}{2}$ when evaluating three-phase torque and input power using the $qs^r$- and $ds^r$-variables.

If, in the case of the three-phase machine, steady-state operation is the focus, the implementation of the transformation can be de-emphasized. Then, the steady-state procedure would be the same as suggested for the two-phase machine with the two exceptions. First, $L_{ss}$ becomes (2.5-30) where the $\frac{3}{2}$ factor times $L_{ms}$ is due to the mutual coupling between stator phase windings. Secondly, torque is given by (2.5-35); here the $\frac{3}{2}$ factor is due to Park's choice of the $\frac{2}{3}$ in the transformation. Otherwise, the equations for a balanced steady-state analysis are the same as for a two-phase machine. A detailed study of Example 2F is recommended.

Although we realize that the $qs^r$- and $ds^r$-voltages equations are going to be essentially those of a two-phase machine, we will still go through the derivation for the sake of completeness. Substituting the change of variables into the stator voltage equations given by (2.5-4) yields

$$(\mathbf{K}_s^r)^{-1}\mathbf{v}_{qd0s}^r = \mathbf{r}_s(\mathbf{K}_s^r)^{-1}\mathbf{i}_{qd0s}^r + p[(\mathbf{K}_s^r)^{-1}\boldsymbol{\lambda}_{qd0s}^r] \tag{2.5-19}$$

Pre-multiplying each side by $\mathbf{K}_s^r$ yields

$$\mathbf{v}_{qd0s}^r = \mathbf{r}_s\mathbf{i}_{qd0s}^r + \omega_r\boldsymbol{\lambda}_{dqs}^r + p\boldsymbol{\lambda}_{qd0s}^r \tag{2.5-20}$$

where

$$(\boldsymbol{\lambda}_{dqs}^r)^T = [\lambda_{ds}^r, \ -\lambda_{qs}^r, \ 0] \tag{2.5-21}$$

If one wishes to work through the steps necessary to go from (2.5-19) to (2.5-20), the trigonometric relations given in Appendix A will be very helpful.

For a magnetically linear system, the stator flux linkages are expressed by (2.5-10). Substituting the change of variables in (2.5-10) yields

$$\left(\mathbf{K}_s^r\right)^{-1}\boldsymbol{\lambda}_{qd0s}^r = \mathbf{L}_s\left(\mathbf{K}_s^r\right)^{-1}\mathbf{i}_{qd0s}^r + \boldsymbol{\lambda}_m' \tag{2.5-22}$$

Pre-multiplying by $\mathbf{K}_s^r$ yields

$$\boldsymbol{\lambda}_{qd0s}^r = \begin{bmatrix} L_{ls} + \frac{3}{2}L_{ms} & 0 & 0 \\ 0 & L_{ls} + \frac{3}{2}L_{ms} & 0 \\ 0 & 0 & L_{ls} \end{bmatrix}\begin{bmatrix} i_{qs}^r \\ i_{ds}^r \\ i_{0s} \end{bmatrix} + \lambda_m'^r\begin{bmatrix} 0 \\ 1 \\ 0 \end{bmatrix} \tag{2.5-23}$$

To be consistent with our previous notation we have added the superscript $r$ to $\lambda_m'$. Note that the $\frac{3}{2}L_{ms}$ has come about due to the mutual coupling of the stator phases and the transformation. In expanded form,

$$v_{qs}^r = r_s i_{qs}^r + \omega_r \lambda_{ds}^r + p\lambda_{ds}^r \tag{2.5-24}$$

$$v_{ds}^r = r_s i_{ds}^r - \omega_r \lambda_{qs}^r + p\lambda_{ds}^r \tag{2.5-25}$$

$$v_{0s} = r_s i_{0s} + p\lambda_{0s} \tag{2.5-26}$$

where

$$\lambda_{qs}^r = L_{ss} i_{qs}^r \tag{2.5-27}$$

$$\lambda_{ds}^r = L_{ss} i_{ds}^r + \lambda'^r_m \tag{2.5-28}$$

$$\lambda_{0s} = L_{ls} i_{0s} \tag{2.5-29}$$

where

$$L_{ss} = L_{ls} + \frac{3}{2} L_{ms} \tag{2.5-30}$$

We must be careful here. When dealing with the two-phase machine, we also used $L_{ss}$; however, there it was $L_{ls} + L_{ms}$ rather than $L_{ls} + \frac{3}{2}L_{ms}$. The difference occurs due to the mutual coupling between the stator windings of a three-phase stator. Although we should probably use a different notation for the two- and three-phase machines, we will not; instead, we will call attention to this difference as we go along.

Substituting (2.5-27) through (2.5-29) into (2.5-24) through (2.5-26) and since $\lambda'^r_m$ is constant, $p\lambda'^r_m = 0$, and we can write

$$v_{qs}^r = (r_s + pL_{ss}) i_{qs}^r + \omega_r L_{ss} i_{ds}^r + \omega_r \lambda'^r_m \tag{2.5-31}$$

$$v_{ds}^r = (r_s + pL_{ss}) i_{ds}^r - \omega_r L_{ss} i_{qs}^r \tag{2.5-32}$$

$$v_{0s} = (r_s + pL_{ls}) i_{0s} \tag{2.5-33}$$

Note that $v_{qs}^r$ and $v_{ds}^r$, (2.5-24) and (2.5-25) or (2.5-31) and (2.5-32), are identical in form to $v_{qs}^r$ and $v_{ds}^r$ for the two-phase machine given by (2.4-17) and (2.4-18) or (2.4-22) and (2.4-23), respectively. The above equations may be written in matrix form as

$$\begin{bmatrix} v_{qs}^r \\ v_{ds}^r \\ v_{0s} \end{bmatrix} = \begin{bmatrix} r_s + pL_{ss} & \omega_r L_{ss} & 0 \\ -\omega_r L_{ss} & r_s + pL_{ss} & 0 \\ 0 & 0 & r_s + pL_{ls} \end{bmatrix} \begin{bmatrix} i_{qs}^r \\ i_{ds}^r \\ i_{0s} \end{bmatrix} + \begin{bmatrix} \omega_r \lambda'^r_m \\ 0 \\ 0 \end{bmatrix} \tag{2.5-34}$$

The expression for electromagnetic torque is

$$T_e = \frac{3}{2}\frac{P}{2} \lambda'^r_m i_{qs}^r \tag{2.5-35}$$

which is positive for motor action. It is important to emphasize again that for the vast majority of operating conditions the rotor-reference frame two-phase voltage equations and torque equation for a three-phase machine are the same as the two-phase machine except for the $\frac{3}{2}$ factor in $L_{ss}$ and in $T_e$ and $P_e$. Moreover, the steady-state phasor equation given by (2.4-61) is valid for a three-phase machine if we use $\frac{3}{2}L_{ms}$. In the case of $L_{ss}$ the $\frac{3}{2}$ factor is due to the mutual coupling of the stator phases and the transformation; in the case of the torque and power, the $\frac{3}{2}$ factor is the transformation. Clearly, the equivalent circuits given in Fig. 2.4-3b for the two-phase machine are valid for the three-phase if $L_{ms}$ is replaced with $\frac{3}{2}L_{ms}$.

As we have mentioned, we will turn this device into a brushless dc machine in Chapter 4 when we consider drive systems. We have also mentioned that the torque-speed characteristic of the permanent-magnet ac machine can be made to closely resemble that of a permanent-magnet dc machine. This will all be done in Chapter 4; however, in Example 2E we took a brief look at the steady-state operation with $\phi_v = 0$, given by (2.4-46), which is the mode of operation for the dc motor look-alike.

**Example 2F. Steady-state operation of a three-phase permanent-magnet machine.** A three-phase four-pole fractional-horsepower permanent-magnet ac machine has the following parameters: $r_s = 3.4\,\Omega$, $L_{ls} = 1.1$ mH, $L_{ms} = 11$ mH, and $\lambda'_m = 0.0826$ V · s/rad. The machine is wye-connected with balanced three-phase applied voltages. The *as*-phase voltage is $V_{as} = \sqrt{2}11.25\cos\omega_e t$. The actual steady-state rotor speed is 750 r/min. Calculate $\omega_e$ and $T_e$, and $\tilde{I}_{as}$ if $\phi_v = 0$ and draw a phasor diagram and show the stator and rotor poles.

The electrical angular velocity of the rotor is from (2.2-34)

$$\begin{aligned}\omega_r &= \frac{P}{2}\omega_{rm}\\ &= \frac{P}{2}\frac{(\mathrm{r/min})\,(\mathrm{rad/r})}{\mathrm{s/min}}\\ &= \frac{4}{2}\frac{(750)\,(2\pi)}{60} = 50\pi\ \mathrm{rad/s}\end{aligned} \tag{2F-1}$$

We are firing the inverter so that the frequency of the voltages applied to the stator windings is equal to the electrical angular velocity of the rotor, therefore, $\omega_e = 50\pi$ rad/s. Now, since the voltage equations are identical for the two- and three-phase machines except for the $\frac{3}{2}L_{ms}$ in $L_{ss}$ and $\frac{3}{2}$ times the torque, we can use (2.4-49). But first, since $\phi_v = 0$ then from (2.4-56)

$$\begin{aligned}\sqrt{2}\tilde{\mathrm{V}}_{as} &= \mathrm{V}^r_{qs} - j\mathrm{V}^r_{ds}\\ &= \sqrt{2}11.25\underline{/0^\circ}\end{aligned} \tag{2F-2}$$

Therefore, $V^r_{ds} = 0$ and from (2.4-49) with the $\frac{3}{2}$ factor included

$$\begin{aligned}T_e &= \left(\frac{3}{2}\right)\left(\frac{P}{2}\right)\frac{r_s\lambda'^r_m}{r_s^2+\omega_r^2L_{ss}^2}\left(V^r_{qs}-\omega_r\lambda'^r_m\right)\\ &= \left(\frac{3}{2}\right)\left(\frac{4}{2}\right)\frac{(3.4)\,(0.0826)\left[\sqrt{2}11.25-50\pi\,(0.026)\right]}{(3.4)^2+(50\pi)^2\left[1.1+\left(\frac{3}{2}\right)11\right]^2\times10^{-6}}\\ &= 0.129\ \mathrm{N\cdot m}\end{aligned} \tag{2F-3}$$

We can use different approaches to calculate $\tilde{I}_{as}$. Perhaps (2.4-61) is the most direct since it is valid for a three-phase machine if we use $\frac{3}{2}L_{ss}$; however, we have calculated the torque and (2.5-35) can be used to calculate $I^r_{qs}$. In particular,

$$I^r_{qs} = \frac{2}{3}\frac{2}{P}\frac{T_e}{\lambda'^r_m} = \left(\frac{2}{3}\right)\left(\frac{2}{4}\right)\frac{(0.129)}{0.0826} = 0.521\text{ A} \tag{2F-4}$$

We can calculate $I^r_{ds}$ from (2.4-48) with $V^r_{ds} = 0$ we have

$$I^r_{ds} = \frac{\omega_r L_{ss}}{r_s} I^r_{qs} = \frac{50\pi\left(1.1 + \frac{3}{2}11\right) \times 10^{-3}(0.521)}{3.4} = 0.424\text{ A} \tag{2F-5}$$

From (2.4-56)

$$\sqrt{2}\tilde{I}_{as} = I^r_{qs} - jI^r_{ds}$$

$$\sqrt{2}\tilde{I}_{as} = 0.521 - j0.424 = 0.671\underline{/-39.1^\circ} \tag{2F-6}$$

Therefore,

$$\tilde{I}_{as} = \frac{1}{\sqrt{2}}(0.671)\underline{/-39.1^\circ} = 0.475\underline{/-39.1^\circ}\text{ A} \tag{2F-7}$$

We need $\tilde{E}_a$, thus from (2.4-60)

$$\tilde{E}_a = \frac{1}{\sqrt{2}}\omega_r\lambda'^r_m\underline{/0^\circ} = \frac{1}{\sqrt{2}}(50\pi)(0.0826)\underline{/0^\circ} = 9.18\underline{/0^\circ}\text{ V} \tag{2F-8}$$

The relationships between the voltage and current phasors, and the resultant north and south poles, are depicted in Fig. 2F-1.

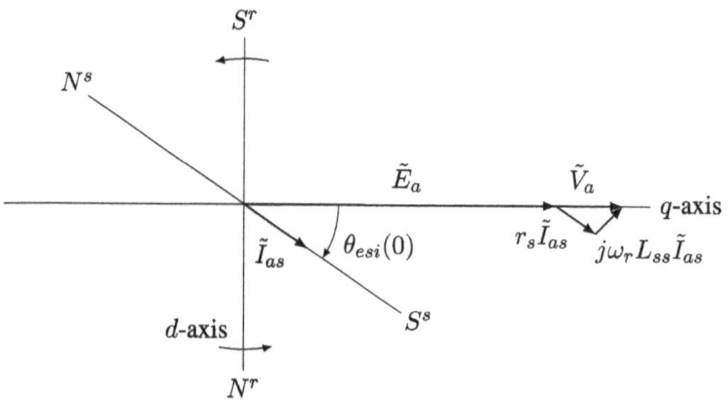


Figure 2F-1 Phasor diagram for steady-state operation of three-phase permanent-magnet ac machine with $\phi_v = 0$.

**SP2.5-1.** A four-pole three-phase permanent-magnet ac motor is operating with $\theta_r(0) = 0$, $V_{as} = 10\cos\omega_r t$, and $I_{as} = \cos(\omega_r t - 20°)$. Calculate $\lambda'^r_m$ if $r_s = 4\,\Omega$, $L_{ss} = 0.01$ H, and $\omega_r = 146$ rad/s. $[\lambda'^r_m = 0.0393\ \text{V}\cdot\text{s/rad}]$

**SP2.5-2.** Repeat SP2.5-1 neglecting $\omega_r^2 L_{ss}^2$. $[\lambda'^r_m = 0.0427\ \text{V}\cdot\text{s/rad}]$

**SP2.5-3.** Calculate (a) the input power $P_e$, (b) $T_e$, and (c) efficiency for the operating conditions given in SP2.5-1. [(a) $P_e = 14.1$ W; (b) $T_e = 0.111$ N · m; (c) eff = 57.4 %]

**SP2.5-4.** The expression for torque of a two-phase, two-pole permanent-magnet ac machine is $T_e = \lambda'^r_m i^r_{qs}$. (a) What is the torque expression for a two-phase multipole permanent-magnet ac machine? (b) Multipole, three-phase machine? [(a) $T_e = \frac{P}{2}\lambda'^r_m i^r_{qs}$, (b) $\frac{3}{2}$ times (a) or (2.5-35)]

**SP2.5-5.** Why is $L_{ss} = L_{ls} + \frac{3}{2}L_{ms}$ for a three-phase machine? [Mutual coupling between phases.]

## REFERENCES


[1] N. Tesla, *My Inventions and Other Writings*, Penguin Books, 2011.

[2] R. H. Park, "Two-Reaction theory of synchronous machines – Generalized method of analysis – Part I," *AIEE Transactions*, vol. 48, July 1929, pp. 716–727.

[3] P. C. Krause, O. Wasynczuk, S. Sudhoff, and S Pekarek, *Analysis of Electric Machinery and Drive Systems*, 3rd edition, Wiley-IEEE Press, 2013.

[4] S. D. Sudhoff, *Power Magnetic Devices*, Wiley-IEEE Press, Hoboken, New Jersey, 2013.


## PROBLEMS

1. For the device shown in Fig. 2.2-7, express the air-gap mmf for balanced steady-state conditions as viewed by an observer running (a) clockwise at $\omega_e$; (b) counterclockwise at $\frac{1}{2}\omega_e$; and (c) counterclockwise at $2\omega_e$. Let $\theta(0) = 0$.

2. Draw the phasor diagram showing the $qs$-axis for Fig. 2.2-8 with $\tilde{V}_{as}$ at zero degrees. Locate the stator poles. Then, redraw Fig. 2.2-8 showing the positions of $\tilde{V}_{as}$ and $\tilde{I}_{as}$.

3. The expression for torque for a two-pole device is

$$T_e(\mathbf{i},\theta_r) = \frac{\partial W_c(\mathbf{i},\theta_r)}{\partial \theta_r}$$

where the coenergy is expressed in terms of currents and the electrical angular displacement. For balanced steady-state conditions the mechanical power output, $P_m$, is the shaft torque multiplied by the shaft angular velocity, $\omega_{rm}$. Derive an expression for $P_m$ in terms of the electrical angular velocity, $\omega_r$ and show that, for balanced steady-state conditions, it is valid for any $P$-pole device.

4. Consider the stator of the four-pole device shown in Fig. 2.2-10. Assume these windings are concentrated with $N_s$ total turns. Plot mmf$_s$ for $0 \le \phi_s \le 2\pi$.

5. Consider the four-pole two-phase permanent-magnet ac machine in Example 2D. The parameters are $r_s = 3.4\ \Omega$, $L_{ss} = 12.1$ mH, and $\lambda'^r_m = 0.0826$ V · s/rad. The stator windings are short circuited and the rotor is driven at $\omega_{rm} = \frac{100}{3}\pi$ rad/s. Calculate $I_{as}$ and $\tilde{I}_{as}$.

6. Suppose the $d$-axis is positioned $\frac{\pi}{2}$ radians counterclockwise from the $q$-axis rather than $\frac{\pi}{2}$ clockwise (i.e., at $\phi_r = \frac{\pi}{2}$). Use the expression for Tesla's rotating field to express $\mathrm{mmf}_s|_{\phi_r=\pi/2}$ and, from this expression, deduce the corresponding transformation to the $d$-axis (a new $\mathbf{K}_s^r$). Show that

$$\mathbf{f}_{qds}^r = \mathbf{K}_s^r \mathbf{f}_{abs}^r$$

where the new $\mathbf{K}_s^r$ is

$$\mathbf{K}_s^r = \begin{bmatrix} \cos\theta_r & \sin\theta_r \\ -\sin\theta_r & \cos\theta_r \end{bmatrix}$$

7. Express $\boldsymbol{\lambda}_{qds}^r$ with the new transformation in Problem 6 for the $ds$-variables obtained for the $d$-axis $\frac{\pi}{2}$ counterclockwise from the $q$-axis.

8. (a) Determine the inverse of the transformation given in Problem 6. (b) Repeat (a) if the change of variables were formulated as

$$\begin{bmatrix} f_{ds}^r \\ f_{qs}^r \end{bmatrix} = \mathbf{K}_s^r \begin{bmatrix} f_{as} \\ f_{bs} \end{bmatrix}$$

where $\mathbf{K}_s^r$ now becomes

$$\mathbf{K}_s^r = \begin{bmatrix} -\sin\theta_r & \cos\theta_r \\ \cos\theta_r & \sin\theta_r \end{bmatrix}$$

9. Starting with (2.4-11) and using $\mathbf{K}_s^r$ and $\left(\mathbf{K}_s^r\right)^{-1}$ obtained in Problem 8(a), derive $\mathbf{v}_{qds}^r$ in a form similar to (2.4-12).

10. In Example 2E a two-phase four-pole permanent-magnet ac machine is operated with $V_s =$ 40 V, $V_{ds}^r = 0$, and $\omega_{rm} = 50$ rad/s. The parameters are $r_s = 4\ \Omega$, $L_{ss} = 20$ mH, and $\theta_r(0) =$ 0. In Example 2E we found $\tilde{E}_a = 35.5\underline{/0°}$. Use the information given above to calculate $\tilde{I}_{as}$ using (2.4-47) and (2.4-48).

11. Calculate the power balance for the mode of operation considered in Example 2F.

CHAPTER 3

# POWER ELECTRONICS

## 3.1 INTRODUCTION

The advent of the silicon-controlled rectifier (SCR) in the 1950s ushered in electric drives, devices which are a major part of the electric power industry. Power electronic circuits, which form the backbone of these drives, have enabled the ever-expanding electrification of consumer electronics, transportation, and wireless technologies. Power electronic-controlled electric machines are impacting industrial and drive systems, propulsion systems in ships and ground vehicles, and aircraft actuator systems. It is not our intent to cover even a small part of the vast power electronics industry in an introductory text. Rather, this chapter attempts to open the door to advanced study by establishing the fundamental principles of switching circuits and then analyzing the basic power electronic converters used in the areas of electric drives and power systems. This chapter is designed to stimulate interest in power electronics and provide a foundation on which to build.

In Section 3.2, we introduce several fundamental concepts relevant to power conversion that can be applied across a broad range of basic and advanced circuits. In Sections 3.3 through 3.5, these concepts are used to introduce dc-dc converters, ac-dc rectifiers, and dc-ac inverters, respectively. Although there are numerous variations of inverters used in electric drive applications, in Section 3.5, we utilize the concepts laid out in Section 3.2 to analyze the basic three-phase inverter. We will use this inverter in Chapter 4 to operate the permanent-magnet ac machine as a brushless dc machine. Finally, in Section 3.6, a common metric for gauging the unavoidable harmonics and distortion that accompany switching power circuits is introduced.

## 3.2 SWITCHING-CIRCUIT FUNDAMENTALS

Because power is the flow of energy with respect to time, we can broadly define "Power Electronics" as the study of electronic circuits and devices that control the flow of energy between sources and loads. Ideally, power electronic circuits accomplish their goal as efficiently as possible so that energy can get where it needs to go without being lost or wasted along the way.

*Introduction to Electric Power and Drive Systems*, First Edition.
Paul Krause, Oleg Wasynczuk, Timothy O'Connell and Maher Hasan.
© 2017 by The Institute of Electrical and Electronics Engineers, Inc. Published 2017 by John Wiley & Sons, Inc.
Companion website: www.wiley.com/go/krause/electricpower

Power electronic devices are everywhere, often doing their jobs silently and diligently behind the scenes while other, more "glamorous" electrical devices get all the attention. In fact, most power converters are designed to be hidden or inconspicuous; those black "bricks" that come with computers are not exactly eye-catching. However, their role in today's world is irreplaceable and cannot be overstated. Because electrical energy is produced and consumed over a wide range of voltages and frequencies, the applications for power electronic devices are countless. For example, transferring energy from an ac source, such as a standard wall outlet, to a dc load, such as the computer on which this text was typed, is called "ac-to-dc conversion," and is accomplished with an ac-dc converter. Thus, you might not be reading this textbook today if it were not for power electronics. Similarly, dc-to-ac, dc-to-dc, and ac-to-ac devices exist for other energy flow control applications.

As you read this chapter, take a look around you. There is a good chance you are surrounded by power electronic devices, many you may not even notice. If you use a cellular phone, a power electronic converter allowed you to charge it by transporting power from the ac grid to the dc battery. The next time you ride in an automobile, you will be surrounded by power electronic devices moving energy from the engine or battery to the car's numerous electrical accessories. And these are just a few of the ways you directly interact with power electronics daily. This chapter will give you a solid understanding of the fundamental concepts that make these circuits work, and we hope it will inspire you to pursue further study of advanced power conversion circuits.

## Power Conversion Principles

Consider the generic power electronic circuit shown in Fig. 3.2-1. This device creates an interface between two circuits, facilitating the flow of power between them when a direct connection is inefficient or impossible. For example, a dc-ac inverter can act as an interface between a dc battery and an ac electric motor. As we have seen in Chapters 1 and 2, most rotating electric machinery requires ac currents and voltages to operate; thus, unless it is a dc machine, a motor cannot operate directly from a battery. An inverter is required and does its job by converting the battery's stored energy from dc to ac that can be used by the motor.

Historically, the connected circuits denoted "Circuit 1" and "Circuit 2" in Fig. 3.2-1 have been characterized as follows: power-producing circuits are called *sources* and power-consuming circuits are called *loads*. However, modern trends in the power field have blurred the line between sources and loads, as many devices are both producers and consumers of power. For example, the battery in a cellular telephone is a source when it is powering the phone and a load when it is being charged. Likewise, the electric machine in a hybrid electric vehicle can act as both a load (motor mode) and source (generator mode) depending on the state of operation of the

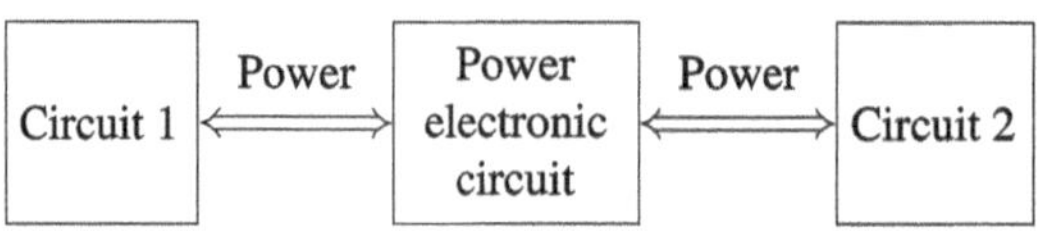


Figure 3.2-1 A generic power electronic circuit.

vehicle. In general, power electronic circuits can, and often do, facilitate bidirectional power flow between circuits, as indicated by the double arrows in Fig. 3.2-1.

The general conceptualization shown in Fig. 3.2-1 directly motivates two key power conversion principles [1]. First, power conversion circuits should consume as little power as possible; an ideal power device is one that is 100% efficient and consumes none of the power flowing through it. Of course, in reality no device is 100% efficient, but real power devices do regularly achieve efficiencies of 95% and above. This directly leads to the second principal: power conversion circuits should transfer much more power than they are rated for. This means that a relatively small power device can interface between two large devices that process a lot of power. In terms of the example mentioned earlier, if a battery and motor are rated to supply/consume 1 kW of electric power during normal operation, a good inverter need only be rated for as little as 20 W; although the inverter can transfer 1 kW from battery to motor, it does this so efficiently that it only consumes 20 W in the process.

One of the first circuits introduced to students that might be considered for power conversion is the resistive voltage divider, shown in Fig. 3.2-2. This circuit creates a dc output voltage that is a fraction of the dc input, and the fraction depends on the values of the resistors in the divider. A voltage divider is thus a candidate for interfacing a battery to a dc circuit that operates at a lower voltage.

Before we continue, we note that throughout this chapter we attempt to be consistent with the common power electronics notation found in texts and literature. In some cases, this may conflict with the notation we have set forth in Chapters 1 and 2; however, rather than modify the notation here, we will adopt the common power electronics notation in order to best prepare the reader for advanced courses in this area. In cases of conflict, we will do our best to clarify which notation is intended. In all other cases, the context should make the distinction clear.

Let us now have a closer look at the resistive voltage divider circuit to see whether it follows the two key power conversion principles. If we assume that $r_{out} \gg r_2$, which is good design practice[1] and is necessary to ensure that the divider creates the expected output voltage, then $r_2//r_{out} \approx r_2$. Thus, the input current, $I_{in}$, is

$$I_{in} = \frac{V_{in}}{r_1 + r_2//r_{out}} \approx \frac{V_{in}}{r_1 + r_2} \tag{3.2-1}$$

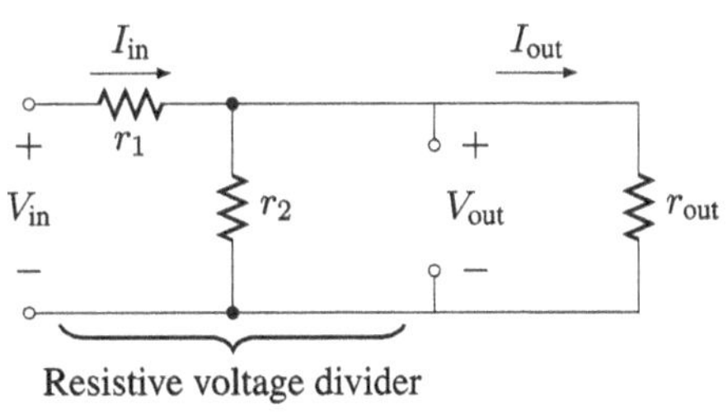


Figure 3.2-2 Resistive voltage divider circuit.

[1] Most modern power loads are not resistive. However, for the purposes of this development we will assume a resistive load. The conclusions about the voltage divider will apply to other load types as well.

and the instantaneous input power, $P_{in}$, is

$$P_{in} = V_{in} I_{in} \approx \frac{V_{in}^2}{r_1 + r_2} \tag{3.2-2}$$

This hints that the voltage divider is not a very good power conversion circuit because most of its input power is consumed by the divider itself in its two resistors.

The output voltage, $V_{out}$, is

$$\begin{aligned} V_{out} &= I_{in}(r_2//r_{out}) \\ &\approx I_{in} r_2 \\ &\approx \frac{V_{in}}{r_1 + r_2} r_2 \end{aligned} \tag{3.2-3}$$

which is the familiar voltage divider relationship with an open-circuit output. If we define the voltage divider's *voltage conversion ratio*, $\alpha$, as

$$\alpha \triangleq \frac{r_2}{r_1 + r_2} \tag{3.2-4}$$

then, from (3.2-3)

$$V_{out} \approx \alpha V_{in} \tag{3.2-5}$$

and

$$P_{in} \approx \frac{\alpha V_{in}^2}{r_2} \tag{3.2-6}$$

In terms of $\alpha$, the output power, $P_{out}$, is

$$\begin{aligned} P_{out} &= \frac{V_{out}^2}{r_{out}} \\ &\approx \frac{\alpha^2 V_{in}^2}{r_{out}} \\ &\approx \alpha P_{in} \left(\frac{r_2}{r_{out}}\right) \end{aligned} \tag{3.2-7}$$

Since we have assumed $r_2 \ll r_{out}$, $\frac{r_2}{r_{out}}$ is exceedingly small. Thus, even if $\alpha$ is close to one, $P_{out}$ is close to zero!

Does a voltage divider qualify as a good power conversion circuit? We have seen that the circuit consumes nearly all of its input power in its resistors, which violates the first principle of power circuits. Its resistors must be rated to handle the full input power, which violates the second principle. Because $r_{out}$ must be large compared to $r_2$ for the divider to function properly, there is minimal available output power. The only way to get more power out would be to greatly reduce $r_{out}$, which would compromise the voltage divider output voltage. It is clear that a voltage divider is NOT a good power conversion circuit. This is not to say that a voltage divider is not a useful circuit; in fact, it is most appropriate in low power (e.g., logic-level) circuits that require minimal current. However, for higher power levels, we need something better.

Let us reconsider the example application suggested earlier, an interface between a dc source and a lower voltage dc load. This requires a dc-dc (pronounced "dc to dc") converter, but one that is more efficient and better able to transfer power than a voltage divider. Essentially, these requirements sound similar to what is provided by a transformer; however, recall from Chapter 1 that a transformer requires ac voltage to function. DC voltage into the transformer's primary winding will not induce a changing magnetic field in the core nor a voltage in its secondary winding. In fact, dc voltage would saturate the transformer's magnetic core, rendering the transformer useless for power transfer. Another method is needed to realize transformer action with dc.

## Fourier Analysis

To better approach this problem, we need to expand our view of the term "dc" from "a signal with constant value" to something broader. Before we proceed, however, it will be useful to review basic Fourier series analysis. Although the motivation may be unclear now, we will see that power conversion circuits often switch between two or more configurations in a periodic manner at a known frequency, and Fourier series are well-suited for analyzing the periodic signals that thus arise.

A periodic function with period $T$ is one for which $f(t) = f(t + T)$. Whereas in other chapters $f$ is a general variable that represents transient voltage, current, or flux linkage, here $f$ is any time-domain periodic function. Although there are some esoteric periodic signals that cannot be represented by a Fourier series, any periodic signal found in a real circuit, including those with step discontinuities, will meet the mathematical conditions necessary. Thus, any periodic signal encountered in a power conversion circuit can be expressed as a sum of sinusoids, also known as the Fourier series:

$$f(t) = a_0 + \sum_{n=1}^{\infty} [a_n \cos(n\omega t) + b_n \sin(n\omega t)] \tag{3.2-8}$$

where $\omega = 2\pi/T$ is the radian frequency, and $a_n$ and $b_n$ are the series coefficients found by computing the following integrals over an interval $[t_0, t_0 + T]$ :

$$a_0 = \frac{1}{T} \int_{t_0}^{t_0+T} f(t)dt \tag{3.2-9}$$

$$a_n = \frac{2}{T} \int_{t_0}^{t_0+T} f(t) \cos(n\omega t)dt \tag{3.2-10}$$

$$b_n = \frac{2}{T} \int_{t_0}^{t_0+T} f(t) \sin(n\omega t)dt \tag{3.2-11}$$

Through algebraic manipulation, the Fourier series can be written in an alternate compact form that is sometimes easier to work with. The sines and cosines in (3.2-8) are combined into phase-shifted cosine terms to give the compact Fourier series:

$$f(t) = a_0 + \sum_{n=1}^{\infty} c_n \cos(n\omega t + \theta_n) \tag{3.2-12}$$

The coefficients of the compact series are related to those of the standard series by

$$c_n = \sqrt{a_n^2 + b_n^2} \tag{3.2-13}$$

and the phase angles are

$$\theta_n = -\tan^{-1}(b_n/a_n) \tag{3.2-14}$$

Recall from calculus that the time average, denoted $\widehat{(\cdot)}$, of a transient signal $f(t)$ over an interval $[t_0, t_0 + T]$ is given by

$$\widehat{f} = \frac{1}{T} \int_{t_0}^{t_0+T} f(t)dt \tag{3.2-15}$$

Comparing this to (3.2-9) we see that $a_0 = \widehat{f}$, which means that the Fourier coefficient $a_0$ is the average value (i.e., the dc component or dc offset) of the periodic signal. To develop a method of dc-dc conversion, let us now consider any periodic signal with a nonzero dc component to be a candidate dc signal. For these, the dc component can be "pulled out" by filtering out the ac components, which are the Fourier series terms with coefficients $a_n$ and $b_n$ (or $c_n$ in the compact series). This broadens the family of dc signals we can consider from "signals that are constant" to "signals with a dc component." This broader classification gives us considerable flexibility as we look for a method of dc-dc conversion.

Although the Fourier series equations can seem daunting when they are first encountered, under certain circumstances that often arise in real power conversion circuits, they can be simplified considerably [2]. For example, if $f(t)$ has no dc component, then $a_0 = 0$:

$$a_0 = 0: \quad \text{when} \quad \widehat{f} = 0 \tag{3.2-16}$$

If, after removing the dc component, the remaining function is odd, then the cosine terms will drop out[2]:

Odd function ⇒ sines only

$$\Rightarrow a_n = 0, \quad \forall\, n \geq 1: \quad \text{when}\, f(t) - a_0 = -[f(-t) - a_0] \tag{3.2-17}$$

Likewise, if the function is even (i.e., it has y-axis symmetry), then the sine terms will drop out:

Even function ⇒ cosines only

$$\Rightarrow b_n = 0, \quad \forall\, n \geq 1: \quad \text{when}\, f(t) = f(-t) \tag{3.2-18}$$

[2] In this chapter, the ∀ symbol stands for "for all."

Finally, if after removing the dc component, the remaining function has half-wave symmetry, it will contain no even harmonics:

$$\begin{aligned}\text{Half-wave symmetry} &\Rightarrow \text{no even harmonics}\\ &\Rightarrow a_n,\ b_n = 0\ \forall\ n \text{ even:}\\ &\quad \text{when } f(t+T/2) - a_0 = -[f(t) - a_0]\end{aligned} \tag{3.2-19}$$

Examples of periodic functions which exhibit some of these properties are shown in Fig. 3.2-3 and described in Example 3A.

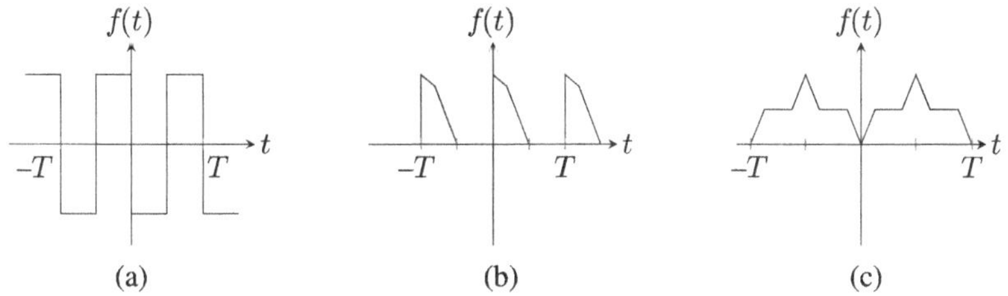


Figure 3.2-3 Periodic functions with various Fourier series properties.

**Example 3A. Fourier series properties of periodic signals.** Describe the general features of each of the periodic functions shown in Fig. 3.2-3 using (3.2-16) through (3.2-19).

(a) This signal has no average value; thus, from (3.2-16) $a_0 = 0$. It is an odd function; thus, from (3.2-17) $a_n = 0, \quad \forall\, n \geq 1$ and the series is made up of sines only. It is half-wave symmetric; thus, from (3.2-19) its Fourier series contains no even harmonics.

(b) The signal has an average value; thus, from (3.2-16) $a_0 \neq 0$. In fact, from (3.2-9) $a_0 = \hat{f}$. The signal is neither even nor odd; thus, from (3.2-17) and (3.2-18) the series contain sines and cosines. The signal has no half-wave symmetry; thus, from (3.2-19) its Fourier series contains even and odd harmonics.

(c) The signal has an average value; thus, from (3.2-16) $a_0 \neq 0$. It is an even function; thus, from (3.2-18) $b_n = 0, \quad \forall\, n \geq 1$ and the series is made up of cosines only. After removing the dc component the signal is half-wave symmetric; thus, from (3.2-19) its Fourier series contains no even harmonics.

## Switches and Switching Functions

A few additional concepts are needed before analyzing a power conversion circuit. The first concept is the *switch*. In power electronics, a switch is a device used to regulate the flow of energy between sources and loads. If a switch is controlled, it is called an *active* device; otherwise, it is a *passive* device. The two most common examples of these two switch types are, respectively, the transistor and the diode. A transistor found in power circuits is typically a bipolar junction transistor (BJT), metal-oxide-semiconductor field-effect transistor (MOSFET), or insulated-gate bipolar transistor (IGBT). Of these, the IGBT is becoming the most popular because of its simple voltage-controlled switching, high power and current capabilities, and fast switching speeds. Diodes are simple p–n junction devices. Although conceptually simpler than transistors, in practice diodes can vary widely in their current-carrying and voltage-blocking capabilities depending on their construction. Every power electronic circuit has at least one switch, and it is the operation and the location of its switches that determine the circuit's behavior.

For this introductory discussion, we approximate a real switch with a concept known as the *ideal switch*, whose circuit representation is shown in Fig. 3.2-4. An ideal switch is lossless, carries no current and maintains an open circuit between its terminals when it is open/off, and has no voltage drop and maintains a short circuit between its terminals when it is closed/on. The real devices described in the previous paragraph are, of course, not ideal. However, they can be approximated well to first order with the ideal assumptions. Thus, to avoid unnecessary complication, we limit our discussion and subsequent analysis to ideal switches.

The ideal switch voltage, $v_{sw}$, and current, $i_{sw}$, behaviors are summarized as follows:

$$\begin{aligned} &\text{Switch open} \Rightarrow\ i_{sw} = 0 \\ &\text{Switch closed} \Rightarrow\ v_{sw} = 0 \end{aligned} \tag{3.2-20}$$

This behavior is described mathematically using a binary switch variable, $q$, as

$$q = \begin{cases} 0, & \text{Switch open} \\ 1, & \text{Switch closed} \end{cases} \tag{3.2-21}$$

Note that this $q$ is a variable and is different from the label of the $q$-axis introduced in Chapter 2. To avoid confusion, we could use a different variable name here; however, in an attempt at consistency with traditional power electronic circuit analysis notation, we will not. Through the remainder of the text it should be clear from the context which $q$ is being used.

Unlike, for example, a light switch, a power electronic switch does not remain in one state for long. Rather, it is opened and closed rapidly in order to transfer power in a desired fashion. Thus, the switch variable $q$ is most often a transient *switching function*, $q(t)$, which opens and closes the switch at a fixed rate known as the *switching frequency*, $f_{sw}$. The inverse of the switching frequency is the *switching period*, $T_{sw}$. The fraction of the switching period that the switch is on is the *duty ratio*, $D$. For example, if a switch is operated at a switching frequency of 40 kHz with a duty ratio of 0.2 (or, 20%), it will be on for 20% of each 25 μs switch period, or 5 μs. The switching frequency is chosen based on many factors, among these are

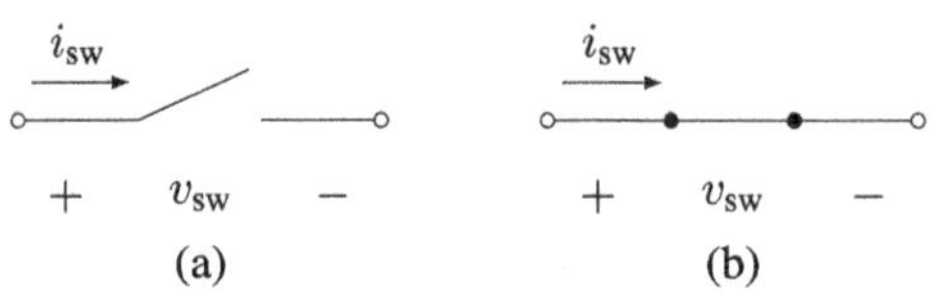


Figure 3.2-4 Ideal switch in (a) open/off and (b) closed/on configurations.

efficiency, circuit ripple and/or distortion requirements, acoustic noise, and device ratings.

A common switching function encountered in the design of power electronic converters is shown in Fig. 3.2-5. By utilizing the even symmetry of the waveform (defined by (3.2-18)) and solving for the Fourier series coefficients using (3.2-9) through (3.2-11), we can write the compact Fourier series of the switching function as

$$q(t) = D + \frac{2}{\pi}\sum_{n=1}^{\infty}\frac{\sin(n\pi D)}{n}\cos(n\omega_{\mathrm{sw}}t - n\theta(0)) \tag{3.2-22}$$

where $\omega_{\mathrm{sw}} = 2\pi f_{\mathrm{sw}}$ and $\theta(0)$ is the phase shift angle of the fundamental component at time zero. Notice that the switching function's average value is its duty ratio, $D$. Also, note that the square wave in Fig. 3.2-5 is an even function centered at $t = 0$, which means that its fundamental has $\theta(0) = 0$. However, the Fourier series in (3.2-22) is general and can easily be adjusted for shifted square waves by setting $\theta(0)$ appropriately.

While switching functions are used to control active switches, passive devices, such as diodes, are not actively controlled. Rather, ideal diodes obey the following:

$$\begin{aligned} &\text{Diode off: } i_{\mathrm{d}} = 0 \Rightarrow v_{\mathrm{d}} < 0 \\ &\text{Diode on: } v_{\mathrm{d}} = 0 \Rightarrow i_{\mathrm{d}} > 0 \end{aligned} \tag{3.2-23}$$

where $v_{\mathrm{d}}$ and $i_{\mathrm{d}}$ are the diode voltage and current, respectively, as shown in Fig. 3.2-6. Note that the non-italicized subscript d here is not indicating *d*-axis variables, as the italicized *d* does in earlier chapters; rather, it simply indicates diode values. In power electronic circuits, active and passive switches are often used in tandem, so that if the active switch is controlled with $q(t)$, the passive switch behavior will be $(1 - q(t))$. When one switch is open, the other is closed, and vice versa. Because a passive switch is not actively controlled, it is simpler to use and often cheaper than an equivalent active device. However, an active device can be used in place of a passive one and controlled with a $(1 - q(t))$ switching function to perform the same function. This

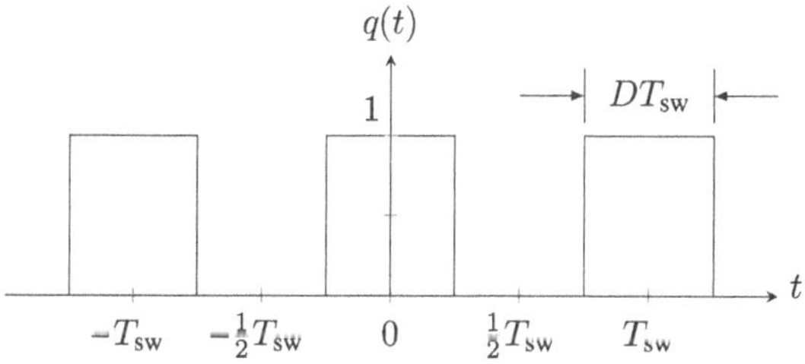

Figure 3.2-5 General periodic switching function $q(t)$.

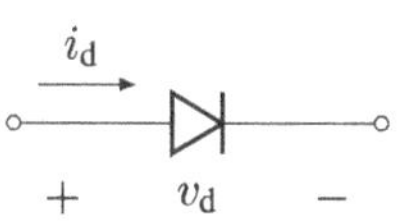

Figure 3.2-6 Ideal diode circuit representation.

is often done in high performance circuits, where precise control of all switches is important.

A final word on switches is in order. Although our analyses in this chapter will assume that all switches are ideal, it is important to understand that this is only an approximation. Although sufficient for our introductory treatment, the switching rules in (3.2-20) do not hold for real devices. In fact, IGBTs, MOSFETs, diodes, and other real switches have finite turn-on and turn-off transition times. During these transitions, which are typically less than 1 μs, both $v_{sw}$ and $i_{sw}$ are nonzero concurrently as they transition to their final state. Because $v_{sw} \times i_{sw} \neq 0$, power is lost in the switch. This switching loss is proportional to $f_{sw}$ and negatively impacts a power converter's efficiency. In addition, due to both solid-state physical effects and ohmic losses, the voltage across a real switch is nonzero when the switch is fully on and conducting current. This conduction loss, which also negatively impacts a converter's efficiency, is proportional to duty cycle – the longer a switch is on, the higher its conduction loss. Because they affect efficiency and generate heat, switching and conduction losses are important considerations in the design of high performance power converters. For a thorough treatment of this topic, interested students should consult a more detailed power electronics text, such as [1] or [3].

**Example 3B. Fourier series of a switching function.** Calculate the Fourier series of the switching function given in Fig. 3B-1. Note that the frequency of the function in this illustrative example is much smaller than would be found in a typical power converter.

Regardless of frequency, the steps to find the Fourier series are the same. The series we seek is given by (3.2-22). We must find the series parameters by comparing Fig. 3B-1 with Fig. 3.2-5. By inspection, we see that

$$D = \frac{\pi/2}{2\pi} = 1/4 \tag{3B-1}$$

and

$$T_{sw} = 2\pi \tag{3B-2}$$

Thus,

$$\omega_{sw} = 2\pi/T_{sw} = 1 \text{ rad/s} \tag{3B-3}$$

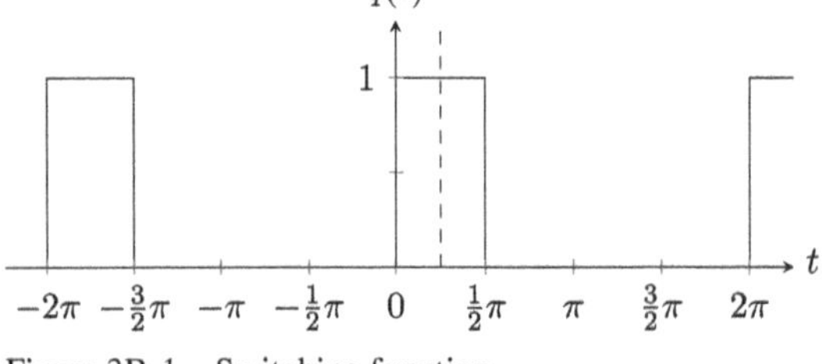


Figure 3B-1 Switching function.

The center of the second pulse in the example, indicated with the dotted vertical line, is at $t = \pi/4$, which is 1/8 of the switching period. Thus,

$$\begin{aligned}\theta(0) &= \omega_{sw} \times \frac{T_{sw}}{8} \\ &= (1\ \text{rad/s}) \times \left(\frac{2\pi}{8}\text{s}\right) = \pi/4\ \text{rad}\end{aligned} \tag{3B-4}$$

We can now write the Fourier series using (3B-1) through (3B-4) and (3.2-22):

$$q(t) = \frac{1}{4} + \frac{2}{\pi}\sum_{n=1}^{\infty} \frac{\sin(n\pi/4)}{n}\cos(nt - n\pi/4) \tag{3B-5}$$

Before concluding this example, note that harmonics that are multiples of four (which is the inverse of $D$) are zero because of the sine coefficient term. This relationship between duty ratio and harmonics is a general one, and it is important in advanced power electronic circuit analysis and design.

## Energy Storage Elements

In addition to switches, power electronic circuits also rely on *energy storage elements*, namely, inductors and capacitors. In fact, with components from just these three fundamental categories – switches, (possibly coupled) inductors, and capacitors – one can construct, in their most basic form, the majority of the important power electronic circuits in common use. In the context of power electronic circuit analysis there are two important concepts, based on the conservation of energy, which will prove invaluable for circuit analysis.

Consider the switched inductor circuit shown in Fig. 3.2-7a. When the switch is closed, the inductor current will ramp up linearly according to $di/dt = V_{dc}/L$. Because the energy stored in the inductor's magnetic field is $1/2Li^2$, it will increase quadratically. Theoretically, if the switch remains closed, the current will ramp up indefinitely and the inductor will continue to store energy at an ever-increasing rate. Of course, in a real inductor the wires will overheat and melt before the current gets too high, but the takeaway is the same: it is a bad idea to close a dc source into an inductor indefinitely. Rather, it is better to keep the average current constant and transfer energy into and out of the inductor cyclically. In the circuit shown, this would require additional circuitry, but we will leave that detail out for now. If the energy put

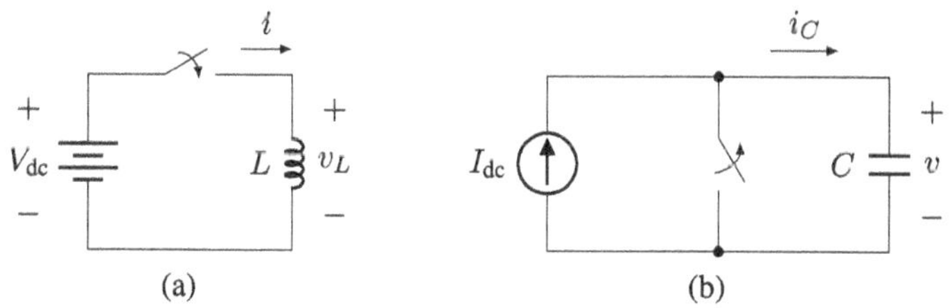


Figure 3.2-7 Switched (a) inductor and (b) capacitor circuits.

into the inductor during charging is always the same as the energy taken out while discharging, the inductor will not overheat. Through simple analysis (see Example 3C), we can show that this steady-state operation is achieved if and only if the average inductor voltage, $\hat{v}_L$, is zero. Thus, the first key principle of energy storage devices in power electronic circuits is

$$\hat{v}_L = 0 \tag{3.2-24}$$

Analogous arguments can be applied to the switched capacitor circuit shown in Fig. 3.2-7b. When the switch is opened, the capacitor voltage will increase linearly according to $dv/dt = I_{dc}/C$. The energy stored in the capacitor's electric field is $1/2Cv^2$, and it will ramp up quadratically with the voltage. Theoretically, the voltage will ramp up indefinitely and the capacitor will continue to store energy at an ever-increasing rate. Of course, as with a real inductor, in a real capacitor physical reality will kick in; the voltage will eventually reach the breakdown voltage of the capacitor separator and arc across the capacitor terminals, creating a short circuit. Thus, it is a bad idea to close a dc current source into a capacitor indefinitely. Instead, energy should be transferred into and out of the capacitor periodically, so that the average voltage remains constant. If the energy put into the capacitor during charging is the same as the energy extracted during discharge, the capacitor will not overcharge and short circuit. One can show that this steady-state operation can be achieved if and only if the average capacitor current, $\hat{i}_C$, is zero. Thus, the second key principle of energy storage devices in power electronic circuits is

$$\hat{i}_C = 0 \tag{3.2-25}$$

We now have all the basic tools to begin an introductory analysis of several power electronic switching circuits. Together with basic circuit analysis techniques, these tools will help us to analyze dc-dc, ac-dc, and dc-ac circuits in the following sections.

**Example 3C. Average inductor voltage for energy balance.** Show that the average inductor voltage must be zero for energy balance in the switched inductor circuit in Fig. 3.2-7a.

The inductor voltage, $v_L$, is expressed as

$$v_L = L\frac{di}{dt} \tag{3C-1}$$

With constant inductance, the average inductor voltage is

$$\hat{v}_L = L\widehat{\frac{di}{dt}} \tag{3C-2}$$

If $\widehat{\frac{di}{dt}} \neq 0$, then the inductor current magnitude will continually increase. Since the stored inductor energy is $1/2Li^2$, this represents an unbalanced energy condition. Therefore, for energy balance

$$\widehat{\frac{di}{dt}} = 0 \tag{3C-3}$$

Combining (3C-2) and (3C-3) for energy balance reveals that

$$\hat{v}_L = 0 \tag{3C-4}$$

---

**SP3.2-1** (a) Find the voltage conversion efficiency $\eta = P_{out}/P_{in}$ for the voltage divider in Fig. 3.2-2 with $r_1 = 1\ \Omega$, $r_2 = 10\ \Omega$, and $r_{out} = 100\ \Omega$. (b) Is this a good power conversion circuit? [(a) $\eta = 0.0817 = 8.17\%$; (b) No, almost all of the input power is consumed in the circuit.]

**SP3.2-2.** Using (3.2-16) through (3.2-19) describe the Fourier series of the square wave switching function defined by $q(t) = 1/2[1 + \text{sign}(\sin(t))]$. [dc component with $a_0 = 1/2$; odd function so no cosines; half-wave symmetry so no even harmonics]

**SP3.2-3.** Show that $\widehat{i_C} = 0$ is necessary for energy balance in a capacitor. [see Example 3C and use analogous arguments]

## 3.3 DC-DC CONVERSION

Consider the ideal dc-dc conversion circuit shown conceptually in Fig. 3.3-1. Power is to be transferred from the dc source, $V_{hi}$, to the lower voltage dc load, $V_{lo}$. Although this motivating example transfers power from a higher voltage source to a lower voltage load, the analysis concepts that follow are equally applicable for low-to-high voltage conversion circuits. Through examining the general Fourier series equations, we have seen that a periodic signal with a dc component can be filtered to give a constant value dc signal. Thus, if we can create a periodic signal from $V_{hi}$ ($V_{in}$ in Fig. 3.2-2) that has an average value of $V_{lo}$, we can filter out the harmonic content and be left with the desired output. A simple periodic signal that meets the requirements is a square wave with magnitude $V_{hi}$ and duty ratio $D = V_{lo}/V_{hi}$. $D$ has a value between 0 and 1, and the length of time, per period, that the square wave is high is $DT_{sw}$, as indicated in Fig. 3.3-2.

This leads us to a dc-dc conversion strategy: transfer power from $V_{hi}$ in square pulses with $D = V_{lo}/V_{hi}$ and filter the result to leave only the dc component, $V_{lo}$.

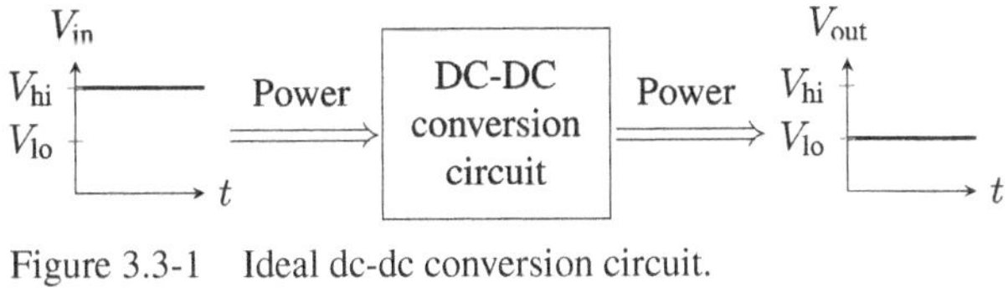


Figure 3.3-1 Ideal dc-dc conversion circuit.

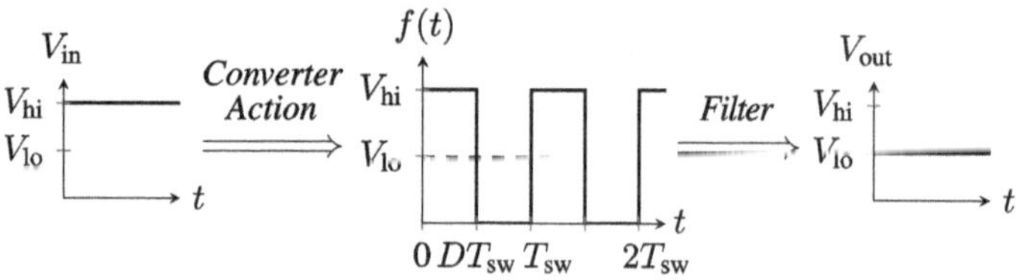


Figure 3.3-2 Filtered square wave voltage dc-dc voltage conversion strategy.

Because a low-pass filter is needed to eliminate all frequencies above dc, the higher the frequency of the pulses, the simpler and more effective this filter can be.

## Buck Converter

A circuit that accomplishes the dc-dc conversion strategy laid out above is the *buck converter*, shown in Fig. 3.3-3. This circuit has one active switch – a transistor – and one passive switch – a diode. Assuming $L$ and $C$ are "large" (we will see what this means a little later) and that the circuit is in steady state, for our initial analysis we can approximate as constants the output voltage, $v_{out}$, and the inductor current, $i_L$. When the active switch is closed, the diode is reverse-biased and does not conduct current; the circuit is in Configuration A, as shown in Fig. 3.3-4, and current flows through the inductor from the input to the output. When the active switch opens, the inductor current cannot change instantaneously and Kirchoff's Current Law (KCL) requires that this current have a path. The inductor current decreases, which creates a negative voltage across the inductor. This forward-biases the diode, turning it on; the circuit switches to Configuration B. An appropriate switching function controls the active switch, switching the circuit between these two configurations. Let us now analyze Configurations A and B in more detail.

In Configuration A, $i_{in} = i_L$, $v_* = V_{in}$, and $v_L = V_{in} - v_{out}$. In Configuration B, the input voltage is detached from the circuit, $v_* = 0$, $i_{in} = 0$, and $v_L = -v_{out}$. If we assume that the active switch is switched very quickly so that the circuit rapidly oscillates between Configurations A and B, an average behavior will emerge. We will first analyze this, and then add the high frequency details afterward. In terms of the switching function,

$$\begin{aligned} v_*(t) &= q(t)V_{in} + (1 - q(t)) \times 0 \\ &= q(t)V_{in} \end{aligned} \tag{3.3-1}$$

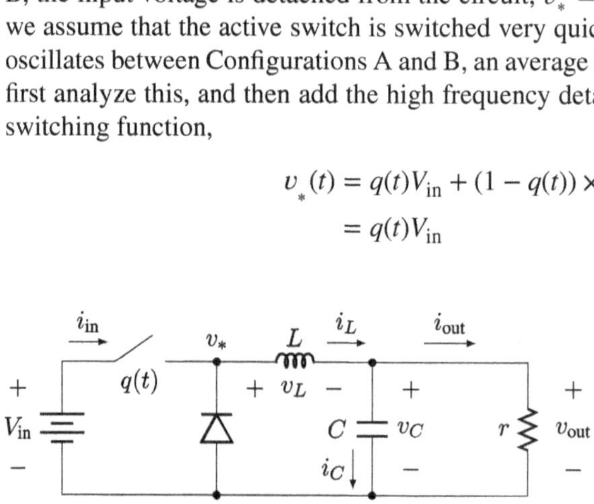


Figure 3.3-3 Buck converter circuit.

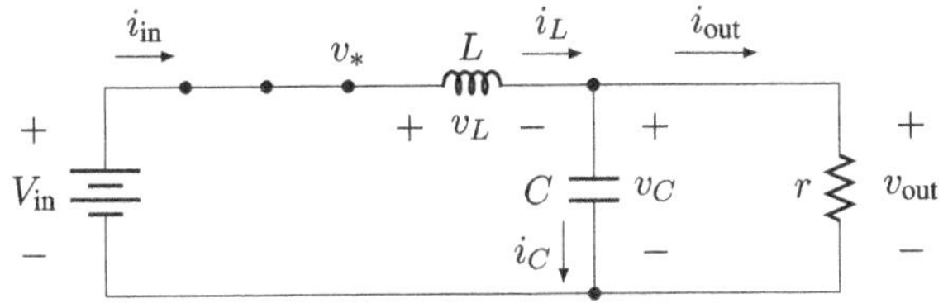


(a)

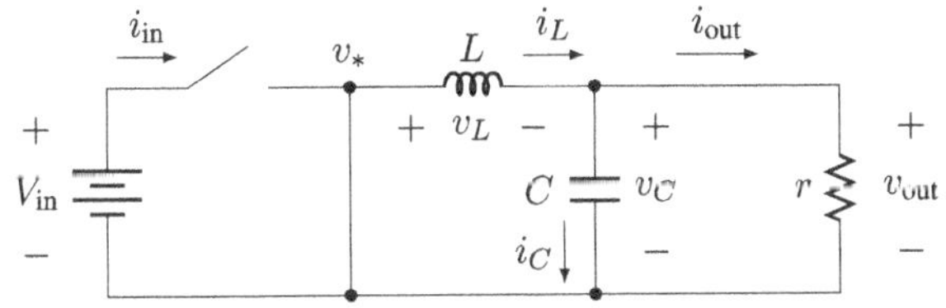


(b)

Figure 3.3-4 Buck converter in (a) Configuration A, with $q(t) = 1$, and (b) Configuration B, with $q(t) = 0$.

The square wave voltage, $v_*$, can be described using (3.2-22) with $\theta(0) = \pi/2$. The time average of $v_*$ over one switching period is

$$\begin{aligned}\hat{v}_* &= \frac{1}{T_{\text{sw}}}\int_0^{T_{\text{sw}}} v_*\,dt \\ &= \frac{1}{T_{\text{sw}}}\left(\int_0^{DT_{\text{sw}}} V_{\text{in}}dt + \int_{DT_{\text{sw}}}^{T_{\text{sw}}} 0 \times dt\right) \\ &= \frac{1}{T_{\text{sw}}}\int_0^{DT_{\text{sw}}} V_{\text{in}}dt = DV_{\text{in}}\end{aligned} \tag{3.3-2}$$

Now, because the average voltage across the inductor is 0 (3.2-24), we have found the average output voltage:

$$\hat{v}_{\text{out}} = \hat{v}_* = DV_{\text{in}} \tag{3.3-3}$$

The buck converter's voltage *conversion ratio* is then

$$\frac{\hat{v}_{\text{out}}}{V_{\text{in}}} = D \tag{3.3-4}$$

Equations (3.3-3) and (3.3-4) equivalently describe the average behavior of the buck converter, showing that the average output voltage is a fraction of the input voltage and is determined by the duty ratio $D$. Look again at Fig. 3.3-2. The equation we have derived encompasses all the features of that figure. The input voltage is first "chopped" by the active switch into the square wave $v_*$. A buck converter is sometimes called a "chopper" because of this. With rapid enough switching and "large" $L$ and $C$, $v_*$ is filtered to leave its average value as $DV_{\text{in}}$. Thus, $\hat{v}_{\text{out}}/V_{\text{in}} = D$ and dc voltage transformation has been accomplished!

Let us now consider the current and power flows in this circuit. Because the average current in the output capacitor is zero (3.2-25),

$$\hat{i}_{\text{out}} = \hat{i}_L \tag{3.3-5}$$

Further, the input current can be expressed in terms of the switching function as

$$\begin{aligned} i_{\text{in}}(t) &= q(t)i_L + (1 - q(t)) \times 0 \\ &= q(t)i_L \end{aligned} \tag{3.3-6}$$

The time-average input current over one switching period is

$$\begin{aligned} \hat{i}_{\text{in}} &= \frac{1}{T_{\text{sw}}} \int_0^{T_{\text{sw}}} i_{\text{in}}(t)dt \\ &= \frac{1}{T_{\text{sw}}} \left( \int_0^{DT_{\text{sw}}} i_L dt + \int_{DT_{\text{sw}}}^{T_{\text{sw}}} 0 \times dt \right) \\ &= \frac{1}{T_{\text{sw}}} \int_0^{DT_{\text{sw}}} i_L dt = D\hat{i}_L \end{aligned} \tag{3.3-7}$$

In the last step we have used our assumption that $i_L$ is approximately constant, and we have set $i_L = \hat{i}_L$. The average power into the converter, $\hat{P}_{\text{in}}$, is

$$\hat{P}_{\text{in}} = V_{\text{in}}\hat{i}_{\text{in}} = V_{\text{in}}D\hat{i}_L \tag{3.3-8}$$

and the average power out of the converter, $\hat{P}_{\text{out}}$, is approximated as

$$\hat{P}_{\text{out}} = \widehat{(v_{\text{out}}i_{\text{out}})} = \hat{v}_{\text{out}}\hat{i}_L \tag{3.3-9}$$

To set the average of a product equal to the product of two averages, which is in general not a valid mathematical step, we have used our assumption that $v_{\text{out}}$ and $i_L$ are constant and equal to $\hat{v}_{\text{out}}$ and $\hat{i}_L$, respectively. Under this assumption (3.3-9) is valid. Substituting the buck converter voltage relationship in (3.3-3) into (3.3-8) and comparing with (3.3-9) yields

$$\hat{P}_{\text{in}} = \hat{v}_{\text{out}}\hat{i}_L = \hat{P}_{\text{out}} \tag{3.3-10}$$

showing that the average power is conserved. Compare this with the voltage divider circuit we studied earlier where $P_{\text{out}}$ was exceedingly small and the majority of the input power was consumed by the device. In contrast, the ideal buck converter transfers all of the average input power to the output and consumes none, making it an exemplary power conversion device.

At this point you may be somewhat unsatisfied with the result, as we have left out a few important details. Namely: what constitutes "large" $L$ and $C$, and what is "fast" switching? We will look at these details now; however, before we do it is worth reviewing the power conversion strategy embodied in Fig. 3.3-2 and the discussion and equations leading to (3.3-3). Similar strategies and concepts are used over and over in power electronic converters, and a firm understanding now will go a long way in your understanding of more advanced devices.

Let us now look at the steady-state operation of the buck converter circuit. We will examine the inductor current and capacitor voltage more closely to gain insight into the relationship between $L$, $C$, and $f_{\mathrm{sw}}$. Thus far, we have assumed constant $v_{\mathrm{out}}$ and $i_L$. To analyze the inductor current in more detail, we will continue to assume that the output capacitance is large so that the capacitor can be approximated as a dc voltage source with voltage $V_{\mathrm{out}}$. Figure 3.3-5 shows the approximate circuit.

The inductor voltage can be expressed using (3.3-1) as

$$\begin{aligned} v_L &= v_* - V_{\mathrm{out}} \\ &= q(t)V_{\mathrm{in}} - V_{\mathrm{out}} \end{aligned} \tag{3.3-11}$$

Because the inductor voltage is also $L di_L/dt$, we have

$$\frac{di_L}{dt} = \frac{q(t)V_{\mathrm{in}} - V_{\mathrm{out}}}{L} \tag{3.3-12}$$

When the active switch is on and the circuit is in Configuration A, the inductor current ramps up linearly with slope $(V_{\mathrm{in}} - V_{\mathrm{out}})/L$; when the active switch is off and the circuit is in Configuration B, the current ramps down linearly with slope $-V_{\mathrm{out}}/L$. The triangular shaped periodic waveform with average $\hat{i}_L = I_L$ shown in Fig. 3.3-6 is the result. The peak-to-peak ripple magnitude of this waveform is $\Delta I$, and its period is $T_{\mathrm{sw}}$.

We can calculate $\Delta I$ by taking the magnitude of the result of integrating (3.3-12) with respect to $t$ either from 0 to $DT_{\mathrm{sw}}$ or from $DT_{\mathrm{sw}}$ to $T_{\mathrm{sw}}$. In either case the result is

$$\begin{aligned} \Delta I &= \frac{V_{\mathrm{out}}}{L}(T_{\mathrm{sw}} - DT_{\mathrm{sw}}) \\ &= \frac{V_{\mathrm{out}}}{L \times f_{\mathrm{sw}}}(1 - D) \end{aligned} \tag{3.3-13}$$

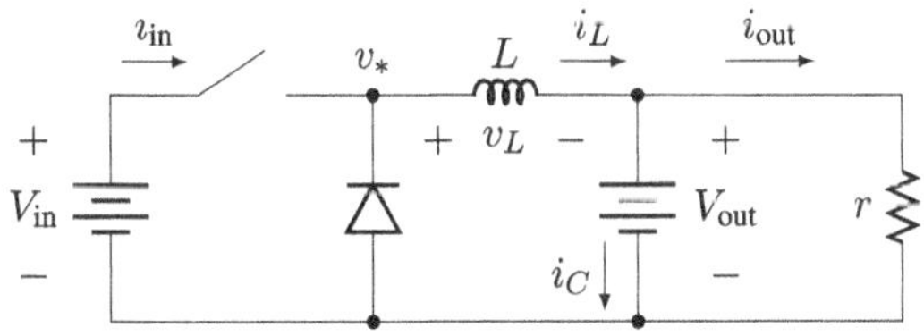


Figure 3.3-5 Buck converter with "large" output capacitance approximated as a voltage source.

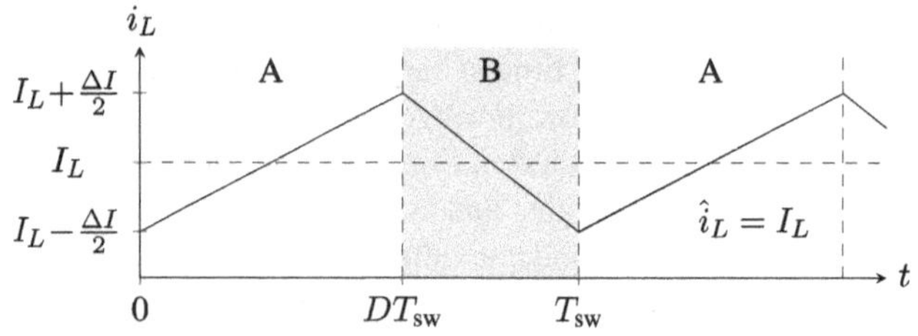


Figure 3.3-6 Buck converter inductor current waveform over one and a half switching periods.

Notice that the ripple magnitude is a function of $L$ and $f_{sw}$. For a given switching frequency, if a maximum ripple $\Delta I_{max}$ is specified, the minimum inductance necessary is

$$L_{min} = \frac{V_{out}(1 - D)}{f_{sw} \times \Delta I_{max}} \tag{3.3-14}$$

According to (3.3-14), if $L \geq L_{min}$, then $\Delta I \leq \Delta I_{max}$.

We have thus derived a minimum threshold for determining a "large" $L$, based on the switching frequency and a specified maximum allowable ripple current. But, what about $C$? Thus far, $C$ has been assumed large so that the capacitor could be considered a dc voltage source. In reality, the capacitor voltage will have an average value with some triangular ripple, similar to the inductor current waveform. To analyze the capacitor, we will use the circuit shown in Fig. 3.3-7. The inductor current is represented as a current source with average $\widehat{i}_L = I_L$ and triangular ripple of magnitude $\pm\Delta I/2$. The capacitor is in parallel with the resistive load. Because according to (3.2-25) the capacitor has no average current in the steady state, $I_L$ flows through the resistor and the ripple current flows through the capacitor. In reality, a small fraction of the ripple current will also flow through the resistive load; however, if the capacitor impedance at the ripple frequency is much smaller than the load impedance, which it will be in normal circumstances, then the majority of the ripple current will flow through the capacitor and our assumption is valid. In fact, in many instances, with a reasonable amount of output capacitance the inductance in the wires from the converter to the load is large enough to validate our assumption, regardless of the load impedance.

The capacitor voltage, $v_C$, will rise and fall according to

$$\frac{dv_C}{dt} = \frac{i_C(t)}{C} \tag{3.3-15}$$

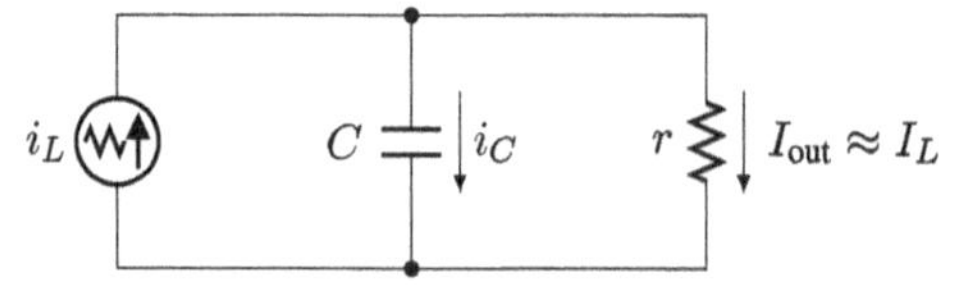


Figure 3.3-7 Buck converter output stage used to calculate $C_{min}$.

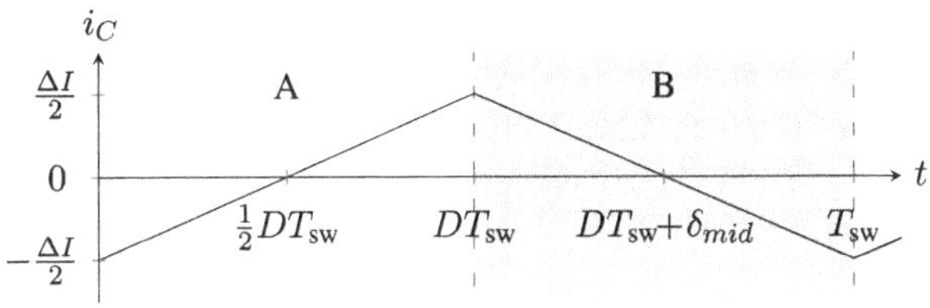


Figure 3.3-8 Buck converter approximate capacitor current.

where $i_C$ is the ripple current from the inductor. Figure 3.3-8, which is similar to Fig. 3.3-6 except that the dc offset has been removed, shows the capacitor current.

The variable $\delta_{mid}$ is introduced to mark the midway point between $DT_{sw}$ and $T_{sw}$:

$$\delta_{mid} \triangleq \frac{T_{sw} - DT_{sw}}{2} \tag{3.3-16}$$

The capacitor charges when $i_C > 0$ and discharges when $i_C < 0$. Thus, the capacitor peak-to-peak voltage ripple $\Delta V$ is found by calculating the voltage rise that occurs between $DT_{sw}/2$ and $DT_{sw} + \delta_{mid}$. Because the ripple is symmetrical about the $t$-axis, its magnitude can also be found by calculating the voltage drop that occurs when $i_C$ is negative; however, because this requires integrating over a split interval, we choose the positive region for simplicity. Specifically,

$$\begin{aligned}
\Delta V &= \int_{DT_{sw}/2}^{DT_{sw}+\delta_{mid}} \frac{dv_C}{dt} dt \\
&= \frac{1}{C} \frac{\left[\left(\frac{DT_{sw}}{2} \times \frac{\Delta I}{2}\right) + \left(\delta_{mid} \times \frac{\Delta I}{2}\right)\right]}{2} \\
&= \frac{\Delta I}{8C}(DT_{sw} + 2\delta_{mid})
\end{aligned} \tag{3.3-17}$$

Using (3.3-16) to simplify, this becomes

$$\begin{aligned}
\Delta V &= \frac{\Delta I \times T_{sw}}{8C} \\
&= \frac{\Delta I}{8C \times f_{sw}}
\end{aligned} \tag{3.3-18}$$

The voltage ripple magnitude is a function of $C, f_{sw}$, and the current ripple magnitude. For a given switching frequency and current ripple, then, the minimum capacitance $C_{min}$ necessary to keep the output voltage ripple below a desired limit $\Delta V_{max}$ is

$$C_{min} = \frac{\Delta I}{8 f_{sw} \Delta V_{max}} \tag{3.3-19}$$

According to (3.3-19), if $C \geq C_{min}$, then $\Delta V \leq \Delta V_{max}$. Thus, we have derived a minimum threshold for determining a "large" $C$, based on the switching frequency, the allowable voltage ripple, and the inductor current ripple.

The last remaining detail is to discuss the switching frequency. Specifically, what is meant by "fast" switching? There are several ways to answer this question; we will only discuss a few of the most important ones here. First, notice from (3.3-14) and (3.3-19) that both $L_{\text{min}}$ and $C_{\text{min}}$ are inversely proportional to $f_{\text{sw}}$. Thus, all else being equal, a higher switching frequency reduces the size of the converter's energy storage components. In a general sense, then, "fast" switching is that which allows the circuit to have reasonably sized components. Next, notice from (3.3-13) and (3.3-18) that $f_{\text{sw}}$ is also inversely proportional to both $\Delta I$ and $\Delta V$. All else being equal, then, faster switching reduces ripple magnitude. Thus, "fast" switching keeps the circuit ripple below a specified threshold. A third way to consider the question is by considering the Fourier series of $v_*$, which, according to (3.3-1) is simply (3.2-22) multiplied by $V_{\text{in}}$. According to our dc-dc conversion process shown in Fig. 3.3-2, we would like to low-pass filter the square wave $v_*$ to leave only its dc component as $V_{\text{out}}$. Any ripple on the output is unwanted. Together, the inductor and capacitor serve as a low-pass filter to remove it. The higher the switching frequency, the farther it will be in the frequency domain from the dc component and the easier it will be for the filter components to remove it and higher harmonics. Thus, "fast" switching is that which allows the low-pass *LC* output filter to work effectively.

Of course, all of these qualitative fast switching descriptions are related and essentially describe the same physical circuit behavior; however, we provide three alternative ways of thinking about the concept in order to facilitate a deeper understanding. Other considerations that influence the switching frequency include audible noise – humans hear frequencies up to around 20 kHz, so switching frequencies are often chosen higher than this to avoid converter "buzzing," – switch hardware limitations, losses – real switches have small losses each time they switch; the higher the frequency, the higher these losses – and switch gate control hardware limits. These advanced topics are considered in later courses.

**Example 3D. Buck converter design.** Design a buck converter to power a 5 V, 25 W resistive load from a 12 V battery. Choose appropriate values for $f_{\text{sw}}$, $D$, $L$, and $C$. The inductor's peak-to-peak current ripple should be no more than 40% of its average current and the output peak-to-peak voltage ripple should be no more than 4% of the nominal value. For the circuit components you choose, calculate the expected $\Delta I$ and $\Delta V$ and compare to the specifications.

The duty ratio is given by (3.3-4) and is

$$D = \hat{v}_{\text{out}}/V_{\text{in}} = 5/12 \tag{3D-1}$$

To avoid the audible range but remain low enough for average switch hardware to accommodate, we will choose $f_{\text{sw}} = 30$ kHz. Once the design is complete, we can verify that our components are reasonably sized. If they are not, we can adjust the switching frequency and component design accordingly. The output power is 25 W; thus

$$\hat{i}_{\text{out}} = 25\text{ W}/5\text{ V} = 5\text{ A} \tag{3D-2}$$

and from (3.3-5) $\hat{i}_L = 5$ A. From the specification,

$$\Delta I_{max} = 0.4 \times \hat{i}_L = 2 \text{ A} \tag{3D-3}$$

thus from (3.3-14)

$$L_{min} = \frac{5 \times (1 - 5/12)}{(30 \times 10^3) \times 2} = 48.61 \ \mu\text{H} \tag{3D-4}$$

Let us choose $L = 50$ μH. Equation (3.3-13) now gives the expected current ripple as

$$\Delta I = \frac{5 \times (1 - 5/12)}{(30 \times 10^3) \times (50 \times 10^{-6})} = 1.94 \text{ A} \tag{3D-5}$$

which is 39% of the nominal current, within the specification. From the specification,

$$\Delta V_{max} = 0.04 \times 5 \text{ V} = 0.2 \text{ V} \tag{3D-6}$$

and from (3.3-19)

$$C_{min} = \frac{1.94}{8 \times (30 \times 10^3) \times 0.2} = 40.4 \ \mu\text{F} \tag{3D-7}$$

Because it is a standard capacitor size, we will choose $C = 47$ μF. Based on this choice, (3.3-18) gives the expected voltage ripple as

$$\Delta V = \frac{1.94}{8 \times (47 \times 10^{-6}) \times (30 \times 10^3)} = 0.17 \text{ V} \tag{3D-8}$$

which is 3.4% of the nominal output voltage, within the specification. The chosen components are reasonably sized for this type of circuit and the circuit achieves its design goals. Thus, the switching frequency does not need to be adjusted.

---

The circuit designed in Example 3D has been simulated and several waveforms are shown in Fig. 3.3-9 to illustrate the behavior. Notice that $v_* = q(t)V_{in}$ is as designed, with $D = 5/12$. The period is 33.33 μs, which corresponds to the 30 kHz switching frequency. The inductor current has an average of 5 A and peak-to-peak ripple of approximately 2 A. The input current is equal to the inductor current when the active switch is on and zero otherwise. The output capacitor voltage has an average of 5 V and peak-to-peak ripple of approximately 0.18 V. Notice that the output voltage ripple is out of phase with the inductor current ripple; this is because the capacitor voltage rises when the inductor current is greater than its average, 5 A, and discharges when it is less than its average. The capacitor and output (resistor) current are plotted together to show that the majority of the ripple in $i_L$ goes through the capacitor while the dc current goes exclusively through the resistor. Finally, the output resistor power is plotted to show the average power is 25 W, as designed, with small peak-to-peak ripple of roughly 2 W, or 8%.

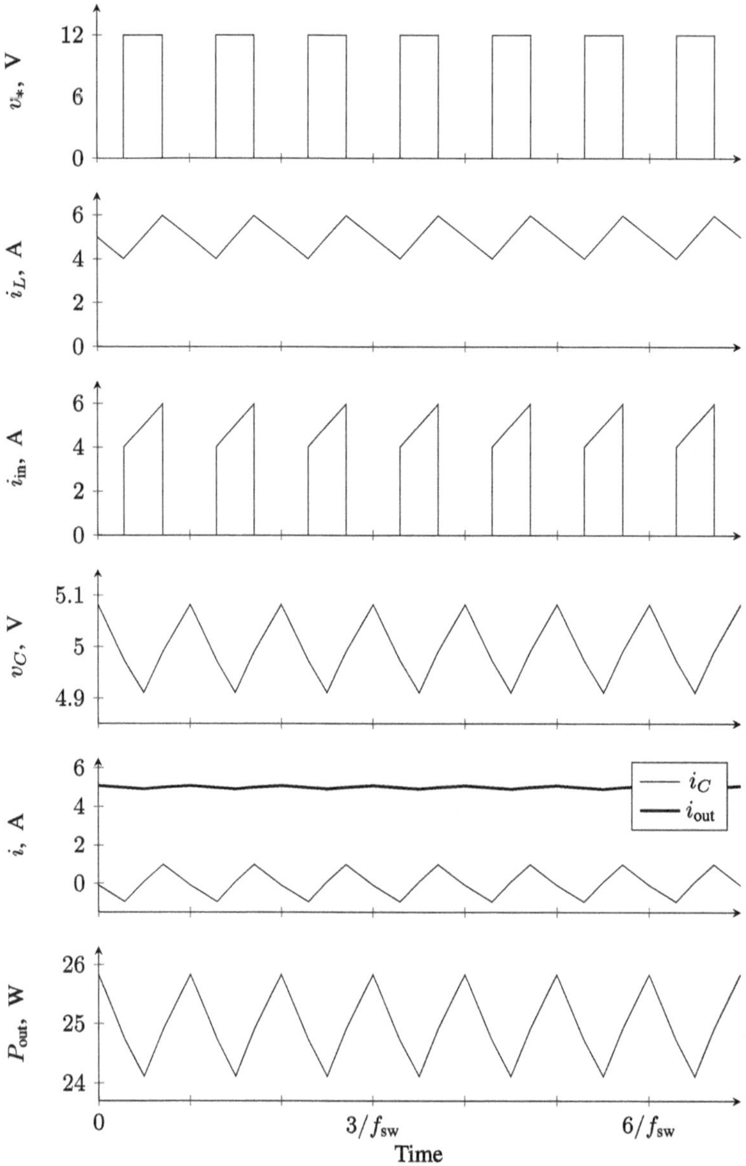


Figure 3.3-9 Circuit waveforms from the 25 W buck converter designed in Example 3D.

## Boost Converter

To highlight the power electronic conversion concepts we have covered thus far, we will briefly introduce a companion circuit to the buck converter: the dc-dc *boost converter*. The basic boost converter circuit is shown in Fig. 3.3-10. Its purpose is the inverse of the buck converter: it transfers power from a lower voltage dc source to a higher voltage dc load. In contrast to the buck converter, in the boost converter the inductor is always connected in series with the input voltage source. By controlling the active switch with an appropriate switching function, the inductor can be charged and discharged while maintaining an average current. This current is alternately disconnected from and connected to the load through the passive (diode) switch. The load current is filtered with the output capacitor. With sufficiently "large" $L$ and $C$, the inductor and capacitor can be approximated as dc current and voltage sources, respectively, to facilitate design and analysis.

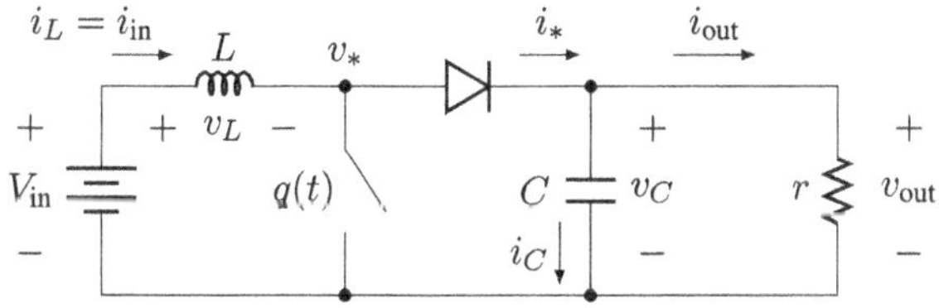


Figure 3.3-10 Boost converter circuit.

Like the buck converter, the boost converter is periodically switched between two circuit configurations. When the active switch is closed, the diode is reverse-biased and turns off. In this circuit configuration, the inductor's voltage is $V_{in}$ and its current is ramped up as energy is transferred from the source to the inductor's magnetic field. The active switch is then opened, which causes the diode to turn on in order to provide a path for the inductor current to the load. Since $\hat{v}_{out} > V_{in}$, the inductor voltage is now negative and its current ramps down as the energy transferred to it from the source is now transferred to the load. The inductor voltage is thus

$$v_L(t) = q(t)V_{in} + (1 - q(t)) \times (V_{in} - \hat{v}_{out}) \tag{3.3-20}$$

Because the average inductor voltage is zero in the steady state (3.2-24), we have

$$\begin{aligned} \hat{v}_L &= \frac{1}{T_{sw}} \int_0^{T_{sw}} [q(t)V_{in} + (1 - q(t)) \times (V_{in} - \hat{v}_{out})]dt \\ &= \frac{1}{T_{sw}} \left[ \int_0^{DT_{sw}} V_{in} dt + \int_{DT_{sw}}^{T_{sw}} (V_{in} - \hat{v}_{out}) dt \right] \\ &= \frac{1}{T_{sw}} [V_{in} \times DT_{sw} + (V_{in} - \hat{v}_{out}) \times (T_{sw} - DT_{sw})] = 0 \end{aligned} \tag{3.3-21}$$

which is simplified to

$$\frac{\hat{v}_{out}}{V_{in}} = \frac{1}{1-D} \tag{3.3-22}$$

Equation (3.3-22) describes the steady-state voltage conversion ratio of the boost converter. It is analogous to (3.3-4) for the buck converter. In the boost converter the average output voltage is greater than or equal to the input voltage, and the conversion ratio is controlled by $D$. When $D = 0$, the active switch is never on and the input and average output voltages are equal. At any other value of $D$, the average output voltage is higher than the input. For example, for $D = 0.5$, $\hat{v}_{out} = 2V_{in}$ and for $D = 0.75$, $\hat{v}_{out} = 4V_{in}$. Note that according to (3.3-22) $\hat{v}_{out}$ tends toward infinity as $D$ approaches unity. This is because as $D$ increases, more energy is transferred to the inductor while the active switch is in the on-state, and the fraction of the switching period for this energy to be transferred to the load when the active switch is off, $(1 - D)$, decreases. In order for this increasing energy to be transferred to the load in a decreasing time, it must be done in short pulses of current at high voltage. In the extreme as $D$ approaches unity, these pulses become infinitesimally small while the voltage becomes infinitely large. Although this description is true in the ideal circuit we have analyzed, in any real converter this is not the case. Series resistance in the inductor will limit the amount of energy that can be transferred to it, bounding the output voltage, as will be examined in Example 3E. More generally, the converter's losses will dominate the behavior at higher duty ratios, bounding the output voltage to a finite value. For example, Fig. 3.3-11 compares the conversion ratio of an ideal boost converter (3.3-22) to that of a boost converter with resistive losses in its inductor equal to 1% of the load resistance. As can be seen, in the lossy circuit the boost ratio peaks at five with $D = 0.9$, then falls off to zero at $D = 1$.

Analysis of the circuit currents and powers in the ideal boost converter reveals that

$$i_{in} = i_L \tag{3.3-23}$$

$$\begin{aligned} \hat{i}_{out} &= (1-D)\hat{i}_L \\ &= (1-D)\hat{i}_{in} \end{aligned} \tag{3.3-24}$$

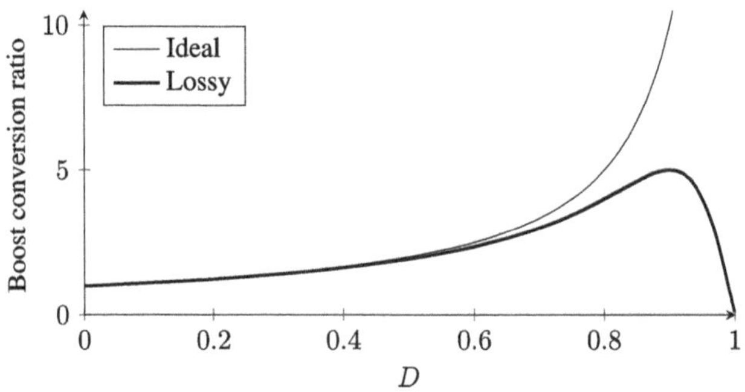


Figure 3.3-11 Boost converter conversion ratio ($\hat{v}_{out}/V_{in}$) for ideal (lossless) and lossy circuits.

$$\begin{aligned}\widehat{P}_{\text{in}} &= V_{\text{in}}\widehat{i}_{\text{in}} \\ &= V_{\text{in}}\frac{\widehat{i}_{\text{out}}}{1-D} = \frac{V_{\text{in}}}{1-D}\widehat{i}_{\text{out}} \\ &= \widehat{v}_{\text{out}}\widehat{i}_{\text{out}}\end{aligned} \tag{3.3-25}$$

Assuming $v_{\text{out}}$ and $i_{\text{out}}$ are approximately constant and equal to $\widehat{v}_{\text{out}}$ and $\widehat{i}_{\text{out}}$, respectively, we have

$$\widehat{P}_{\text{out}} = (\widehat{v_{\text{out}}i_{\text{out}}}) = \widehat{v}_{\text{out}}\widehat{i}_{\text{out}} \tag{3.3-26}$$

Thus, $\widehat{P}_{\text{in}} = \widehat{P}_{\text{out}}$ and power is conserved.

The boost converter power conversion strategy is hinted at by (3.3-24), and is summarized as follows. In the steady state, the inductor can be approximated as a dc current source; we ignore small ripple current in this discussion. Because $\widehat{P}_{\text{in}} = \widehat{P}_{\text{out}}$ and $\widehat{v}_{\text{out}} > V_{\text{in}}$, this requires that $\widehat{i}_{\text{out}} < \widehat{i}_{\text{in}}$. This dc current reduction is accomplished by first "chopping" the input current to create the square wave diode current $i_*$. The dc value of $i_*$ is then obtained by filtering; the whole process being analogous to the voltage reduction strategy used in the buck converter and shown in Fig. 3.3-2. In the boost converter, the circuit reduces dc current the same way the buck converter reduces dc voltage. In both circuits, because power is conserved, when output voltage or current decreases, its corresponding current or voltage must increase.

Circuit waveforms from a 25 W, 5 V to 12 V boost converter with the same inductor and capacitor values as the buck converter in Example 3D are shown in Fig. 3.3-12. The waveforms are arranged so that when compared side-by-side with Fig. 3.3-9, the analogous conversion strategies are evident. For example, the top boost waveform, $i_*$, is analogous to the top buck waveform, $v_*$. Each of these is the "chopped" version of the related dc input value.

Choosing appropriate $L$, $C$, and $f_{\text{sw}}$ in boost converter design requires similar analysis as was done for the buck converter. Inductor current ripple, capacitor voltage ripple, and switch frequency considerations are all used to choose appropriate values. Although we will not explicitly cover this here, you should now have the necessary knowledge and tools to accomplish the task.

**Example 3E. Boost converter with resistive losses.** Derive the boost conversion ratio for the non-ideal boost converter shown in Fig. 3E-1.

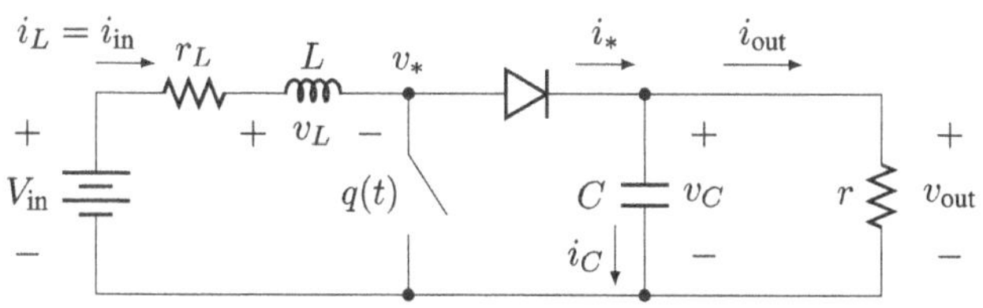


Figure 3E-1 Boost converter circuit with resistive losses in the inductor.

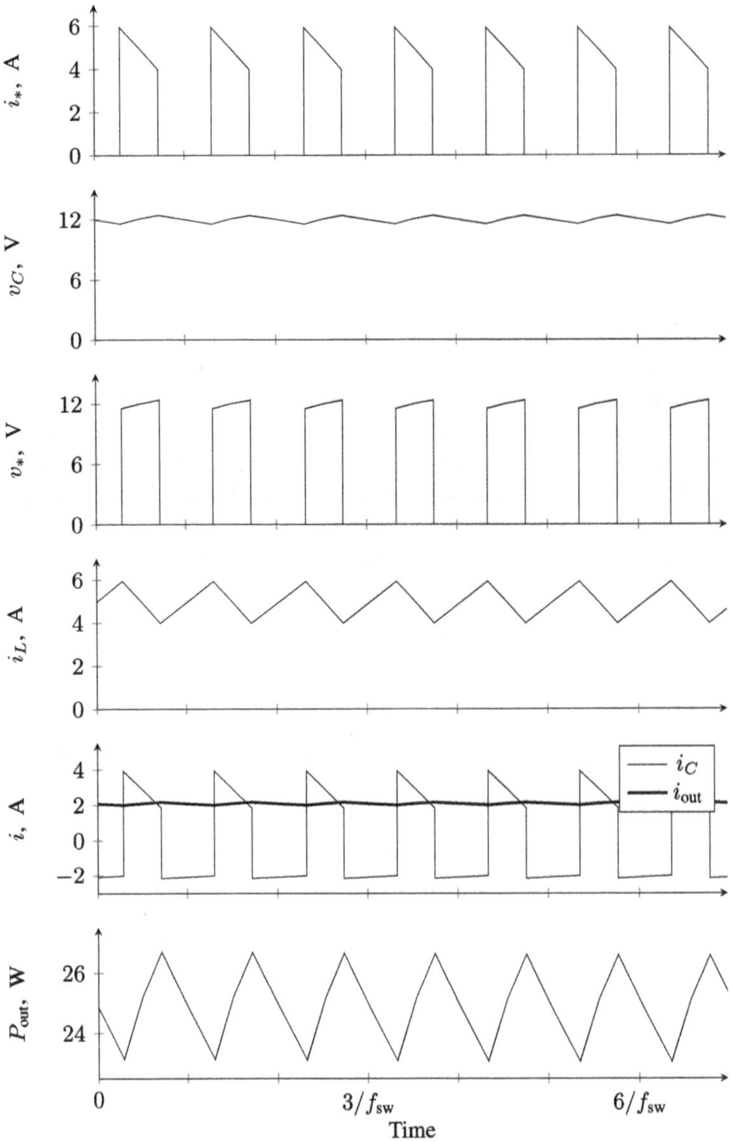


Figure 3.3-12 Circuit waveforms from ideal 25 W boost converter with $L = 50\ \mu H$, $C = 47\ \mu F$, $f_{sw} = 30$ kHz, $V_{in} = 5$ V, $\hat{v}_{out} = 12$ V, and $D = 7/12$.

With the inductor resistive losses, the power balance for the circuit can be expressed

$$\widehat{P}_{\text{in}} = \widehat{P}_{\text{out}} + \widehat{P}_{\text{loss}} \tag{3E-1}$$

where $\widehat{P}_{\text{loss}}$ is the average resistive loss in the inductor. By inspection

$$\widehat{P}_{\text{in}} = V_{\text{in}}\widehat{i}_{\text{in}} = V_{\text{in}}\widehat{i}_L \tag{3E-2}$$

and

$$\widehat{P}_{\text{out}} = \widehat{i^2_{\text{out}}}r = \widehat{i}^2_{\text{out}}r \tag{3E-3}$$

The last equality is true under the assumption that $i_{\text{out}}$ is approximately constant and equal to $\widehat{i}_{\text{out}}$. The added resistance does not affect the relationship between the average input and output current (3.3-24); thus,

$$\widehat{P}_{\text{out}} = ((1-D)\widehat{i}_L)^2 r \tag{3E-4}$$

Assuming $i_L$ is approximately constant and equal to $\widehat{i}_L$, the resistive losses are

$$\widehat{P}_{\text{loss}} = \widehat{\left(i^2_L\right)} r_L = \widehat{i}^2_L r_L \tag{3E-5}$$

Combining (3E-1), (3E-4), and (3E-5) gives

$$\widehat{P}_{\text{in}} = (1-D)^2\widehat{i}^2_L r + \widehat{i}^2_L r_L \tag{3E-6}$$

Now, equating (3E-6) to (3E-2) and solving for $V_{\text{in}}$ gives

$$V_{\text{in}} = (1-D)^2 i_L r + i_L r_L \tag{3E-7}$$

We must now find an equation for $\widehat{v}_{\text{out}}$. By inspection and (3.3-24) we have

$$\widehat{v}_{\text{out}} = (1-D)\widehat{i}_L r \tag{3E-8}$$

Solving (3E-8) for $\widehat{i}_L$ substituting this expression into (3E-7) gives

$$V_{\text{in}} = (1-D)\widehat{v}_{\text{out}} + \frac{\widehat{v}_{\text{out}} r_L}{(1-D)r} \tag{3E-9}$$

Some basic algebra will now lead to the desired boost conversion ratio:

$$\frac{\widehat{v}_{\text{out}}}{V_{\text{in}}} = \left(\frac{1}{1-D}\right) \times \left(1 + \frac{r_L}{(1-D)^2 r}\right)^{-1} \tag{3E-10}$$

Comparing (3E-10) with (3.3-22) shows that an additional $r_L$-dependent term modifies the ideal boost conversion ratio. As $r_L$ increases, the conversion ratio decreases relative to the ideal case. Further algebraic manipulation reduces the conversion ratio to

$$\frac{\widehat{v}_{\text{out}}}{V_{\text{in}}} = \frac{(1-D)r}{(1-D)^2 r + r_L} \tag{3E-11}$$

This form clearly shows that, because of the $r_L$ term in the denominator, the boost conversion ratio is zero when $D = 1$. The conversion ratio in (3E-11) with $r = 1\ \Omega$ and $r_L = 0.01\ \Omega$ is what is plotted in Fig. 3.3-11.

### Advanced Circuit Topologies

Numerous other dc-dc converter topologies exist. Buck-boost and boost-buck converters combine elements of the buck and boost converters to allow both voltage increase and decrease in a single converter. A flyback converter also allows voltage increase and decrease, and uses a coupled inductor to achieve isolation and to add flexibility in choosing the output voltage polarity. Multi-stage series and parallel configurations are also common when efficiency, ripple, or other design specifications are restrictive. Although the topologies of these circuits are more elaborate than the simple buck and boost circuits introduced here, the basic concepts we have introduced can be applied to their analysis.

---

**SP3.3-1.** For a buck converter with $f_{sw} = 25$ kHz, what is $T_{sw}$? [ $T_{sw} = 40$ μs]

**SP3.3-2.** For the buck converter in SP3.3-1, if the active switch is on for 10 μs of each switch period, what is the duty ratio, $D$? [ $D = 0.25$]

**SP3.3-3.** For the buck converter in SP3.3-1 and SP3.3-2, (a) what is the average output voltage, $\hat{v}_{out}$, if the input voltage, $V_{in} = 28$ V? (b) What is the buck voltage conversion ratio? [(a) $\hat{v}_{out} = 7$ V; (b) $\frac{\hat{v}_{out}}{V_{in}} = D = 0.25$]

**SP3.3-4.** (a) Use the result in Example 3E to derive the maximum boost conversion ratio for the lossy boost converter shown in Fig. 3E-1. Assume $r = 1\ \Omega$ and $r_L = 0.01\ \Omega$. (b) What is $D$ at this point? Compare your results to the plot in Fig. 3.3-11. [(a) $\hat{v}_{out}/V_{in} = 5$; (b) $D_{max} = 0.9$]

## 3.4 AC-DC CONVERSION

We will now shift gears, moving from dc-dc converters to ac-dc conversion circuits, commonly called *rectifiers*. We will look at two common and simple rectifiers in order to introduce the most important concepts. With a firm understanding of the concepts introduced in this section, analysis of more advanced rectifier circuits is straightforward.

### Half-Wave Rectifier

The first circuit we will study is the single-phase, *half-wave rectifier* shown in Fig. 3.4-1. An ac voltage source $V_{in}(t) = V_p\sin(\omega_{in}t)$, where $\omega_{in} = 2\pi f_{in}$ is the radian frequency, is connected in series with a diode and a load. In the steady state, the circuit current is $i_{load}$ and the diode and load voltages are $v_d$ and $v_{load}$, respectively, as shown.

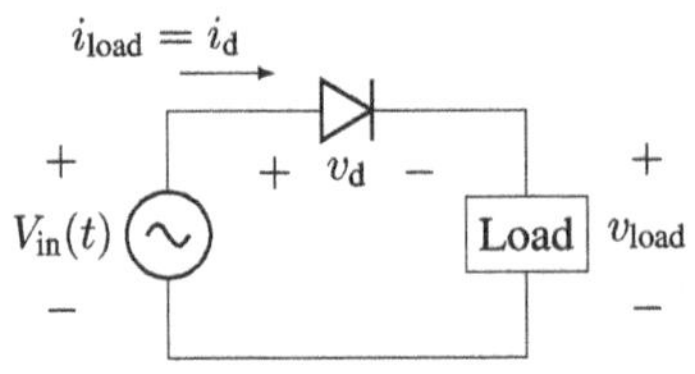


Figure 3.4-1 Half-wave rectifier circuit

In our study, the load will take on various forms consisting of passive elements. The diode obeys the rules given in (3.2-23).

To analyze a diode circuit such as this, an effective strategy is the *trial method* [1], which is described as follows. At each instant of time, assume a diode state (on or off); if the diode is on, set $v_d = 0$ and analyze the circuit to determine the direction of the diode current $i_d$; if the diode is off, set $i_d = 0$ and analyze the circuit to determine the polarity of $v_d$. Except at the instants when the diode switches state, one and only one diode state will be self-consistent, as described by (3.2-23); the other state will be impossible. Once the diode state is known, the circuit can be solved. Formally, this process must be applied at each instant of time to calculate the circuit's full transient behavior. However, in practice the behavior can usually be bracketed into a few discrete time regions. For example, with sinusoidal input the time domain can be divided into three regions, where the input is positive, negative, and zero, respectively. The diode state will remain constant within each of these regions.

Now assume a resistive load $r$. We can apply the diode trial method and solve the circuit in Fig. 3.4-1. Within the interval $t \in (0, T/2)$, $V_{in} > 0$. We first assume the diode is off and set $i_d = i_{load} = 0$. With $i_{load} = 0$, $v_{load} = i_{load} \times r = 0$ and $v_d = V_{in} - v_{load} = V_{in} > 0$. But, if the diode is off, $v_d$ must be negative; thus, this state is inconsistent and cannot be true. Thus, the diode is on. To see this, set $v_d = 0$ and solve the circuit. In this case, $v_{load} = V_{in} > 0$, which means that $i_{load} = v_{load}/r > 0$. $i_{load} > 0$ is consistent with the diode being on, so this state is consistent. Following this same procedure for the interval $t \in (T/2, T)$ when $V_{in} < 0$, we find that the only consistent diode state is the off state. Finally, the instants when $V_{in} = 0$ must be examined. Unfortunately, the trial method tells us noting about the diode state in these instants, because both the on and off states reveal inconsistent results. This simply means that the diode is changing states at these instants, going from on to off at $t = T/2$ and from off to on at $t = T$. Because the diode is transitioning at these points, we cannot assign it a definitive state. The full diode trial method analysis procedure is summarized in Table 3-1.

From Table 3-1, the operation of the circuit is clear. When the input voltage is positive the diode is on, the load voltage equals the input voltage, and positive current flows through the load. As the input voltage crosses zero, the diode switches off. When the input voltage is negative, the diode is off, no current flows, and the load voltage is zero. The diode switches back on as the voltage crosses zero again one half-period later, and the process continues periodically. Half-wave rectifier input and output voltage waveforms for $f_{in} = 60$ Hz are shown in Fig. 3.4-2. This plot shows why the circuit is named the way it is: half of the input sine wave is allowed to pass

**TABLE 3-1 Summary of diode trial method for half-wave rectifier with resistive load**

| $t$ | $V_{in}$ | Diode state assumed | Test | Result | OK? |
|---|---|---|---|---|---|
| $(0, T/2)$ | $> 0$ | on | $v_d = 0$ | $i_d > 0$ | YES |
| | | off | $i_d = 0$ | $v_d > 0$ | NO |
| $(T/2, T)$ | $< 0$ | on | $v_d = 0$ | $i_d < 0$ | NO |
| | | off | $i_d = 0$ | $v_d < 0$ | YES |
| $T/2$ or $T$ | $= 0$ | on | $v_d = 0$ | $i_d = 0$ | NO |
| | | off | $i_d = 0$ | $v_d = 0$ | NO |

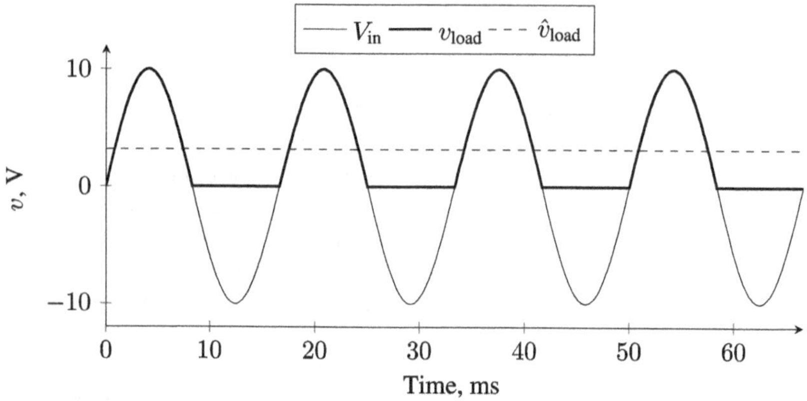


Figure 3.4-2 Voltage waveforms from a half-wave rectifier with resistive load; $V_p = 10$ V, $r = 1\ \Omega$, and $f_{in} = 60$ Hz.

through to the output; the negative half of the wave is blocked. The average value of this rectified signal is easily calculated as

$$\begin{aligned}\hat{v}_{load} &= \frac{1}{T}\int_0^{T/2} V_p \sin(\omega_{in} t)dt \\ &= \frac{-V_p}{\omega_{in} T}\left[\cos\left(\frac{\omega_{in} T}{2}\right) - \cos(0)\right] \\ &= \frac{V_p}{\pi} \end{aligned} \tag{3.4-1}$$

For the example in Fig. 3.4-2, $\hat{v}_{load} = 10/\pi \approx 3.18$ V, as plotted. Although the load voltage of the half-wave rectifier does not look the part, it is, in fact, a dc signal because it fits our expanded definition: it has a nonzero average value. Accordingly, a Fourier series can be constructed for this waveform, and its dc component is $V_p/\pi$. The half-wave rectifier accomplishes the goal of converting the ac $V_{in}(t)$ to the dc $v_{load}$, with average value $\hat{v}_{load} = V_p/\pi$. Additional waveforms for circuit current and load power are plotted in Fig. 3.4-3. Notice that current and power only flow to the load when the input voltage is positive and the diode is on.

The analysis is complicated somewhat when the load contains inductors and capacitors, because now the load can maintain a current and/or voltage independent of the source. This changes the diode switching behavior compared to the resistive load case. However, by adjusting the results of applying the diode trial method to the resistive load and using some basic circuit analysis, we can meet the challenge.

First, let us examine a parallel $rC$ load. If the capacitance is "large" so that the load time constant is much greater than the input voltage period (i.e., $\tau = rC \gg T$), the capacitor will charge up when the diode is on, but will maintain its charge and slowly discharge approximately linearly through the resistor when the diode is off. Here, we only examine the large capacitance case, because if $C$ is small the capacitor will discharge quickly when the diode is off and the load behavior will not differ significantly from the resistive case. With large output capacitance the diode will be reverse-biased and remain off until the input voltage is greater than the

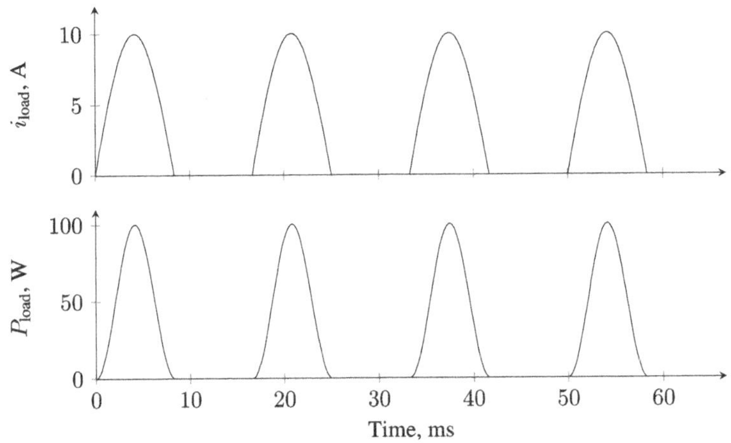


Figure 3.4-3 Load current and power waveforms from a half-wave rectifier with $V_p = 10$ V, $r = 1\ \Omega$, and $f_{\text{in}} = 60$ Hz.

decaying capacitor voltage, at which point the diode will turn on again. In essence, with respect to the resistive load circuit, the introduction of the capacitor delays the turn-on time of the diode until the input voltage is greater than the capacitor voltage and accelerates the turn-off time to when the diode current crosses zero (when the capacitor and resistor currents balance exactly), which occurs while the input voltage is still positive. This is another way of saying that, because the source current leads the source voltage in this capacitive circuit, the diode current crosses zero before the source voltage.

Calculating the diode turn-on and turn-off times $t_{\text{on}}$ and $t_{\text{off}}$, respectively, requires some calculus [1], which we will not explicitly show here. However, it suffices to understand conceptually the differences between the resistive and $rC$ load circuits. Waveforms for a half-wave rectifier with a parallel $rC$ load with $r = 1\ \Omega$ and $C = 47$ mF ($\tau = 47$ ms) and source parameters $V_p = 10$ V and $f_{\text{in}} = 60$ Hz ($T \approx 16.67$ ms) are given in Figs. 3.4-4 and 3.4-5. For this circuit, $\tau/T \approx 2.82$. Comparing the voltage waveforms in Fig. 3.4-4 to those in Fig. 3.4-2 shows the effect of the capacitor. $\hat{v}_{\text{load}}$ is much higher and the ripple about the average is much lower.

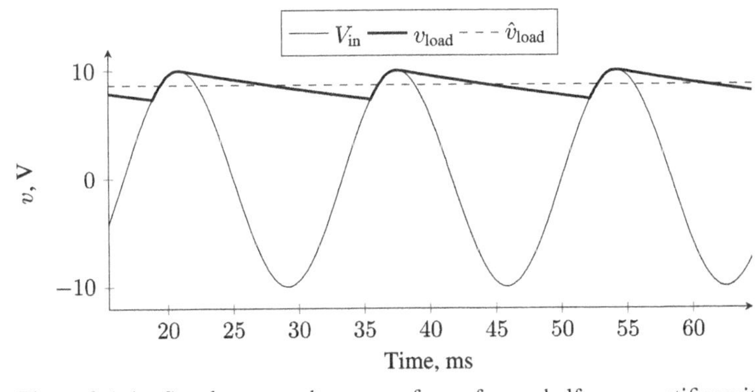


Figure 3.4-4 Steady-state voltage waveforms from a half-wave rectifier with parallel $rC$ load; $V_p = 10$ V, $r = 1\ \Omega$, $C = 47$ mF, and $f_{\text{in}} = 60$ Hz.

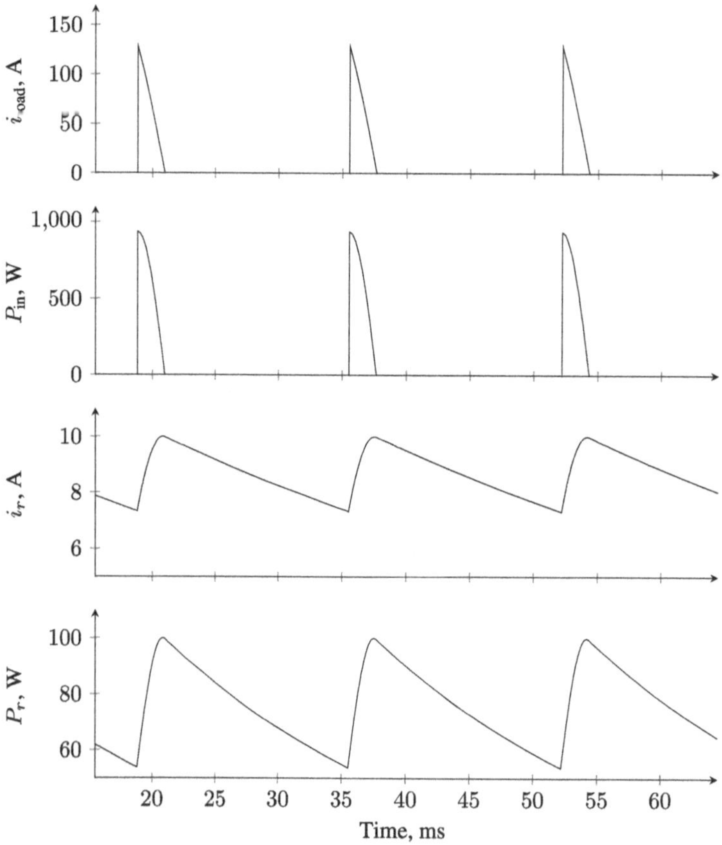


Figure 3.4-5 Source current and power, and load (resistor) current and power waveforms from a half-wave rectifier with parallel $rC$ load; $V_p = 10$ V, $r = 1\ \Omega$, $C = 47$ mF, and $f_{in} = 60$ Hz.

The capacitor acts as both a low-pass filter and a charge storage device increasing and smoothing the resistor voltage. Figure 3.4-5 shows that the RC load draws current in short, large bursts when the diode is on. Compared to Fig. 3.4-3, Fig. 3.4-5 shows both current and power pulses from the source over ten times larger than in the resistive load circuit. This is typical of diode capacitor circuits, and it has important consequences for the power grid supplying the power, as will be mentioned again at the end of this section.

Now, let us consider a series $rL$ load. If the inductance is "small" so that the load time constant is much smaller than the input voltage period (i.e., $\tau = L/R \ll T$), the source current will lag the voltage by a small angle much less than 90°. The current zero-crossing, and the corresponding diode turn-off, will lag behind the voltage zero-crossing. We will only consider the small inductance case, because with large $L$ the inductor current will not go to zero during the half-cycle when the input voltage is negative; thus, the diode will never turn off, and rectification will not occur. For a short time during each cycle the source voltage will be negative while the current is positive: energy will flow from the load to the source. Once the diode turns off, the load conditions are identical to the off-state conditions of the resistive load circuit with zero load current and voltage; thus, the diode will turn on at the same times. However, the turn-off will be delayed, which will affect the rectified voltage across the load.

As with the $rC$ load, explicitly calculating the diode turn-off times requires nontrivial calculations that will not be shown here. However, a firm conceptual understanding of the impact of the inductor's addition is our goal. This is achieved through an understanding of the previous discussion and by studying Figs. 3.4-6 and 3.4-7, which plot waveforms from a half-wave rectifier with series $rL$ load, the same parameters as the resistive circuit and $L = 3$ mH ($\tau = 3$ ms). For this circuit, $\tau/T \approx 0.18$. Comparing Fig. 3.4-6 with its resistive (Fig. 3.4-2) and parallel $rC$ (Fig. 3.4-4) counterparts, and comparing Fig. 3.4-7 with Figs. 3.4-3 and 3.4-5 shows how the addition of energy storage elements impacts the performance of the half-wave rectifier.

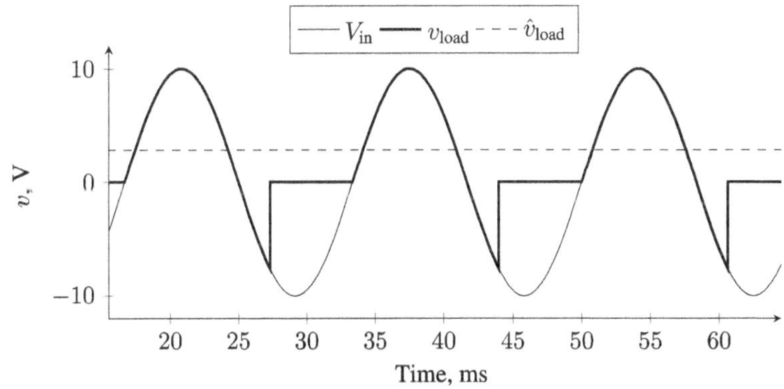


Figure 3.4-6 Steady-state voltage waveforms from a half-wave rectifier with series $rL$ load; $V_p = 10$ V, $r = 1\ \Omega$, $L = 3$ mH, and $f_{in} = 60$ Hz.

**Example 3F. Average power in a half-wave rectifier.** Calculate the average power delivered by the half-wave rectifier to a resistive load, $r$.

The relevant circuit waveforms are shown in Figs. 3.4-2 and 3.4-3. The load power, $P_{load}$, is expressed as

$$P_{load} = \frac{v_{load}^2}{r} \tag{3F-1}$$

The average power delivered to the load is

$$\hat{P}_{load} = \frac{\widehat{\left(v_{load}^2\right)}}{r} \tag{3F-2}$$

which is calculated as

$$\begin{aligned}\hat{P}_{load} &= \frac{1}{r}\frac{1}{T_{in}}\int_0^{T_{in}/2} V_p^2 \sin^2(\omega_{in}t)dt \\ &= \frac{1}{r}\frac{V_p^2}{T_{in}}\int_0^{T_{in}/2} \sin^2(\omega_{in}t)dt \end{aligned} \tag{3F-3}$$

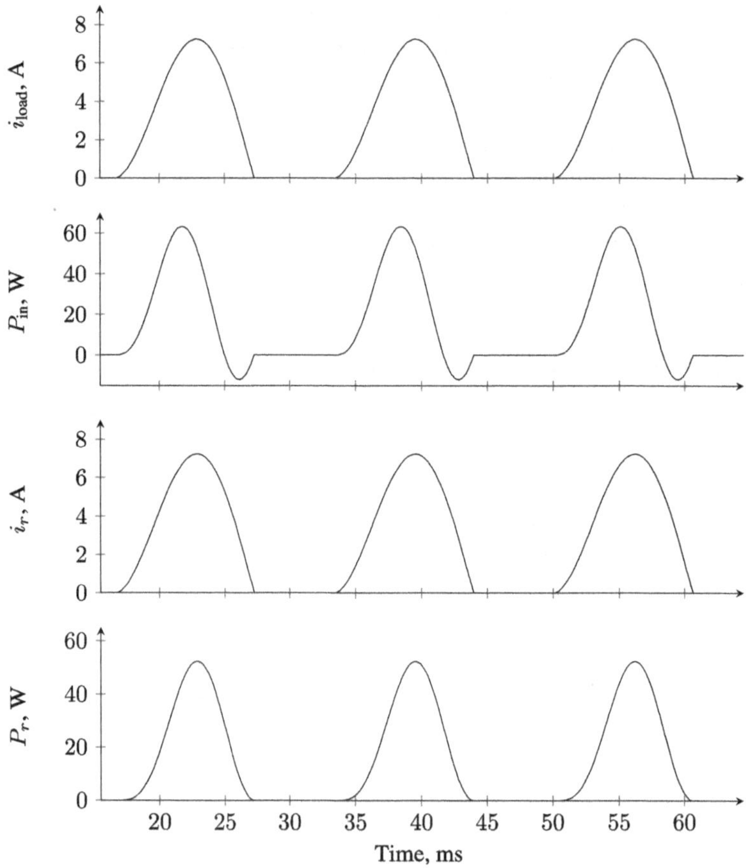


Figure 3.4-7 Source current and power, and load (resistor) current and power waveforms from a half-wave rectifier with series $rL$ load; $V_p = 10$ V, $r = 1\ \Omega$, $L = 3$ mH, and $f_{in} = 60$ Hz.

We use the identity

$$\sin^2(\omega_{in}t) = \frac{1}{2}[1 - \cos(2\omega_{in}t)] \tag{3F-4}$$

to integrate (3F-3), giving

$$\widehat{P}_{load} = \frac{V_p^2}{4r} \tag{3F-5}$$

For comparison, note that from (3.4-1)

$$\begin{aligned}\frac{(\widehat{v}_{load})^2}{r} &= \frac{(V_p/\pi)^2}{r}\\ &= \frac{V_p^2}{\pi^2 r} \neq \widehat{P}_{load}\end{aligned} \tag{3F-6}$$

This illustrates an important fact about averaging transient signals. The average of a product of two signals (i.e., $(\widehat{v_{\text{load}}^2})$) is not necessarily equal to the product of the averages of the two signals (i.e., $(\hat{v}_{\text{load}})^2$). This is important to remember when calculating average power.

## Full-Wave Rectifier

The single diode half-wave rectifier converts ac to dc by clipping the input ac voltage and leaving only the positive "humps" of the sine wave, which have a nonzero average value. Looking at Fig. 3.4-2, we see that we can double the average load voltage obtained from the half-wave rectifier by flipping the clipped negative humps, making them positive, and applying them to the load as well. In other words, we would like to obtain $v_{\text{load}} = \text{abs}(V_{\text{in}})$. One of the most common ways to accomplish this is by adding three diodes and forming a *full-wave rectifier* (also called a *full-bridge rectifier*). The circuit is shown in Fig. 3.4-8.

Although the circuit contains three more diodes than the half-wave rectifier, we can analyze it in the same way using the trial method. We assume a resistive load and analyze all the diode state combinations, both for positive, negative, and zero voltage. Because there are now four diodes, there are $2^4 = 16$ switch combinations for each of the three input voltage regions, leaving 48 combinations to analyze. Although this seems daunting at first, many of the combinations can be eliminated quickly.

Similarly to the half-wave rectifier, when $V_{\text{in}} = 0$ none of the 16 diode states are self-consistent. These are the instants when the diodes change state; each diode will switch from its state before the zero-crossing to its new state after the zero-crossing. Of the remaining 32 configurations, several are quickly eliminated due to Kirchoff's Voltage Law (KVL) and KCL violations. Any configuration with only one diode on violates KCL. For example, if diode 1 is on, $v_{\text{d1}} = 0$; however, in order for $i_{\text{d1}}$ to be positive, as required, at least one other diode must be on to give the current a path. Thus, there is a KCL violation. This logic eliminates eight configurations (four with $V_{\text{in}} > 0$ and four with $V_{\text{in}} < 0$). Likewise, any configuration with diodes 1 and 3 on simultaneously and/or diodes 2 and 4 on simultaneously violates KVL, because these configurations short the input source and also require negative current to flow through one of the diodes. This eliminates 14 more configurations. In the two cases with all

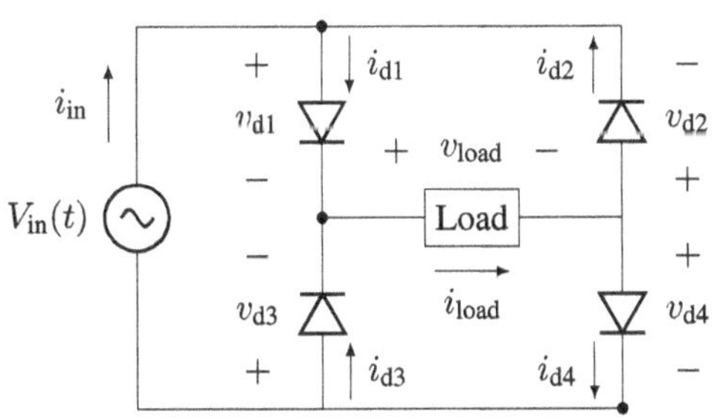


Figure 3.4-8 Full-wave rectifier circuit.

**TABLE 3-2 Summary of diode trial method for a full-wave rectifier with resistive load**

| Case | $t$ | $V_{in}$ | Diode state | Test result d1 | d2 | d3 | d4 | OK? | Notes |
|---|---|---|---|---|---|---|---|---|---|
| 0 | | | 0000 | $v_{d1} = -v_{d2}$ | | $v_{d3} = -v_{d4}$ | | NO | Inconsistent |
| 1 | | | 0001 | - | - | - | - | NO | KCL |
| 2 | | | 0010 | - | - | - | - | NO | KCL |
| 3 | | | 0011 | $v_{d1} > 0$ | $v_{d2} < 0$ | $i_{d3} = 0$ | $i_{d4} = 0$ | NO | Inconsistent |
| 4 | | | 0100 | - | - | - | - | NO | KCL |
| 5 | | | 0101 | - | - | - | - | NO | KVL |
| 6 | | | 0110 | $v_{d1} > 0$ | $i_{d2} < 0$ | $i_{d3} < 0$ | $v_{d4} > 0$ | NO | Inconsistent |
| 7 | $\left(0, \frac{T}{2}\right)$ | $> 0$ | 0111 | - | - | - | - | NO | KVL |
| 8 | | | 1000 | - | - | - | - | NO | KCL |
| 9 | | | 1001 | $i_{d1} > 0$ | $v_{d2} < 0$ | $v_{d3} < 0$ | $i_{d4} > 0$ | YES | |
| 10 | | | 1010 | - | - | - | - | NO | KVL |
| 11 | | | 1011 | - | - | - | - | NO | KVL |
| 12 | | | 1100 | $i_{d1} = 0$ | $i_{d2} = 0$ | $v_{d3} < 0$ | $v_{d4} > 0$ | NO | Inconsistent |
| 13 | | | 1101 | - | - | - | - | NO | KVL |
| 14 | | | 1110 | - | - | - | - | NO | KVL |
| 15 | | | 1111 | - | - | - | - | NO | KVL |
| 16 | | | 0000 | $v_{d1} = -v_{d2}$ | | $v_{d3} = -v_{d4}$ | | NO | Inconsistent |
| 17 | | | 0001 | - | - | - | - | NO | KCL |
| 18 | | | 0010 | - | - | - | - | NO | KCL |
| 19 | | | 0011 | $v_{d1} < 0$ | $v_{d2} > 0$ | $i_{d3} = 0$ | $i_{d4} = 0$ | NO | Inconsistent |
| 20 | | | 0100 | - | - | - | - | NO | KCL |
| 21 | | | 0101 | - | - | - | - | NO | KVL |
| 22 | | | 0110 | $v_{d1} < 0$ | $i_{d2} > 0$ | $i_{d3} > 0$ | $v_{d4} < 0$ | YES | |
| 23 | $\left(\frac{T}{2}, T\right)$ | $< 0$ | 0111 | - | - | - | - | NO | KVL |
| 24 | | | 1000 | - | - | - | - | NO | KCL |
| 25 | | | 1001 | $i_{d1} < 0$ | $v_{d2} > 0$ | $v_{d3} > 0$ | $i_{d4} < 0$ | NO | Inconsistent |
| 26 | | | 1010 | - | - | - | - | NO | KVL |
| 27 | | | 1011 | - | - | - | - | NO | KVL |
| 28 | | | 1100 | $i_{d1} = 0$ | $i_{d2} = 0$ | $v_{d3} > 0$ | $v_{d4} < 0$ | NO | Inconsistent |
| 29 | | | 1101 | - | - | - | - | NO | KVL |
| 30 | | | 1110 | - | - | - | - | NO | KVL |
| 31 | | | 1111 | - | - | - | - | NO | KVL |
| 32-47 | $\frac{T}{2}$ or $T$ | $= 0$ | Indeterminate diode transition state. | | | | | | |

four diodes off, no load current can flow, which means that $v_{d1} = -v_{d2}$ and $v_{d3} = -v_{d4}$. This is inconsistent with all off-diode voltages being negative, eliminating these two cases. There remain only eight configurations to examine.

The full results of the trial method for the full-wave rectifier are given in Table 3-2. For simplicity, the collective diode state assumed is represented by a four-digit binary number, where 1 = on and 0 = off. For example, the diode state with diodes 2 and 3 on and diodes 1 and 4 off is represented as 0110. The result of each case's test assumption for each diode is listed under "Test Result": if diode 1 is assumed on ($v_{d1} = 0$), its test result will show whether $i_{d1} > 0$; if it is assumed off ($i_{d1} = 0$), its test result will show whether $v_{d1} < 0$. If all four diode tests are consistent, the case is a viable circuit configuration and is marked YES in the "OK" column. After applying the trial method to the remaining eight cases, only Case 9, where $V_{in} > 0$ and diodes 1 and 4 are on, and Case 22, where $V_{in} < 0$ and diodes 2 and 3 are on, are allowable circuit configurations.

The full-wave rectifier operation is now evident. As time increases from 0 to $T/2$ the circuit is in diode state 1001, the positive half-cycle of the sinusoid is applied across the load and current flows through the load and diodes 1 and 3. The diodes switch state as the voltage crosses zero at $t = T/2$. The input voltage sinusoid is negative from $T/2$ to $T$, and $-V_{in}$ (which is positive) is applied across the load. Current flows through the load and diodes 3 and 4. As the input voltage crosses zero again, the diodes switch back to their original state, and the cycle continues.

The diode trial method thus far presented may seem tedious; however, once you work through it for a few circuits, you will develop an intuition for how diode circuits work. The tedium of checking each and every switch state can then be bypassed as you quickly identify the acceptable configurations. Many introductory texts bypass any discussion of this rigorous method in favor of simply stating how the full-wave rectifier operates; however, a thorough understanding of the trial method is very useful when you encounter a complicated circuit containing diodes. Thus, time invested now will pay dividends down the road.

Input, load, and average voltage waveforms for the full-wave rectifier with resistive load are shown in Fig. 3.4-9. Performing the averaging calculation in (3.4-1) on $v_{load}$ gives

$$\hat{v}_{load} = \frac{2V_p}{\pi} \tag{3.4-2}$$

which is twice the average value of the half-wave rectifier output. Figure 3.4-10 shows load current and power waveforms. Figures 3.4-9 and 3.4-10, when compared to

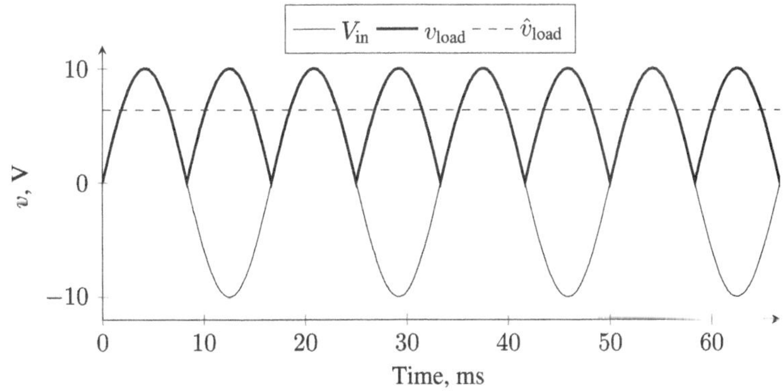


Figure 3.4-9 Voltage waveforms from a full-wave rectifier with resistive load; $V_p = 10$ V, $r = 1\ \Omega$, and $f_{in} = 60$ Hz.

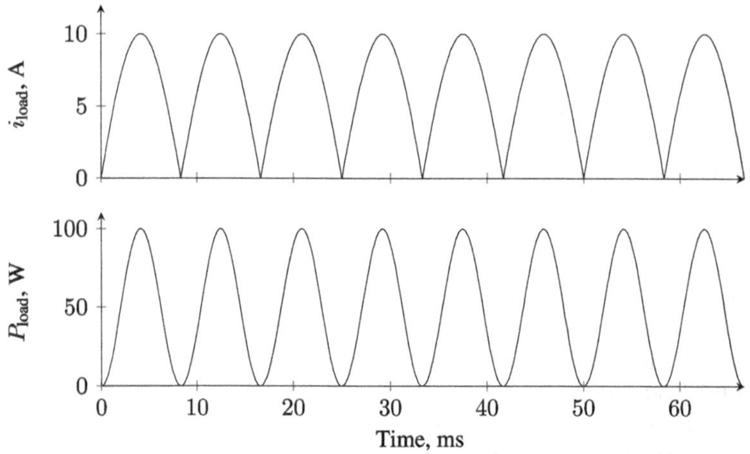


Figure 3.4-10 Load current and power waveforms from a full-wave rectifier with resistive load; $V_p = 10$ V, $r = 1\ \Omega$, and $f_{in} = 60$ Hz.

Figs. 3.4-2 and 3.4-3, respectively, show that the full-wave rectifier output waveforms have double the frequency of the half-wave rectifier. The compact Fourier series representation of the full-wave rectified waveform shown in Fig. 3.4-9, labeled $v_{rec}(t)$, can be calculated using (3.2-8) through (3.2-14) and the fact that the waveform is even to give

$$v_{rec}(t) = \frac{2V_p}{\pi} + \frac{4V_p}{\pi}\sum_{n=1}^{\infty}\frac{\cos(n\pi)}{1-4n^2}\cos[n(2\omega_{in}t - 2\theta(0))] \tag{3.4-3}$$

where $V_p = 10$ V, $\omega_{in} = 2\pi \times 60$ rad/s, and $\theta(0) = \pi/2$. Notice that the dc component of the Fourier series matches the average value in (3.4-2), and the series' fundamental frequency is twice the frequency of $V_{in}$. When compared to the half-wave rectifier output, the double frequency harmonics of the full-wave rectifier are relatively easier to filter.

The effect of adding parallel capacitance to the full-wave rectifier output is similar to the half-wave rectifier. However, since the full-wave rectified voltage has twice the frequency of the half-wave, the filtered voltage has a higher average and reduced ripple magnitude. With a capacitor in parallel with the load, the circuit is often called a *classical rectifier*, because this is a classical (i.e., common) charging circuit for batteries. Waveforms for the full-wave rectifier shown in Fig. 3.4-8 with 47 mF parallel load capacitance are shown in Figs. 3.4-11 and 3.4-12. Comparing these figures to the half-wave rectifier waveforms in Figs. 3.4-4 and 3.4-5, respectively, clearly shows the double frequency and reduced load ripple.

Adding inductance in series with the full-wave rectifier output has effects similar to the half-wave rectifier, with one key difference. Because there are four switches instead of one, with an inductive load the full-wave rectifier always provides a path for the current, so there are no diode switching delays and the current need not decay to zero for the diodes to switch states. This is illustrated in Figs. 3.4-13 and 3.4-14.

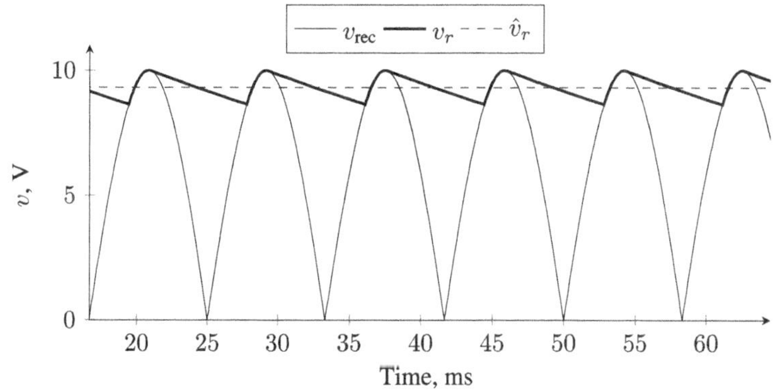


Figure 3.4-11 Voltage waveforms from a full-wave rectifier with parallel $rC$ load; $V_p = 10$ V, $r = 1\ \Omega$, $C = 47$ mF, and $f_{in} = 60$ Hz.

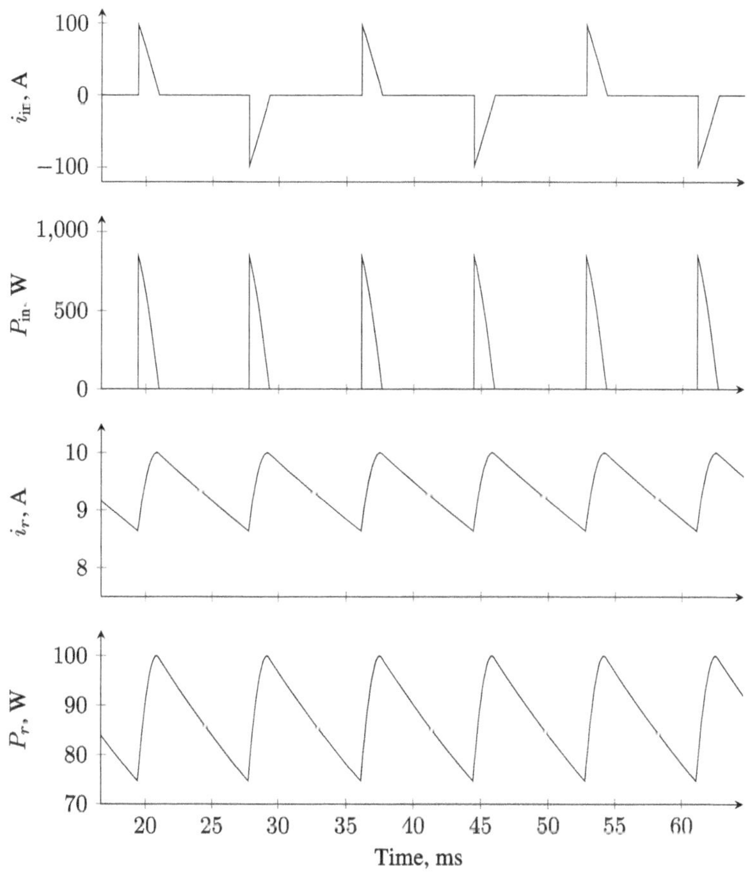


Figure 3.4-12 Source current and power, and load (resistor) current and power waveforms from a full-wave rectifier with parallel $rC$ load; $V_p = 10$ V, $r = 1\ \Omega$, $C = 47$ mF, and $f_{in} = 60$ Hz.

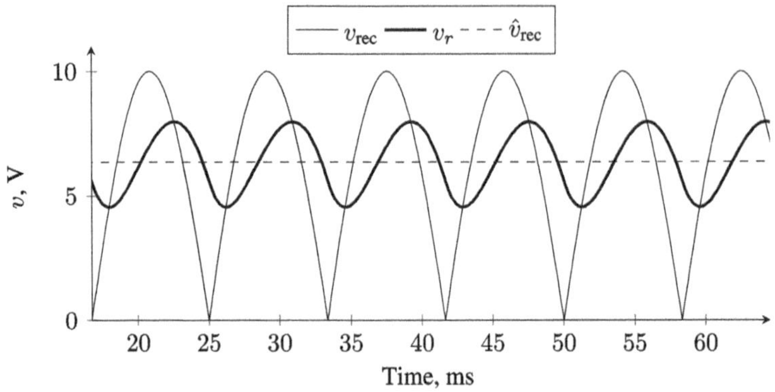


Figure 3.4-13 Steady-state voltage waveforms from a full-wave rectifier with series $rL$ load; $V_p = 10$ V, $r = 1\ \Omega$, $L = 3$ mH, and $f_{in} = 60$ Hz.

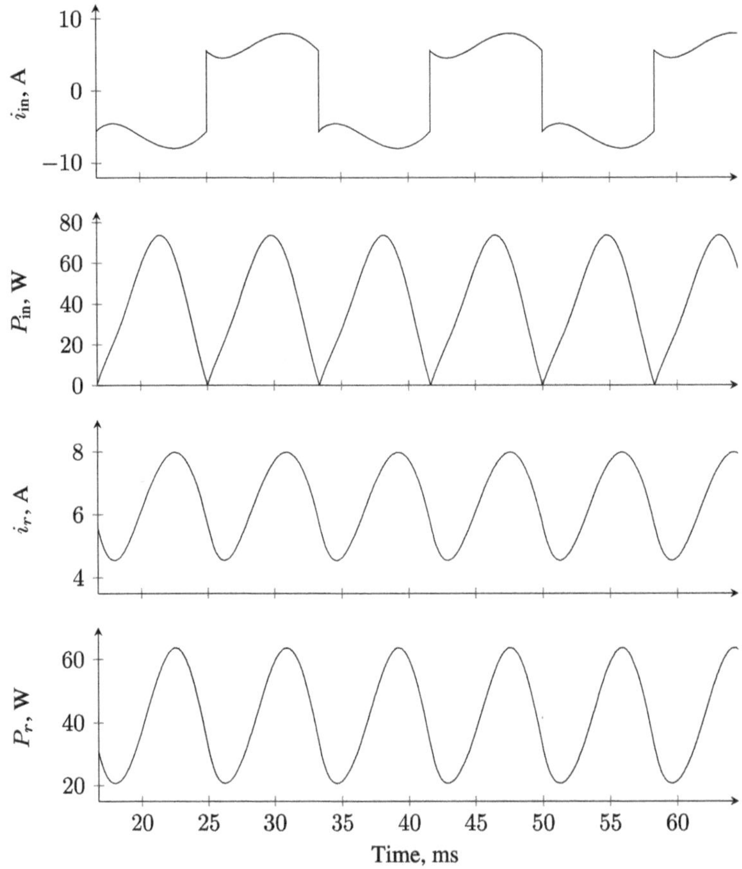


Figure 3.4-14 Source current and power, and load (resistor) current and power waveforms from a full-wave rectifier with series $rL$ load; $V_p = 10$ V, $r = 1\ \Omega$, $L = 3$ mH, and $f_{in} = 60$ Hz.

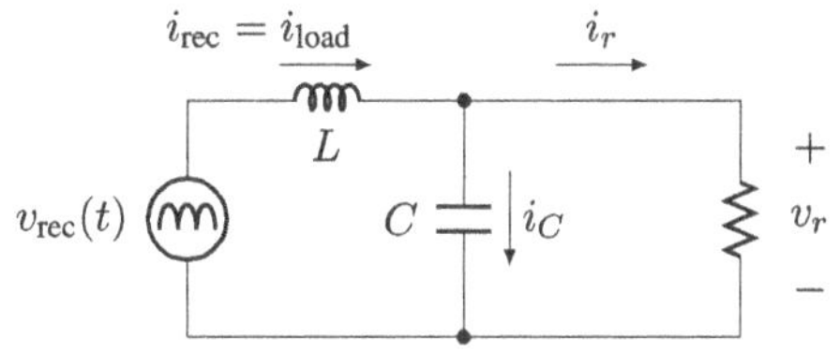


Figure 3.4-15 Full-wave rectifier with $LC$ output filter.

Comparing these to Figs. 3.4-6 and 3.4-7, respectively, shows that the rectified voltage is not altered by diode switching delays, the input power is always non-negative, and the load current remains positive. The inductor acts as a low-pass filter that smooths the resistor voltage.

As we have seen, unlike the half-wave rectifier, which contains only one diode, the full-wave rectifier contains four diodes and provides a path for the load current in each of its allowed diode switch states. This facilitates the implementation of more sophisticated load filtering to achieve the desired load power quality. To see this, consider the $LC$ low-pass filter shown in Fig. 3.4-15. The input to the filter is $v_{rec}(t) = \text{abs}(V_{in}(t))$, which consists of a dc component equal to $2V_p/\pi$ and harmonics at multiples of $2\omega_{in}$. To achieve true ac-to-dc conversion, all the harmonics above dc should be filtered out. A straightforward steady-state analysis of the circuit (this is left to the reader in SP3.4-2) reveals that the filter transfer function $G_{filt}(\omega)$ is

$$G_{filt}(\omega) = \frac{v_r(\omega)}{v_{rec}(\omega)} = \frac{r}{j\omega L + r(1 - \omega^2 LC)} \tag{3.4-4}$$

The magnitude of the complex $G_{filt}$, notated as $\|G_{filt}\|$, is plotted in Fig. 3.4-16. Also plotted are the compact Fourier series harmonic coefficients (i.e., the $c_n$ values of

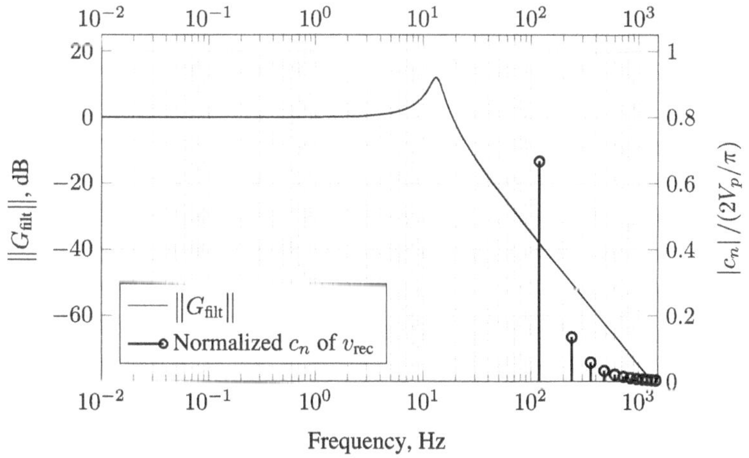


Figure 3.4-16 $\|G_{filt}\|$ and normalized compact Fourier series harmonic coefficients of $v_{rec}$ versus frequency with $V_p = 10$ V, $L = 3$ mH, $C = 47$ mF, and $r = 1\ \Omega$.

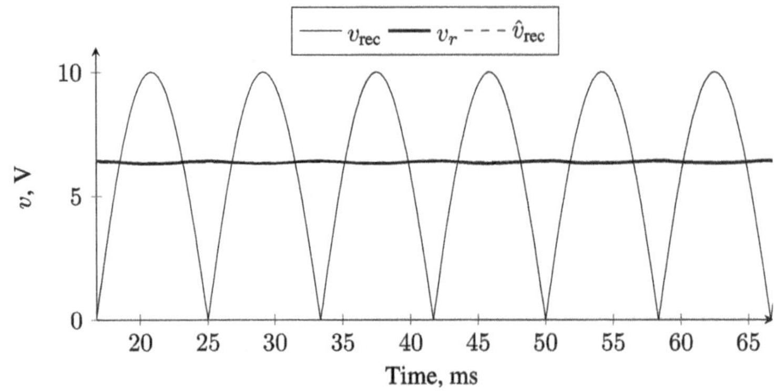


Figure 3.4-17 Steady-state voltage waveforms from a full-wave rectifier with $rLC$ load; $V_p = 10$ V, $r = 1\ \Omega$, $L = 3$ mH, $C = 47$ mF, and $f_{in} = 60$ Hz.

$v_{rec}(t)$ in (3.4-3)), normalized to the dc component $\frac{2V_p}{\pi}$. With $L = 3$ mH and $C = 47$ mF, the harmonics of $v_{rec}(t)$ are well to the right of the resonance frequency, $f_{res} = \frac{1}{2\pi\sqrt{LC}} \approx 13.4$ Hz, in the filter's capacitive region. The harmonics are thus filtered out effectively, leaving the dc component to dominate the filter output.

Voltage waveforms for the filtered full-wave rectifier are shown in Fig. 3.4-17. The load voltage $v_r$ closely matches $\hat{v}_{rec}$, highlighting the $LC$ low-pass filter effects and clearly showing the ac-dc converter action of this circuit. In general, $L$ and $C$ should be chosen so that the harmonics of the rectified waveform are located well into the capacitive region of the filter to the right of the resonance frequency. If the harmonics fall to the left of the resonance frequency, the filter's attenuation will be minimal. This is the essence of power electronics, as we have also seen with the dc-dc converters earlier in this chapter. Power conversion is achieved by chopping up an input signal using switches and filtering the chopped up signal before applying to a load. In the buck converter, the input voltage was chopped into a square wave and filtered; in the boost converter, the input current was chopped into a square wave and filtered; in the full- and half-wave rectifiers, the input voltage sinusoid is chopped into half-waves with diodes, and the waves are filtered and applied to a dc load.

---

**SP3.4-1.** (a) Calculate the average power delivered by the full-wave rectifier to a resistive load. The input voltage is $V_{in}(t) = V_p \sin(\omega_{in} t)$, where $T_{in} = 2\pi/\omega_{in}$ is the input period. Relevant circuit waveforms are shown in Figs. 3.4-9 and 3.4-10. (b) Compare your answer in (a) to the result in Example 3F. [(a)$\widehat{P}_{load} = V_p^2/2r$; (b) The average power in the full-wave rectifier is twice that in the half-wave rectifier]

**SP3.4-2.** Derive (3.4-4) using Fig. 3.4-15.

**SP3.4-3.** Why does the Fourier series of $v_{rec}$ in (3.4-3) have a fundamental frequency of $2\omega_{in}$? [$|V_{in}|$ has twice the frequency of $V_{in}$ (the absolute value of a sinusoid has twice the frequency of the sinusoid itself)]

## 3.5 DC-AC CONVERSION

Now that we have covered the fundamentals of dc-dc conversion and ac-dc rectification, we conclude our study of power electronic circuits with dc-ac conversion circuits, called *inverters*. We will first analyze the single-phase ac-to-dc inverter before concluding this section with a study of the three-phase inverter.

### Single-Phase Inverter

The single-phase dc-ac inverter we will study is shown in Fig. 3.5-1. This circuit converts energy from the dc source, $V_{in}$, to the ac load using a bridge of four active transistor switches, notated with the capital letter T. There are numerous ways of controlling these switches to accomplish the task. We will cover one method here and leave the others to an advanced course.

It is fairly straightforward to identify the allowed and useful switch states. First, switches 1 and 3 and 2 and 4, respectively, can never be on simultaneously, lest they short the source. Thus, their respective switching functions obey $q_1(t) = (1 - q_3(t))$ and $q_2(t) = (1 - q_4(t))$. Second, if switches 1 and 2 or 3 and 4, respectively, are on simultaneously, the load is essentially short-circuited and no power will be transferred. Further, if only one (or no) switch is on, there will not be a continuous path for current to flow through the load. Thus, the only allowable switch states that transfer power from the source to the load are $(q_1, q_2, q_3, q_4) = (1, 0, 0, 1)$ and $(0, 1, 1, 0)$. If the circuit is switched between these two states alternately with equal time in each state, the square wave voltage shown in Fig. 3.5-2 will be applied across the load.

Although typically when we think of an ac waveform we think of a sinusoid, the inverter voltage square wave shown in Fig. 3.5-2 also qualifies. In fact, the Fourier analysis we presented earlier showed that this periodic waveform can be expressed as a sum of sinusoids: the fundamental and its harmonics. The series in (3.2-22) can be easily modified to represent the inverter square wave (see SP3.5-1). If we apply appropriate filtering, we can remove the harmonics and "pull out" the fundamental frequency of the square wave, leaving a pure sinusoid. In practice, such filtering is difficult to achieve because the harmonics are relatively close to the fundamental in the frequency domain. In later courses, more advanced switching methods based on *pulse width modulation* (PWM) are used to separate the unwanted harmonics from the

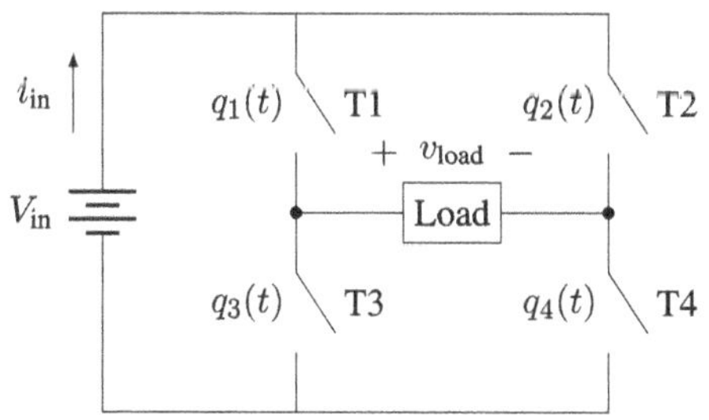


Figure 3.5-1 Single-phase dc-ac inverter circuit.

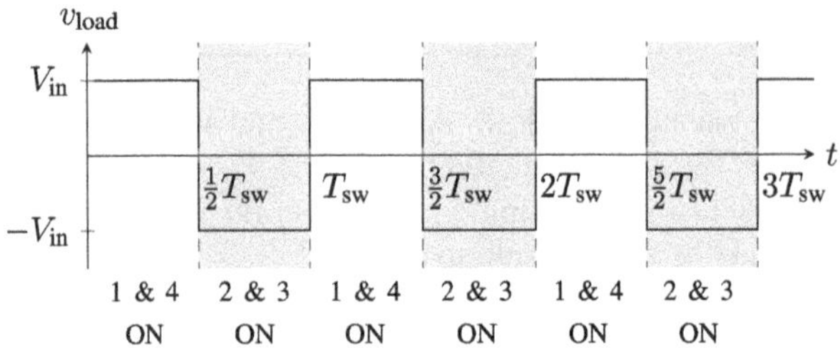


Figure 3.5-2 Single-phase inverter voltage.

fundamental and make it easier to filter them. But for the purposes of our introductory analysis, we will limit ourselves to the square wave switching scheme and analyze the circuit with both inductive and resonant $LC$ load filters applied.

Figure 3.5-3 shows voltage, current, and power waveforms for the single-phase inverter with a series $rL$ load and $V_{in} = 10$ V, $r = 1\ \Omega$, $L = 3$ mH, and $f_{in} = 60$ Hz. The load $rL$ time constant is $\tau = 3$ ms, which is about one sixth of the switching period $T_{sw} = 16.7$ ms, and the response is as expected: a series of $rL$ charges and decays. The load current is slightly more sinusoidal than the square wave current that would result without the inductive filtering, but we are still far from a sinusoid. Luckily, we can do better.

A load waveform closer to a sinusoid can be formed by adding a capacitor in series with the series $rL$ load and choosing $L$ and $C$ to resonate at the inverter

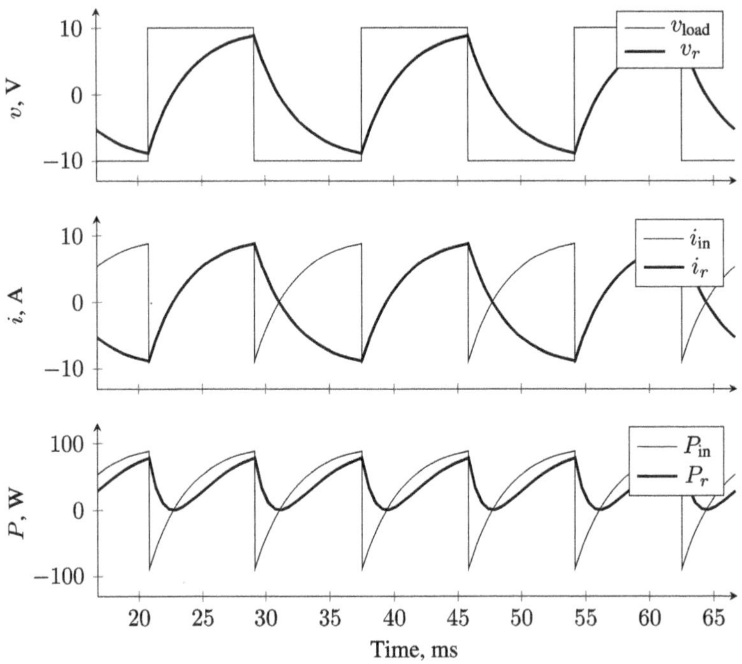


Figure 3.5-3 Steady-state voltage, current, and power waveforms from a single-phase inverter with series $rL$ load; $V_{in} = 10$ V, $r = 1\ \Omega$, $L = 3$ mH, and $f_{in} = 60$ Hz.

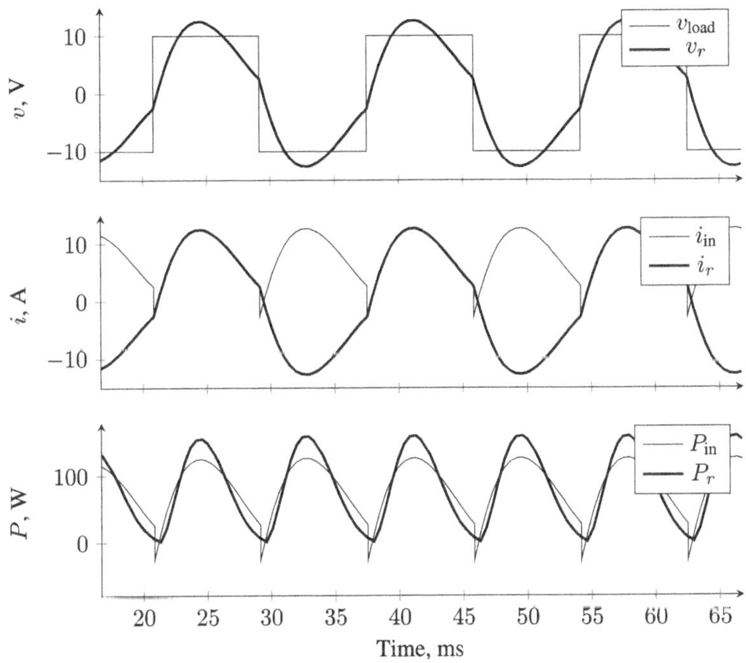


Figure 3.5-4 Steady-state voltage, current, and power waveforms from a single-phase inverter with series resonant *rLC* load; $V_{in} = 10$ V, $r = 1\ \Omega$, $L = 3$ mH, $C = 2.345$ mF, and $f_{in} = 60$ Hz.

switch frequency. This creates a series *resonant filter*. For example, by adding series capacitance $C$ to the circuit characterized by the waveforms in Fig. 3.5-3 such that $\frac{1}{2\pi\sqrt{LC}} = 60$ Hz, the waveforms in Fig. 3.5-4 result. Although the shapes of these load voltage and current waveforms are more sinusoidal than their counterparts in Fig. 3.5-3, they are not strictly sinusoidal. Luckily, though, in many applications these are acceptable ac waveforms. For example, you may be surprised to know that numerous low cost consumer single-phase dc-ac inverters create similar waveforms. If you were to probe the output voltage of one of these devices, its shape would likely be similar to that of $v_r$ in Fig. 3.5-4.

**Example 3G. Power calculation for single-phase inverter.** Calculate the average power delivered to a resistive load by the single-phase inverter in Fig. 3.5-1. Because the load is resistive, the load current is simply the square wave in Fig. 3.5-2 scaled by the load resistance. The power delivered to the load is calculated as

$$P_{load} = i_{load}^2 \times r \tag{3G-1}$$

The average power is

$$\begin{aligned} \widehat{P}_{load} &= \widehat{\left(i_{load}^2 \times r\right)} \\ &= \widehat{\left(i_{load}^2\right)} \times r \end{aligned} \tag{3G-2}$$

Notice that this is *not* equivalent to $(\widehat{i}_{load})^2$, which is zero. We can calculate the load current's mean square, denoted $(\cdot)_{ms}$, as

$$
\begin{aligned}
\widehat{P}_{load} &= r \times (i_{load})_{ms} \\
&= \frac{r}{T_{sw}} \int_0^{T_{sw}} i_{load}^2 dt \\
&= \frac{r}{T_{sw}} \int_0^{T_{sw}} \left(\frac{V_{in}}{r}\right)^2 dt \\
&= \frac{V_{in}^2}{rT_{sw}} \int_0^{T_{sw}} dt = \frac{V_{in}^2}{r}
\end{aligned}
\tag{3G-3}
$$

## Three-Phase Inverter

The single-phase inverter can be expanded to a three-phase inverter, a device which is common in electric drives. In Chapter 4, this inverter is used to drive the permanent-magnet ac machine as a brushless dc machine. Our purposes here are to introduce this inverter, set the stage for the discussion in Chapter 4, and add justification for portraying the output voltages of this inverter as sinusoidal variations corresponding to the switching frequency as we did in Chapter 2 in the analysis of the permanent-magnet ac machine.

The three-phase inverter of interest, which is often referred to as a "six-step" inverter, is shown in Fig. 3.5-5a. It consists of six transistors, each with an anti-parallel diode, supplied from a dc source $V_{in}$. The logic (switching) signals for the transistors are shown in Fig. 3.5-5b. It is assumed that the forward resistance of the diodes and transistors is negligibly small and the turn-on and turn-off times are neglected, whereupon the transistors and diodes are considered to be ideal switches. With these assumptions, at the instant when the "top-rail" transistor in a phase is turned off, its corresponding "bottom-rail" transistor is turned on. In other words, each phase is either connected to the top or bottom rail; this is referred to as continuous-current or 180° mode of operation.

Let us assume that the inverter is supplying a symmetrical wye-connected three-phase $rL$ load, thus,

$$v_{as} = r_s i_{as} + L_s \frac{di_{as}}{dt} \tag{3.5-1}$$

$$v_{bs} = r_s i_{bs} + L_s \frac{di_{bs}}{dt} \tag{3.5-2}$$

$$v_{cs} = r_s i_{cs} + L_s \frac{di_{cs}}{dt} \tag{3.5-3}$$

Since the load is three-wire and wye-connected, the sum of the currents is zero and since the load is symmetrical the sum of the phase voltages, which is $v_{0s}$, is zero.

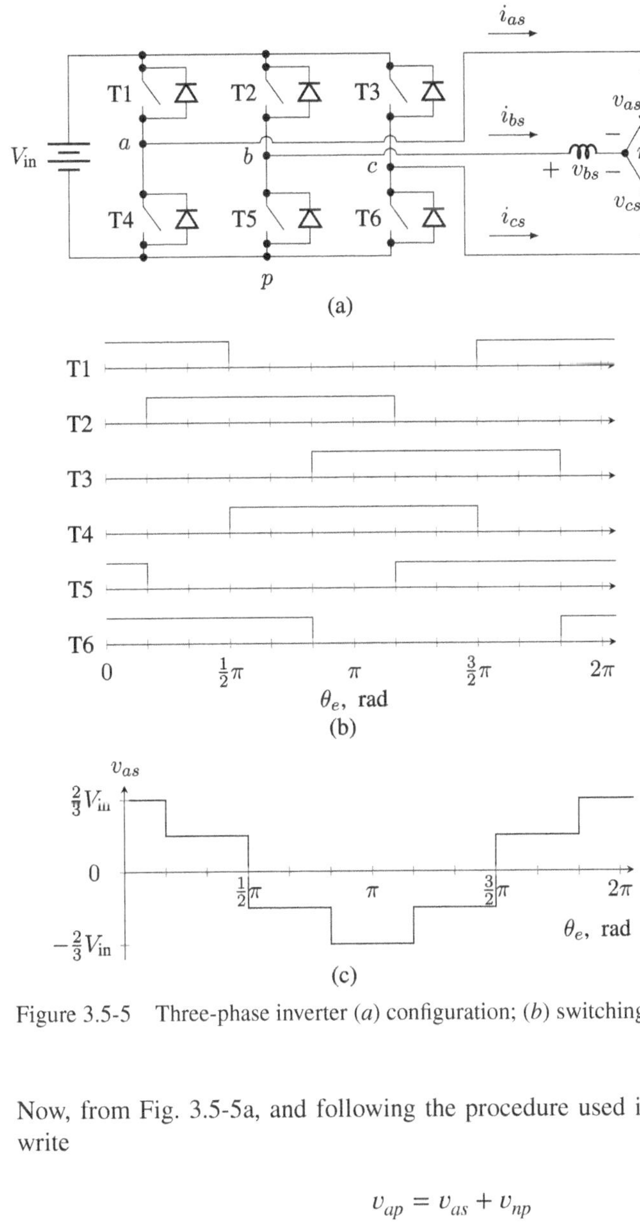


Figure 3.5-5 Three-phase inverter (*a*) configuration; (*b*) switching logic; (*c*) plot of $v_{as}$.

Now, from Fig. 3.5-5a, and following the procedure used in Example 1G, we can write

$$v_{ap} = v_{as} + v_{np} \tag{3.5-4}$$

$$v_{bp} = v_{bs} + v_{np} \tag{3.5-5}$$

$$v_{cp} = v_{cs} + v_{np} \tag{3.5-6}$$

Adding (3.5-4) through (3.5-6) and since for a symmetrical load $v_{as} + v_{bs} + v_{cs} = 0$, we can write

$$v_{np} = \frac{1}{3}(v_{ap} + v_{bp} + v_{cp}) \tag{3.5-7}$$

Substituting (3.5-7) into (3.5-4) through (3.5-6) and then solving for $v_{as}$, $v_{bs}$, and $v_{cs}$ yields

$$v_{as} = \frac{2}{3}v_{ap} - \frac{1}{3}(v_{bp} + v_{cp}) \tag{3.5-8}$$

$$v_{bs} = \frac{2}{3}v_{bp} - \frac{1}{3}(v_{cp} + v_{ap}) \tag{3.5-9}$$

$$v_{cs} = \frac{2}{3}v_{cp} - \frac{1}{3}(v_{ap} + v_{bp}) \tag{3.5-10}$$

where $v_{ap}$, $v_{bp}$, and $v_{cp}$ are either $V_{\text{in}}$ or zero depending upon the state of the transistors. A plot of $v_{as}$ is shown in Fig. 3.5-5c. The voltages $v_{bs}$ and $v_{cs}$ are of the same waveform as $v_{as}$ but lag by 120° and 240°, respectively, for an *abc* sequence.

In Chapter 4 the three-phase inverter shown in Fig. 3.5-5a is used to drive a permanent-magnet ac machine and the firing of the inverter is controlled so that the fundamental frequency of the applied voltages ($v_{as}$, $v_{bs}$, and $v_{cs}$) corresponds to the rotor speed $\omega_r$ of the permanent-magnet ac machine. Moreover, we will find that with appropriate control of the phase of the applied voltages, the operating characteristics of the inverter-machine combination can be made to resemble that of a permanent-magnet dc machine.

In order to analytically "connect" the inverter to the machine equations, we need to transform $v_{as}$, $v_{bs}$, and $v_{cs}$ to $v_{qs}^r$ and $v_{ds}^r$ since Park's equations are used to analyze the permanent-magnet ac machine in Chapter 2. Using (3.2-8) through (3.2-11) and the rules given by (3.2-16) through (3.2-19), a Fourier expansion of $v_{as}$ shown in Fig. 3.5-5c, which is even, has no dc component, and is half-wave symmetric, may be written as

$$v_{as} = \frac{2V_{\text{in}}}{\pi}\left(\cos\omega_r t + \frac{1}{5}\cos 5\omega_r t - \frac{1}{7}\cos 7\omega_r t + \cdots\right) \tag{3.5-11}$$

The question arises as to why are we using $\omega_r$ rather than $\omega_e$ in (3.5-11). We have two reasons; neither of which would stand up in court. First, we are using Park's transformation which involves $\theta_r(\omega_r)$ and second, in Chapter 4 we will control the inverter such that the frequency of $v_{as}$, $v_{bs}$, and $v_{cs}$ correspond to $\omega_r$, the electrical angular velocity of the rotor.

It should be noted that the maximum value of the Fourier expansion of $v_{as}$ given in Fig. 3.5-5c is $\frac{2}{3}V_{\text{in}}$; however, the peak value of the fundamental component of $v_{as}$ given in (3.5-11) is $\frac{2}{\pi}V_{\text{in}}$. The voltages $v_{bs}$ and $v_{cs}$ may be expressed, for an *abc* sequence, by substituting $\omega_r t - \frac{2}{3}\pi$ and $\omega_r t + \frac{2}{3}\pi$, respectively, for $\omega_r t$ in (3.5-11). If we use (2.5.17) to transform $v_{as}$, $v_{bs}$, and $v_{cs}$, where $v_{as}$ is given by (3.5-11), to the $\omega_r$ reference frame we obtain

$$v_{qs}^r = \frac{2V_{\text{in}}}{\pi}\left(1 + \frac{2}{35}\cos 6\omega_r t - \frac{2}{143}\cos 12\omega_r t + \cdots\right) \tag{3.5-12}$$

$$v_{ds}^r = \frac{2V_{\text{in}}}{\pi}\left(\frac{12}{35}\sin 6\omega_r t - \frac{24}{143}\sin 12\omega_r t + \cdots\right) \tag{3.5-13}$$

In Chapter 2 when we transformed the fundamental component of the applied voltages (two- or three-phase) with $v_{as} = \sqrt{2}V_s\cos\omega_r t$, we obtained $\sqrt{2}V_s$ for $v_{qs}^r$ and zero for $v_{ds}^r$. This is similar to (3.5-12) and (3.5-13), respectively, with the harmonics neglected. We are not surprised by this; however, it is interesting that the frequencies of the harmonics have changed. The 5th and 7th of the three-phase voltages have become the 6th harmonic and the 11th and 13th have become the 12th in the $qs^r$- and $ds^r$-voltages. You are asked to show this for the 5th and 7th harmonics in a problem at the end of the chapter. Plots of $v_{qs}^r$ and $v_{ds}^r$ with $v_{as}$ are shown in Fig. 3.5-6 with $V_{\text{in}} = 25$ V.

Before leaving this section let us consider the operation of the three-phase inverter shown in Fig. 3.5-5 for three different switching frequencies, which are shown in Figs. 3.5-7 through 3.5-9. The parameters of the symmetrical three-phase load are $r_s = 3.4\ \Omega$, $L_s = 12.1$ mH, and $V_{\text{in}} = 25$ V. In Fig. 3.5-7 the switching frequency $f_{\text{sw}} = 1.58$ Hz ($\omega_r = 10$ rad/s); in Fig. 3.5-8, $f_{\text{sw}} = 15.8$ Hz ($\omega_r = 100$ rad/s); and $f_{\text{sw}} = 28.9$ Hz ($\omega_r = 183$ rad/s) in Fig. 3.5-9. It should also be noted that in order to display this range of switching frequencies it is necessary to use a different time scale in each figure. Also, $v_{as}$, $i_{as}$, $i_{a\text{T}1}$, $-i_{ad1}$, $-i_{a\text{T}4}$, and $i_{ad4}$ are plotted where $i_{ad1}$ and $i_{ad4}$ are the anti-parallel diode currents. Also $I_{as}$, the fundamental component of $i_{as}$ obtained by a Fourier expansion, is superimposed on each plot of $i_{as}$.

It is helpful to go through a switching cycle of the $as$-phase of the inverter. With the logic signal T1 high, current is flowing through transistor T1 (Fig. 3.5-5a) and $i_{a\text{T}1}$ is positive and equal to $i_{as}$ (Figs. 3.5-7 through 3.5-9). When T1 is "turned off" and T4 is "turned on," $i_{a\text{T}1}$ is diverted to the anti-parallel diode d4. Since, at the instant of switching, $v_{as}$ is stepped from positive to negative, this negative voltage drives $i_{ad4}$ ($i_{as}$) to zero. Whereupon d4 shuts off and T4 conducts the negative current. Note

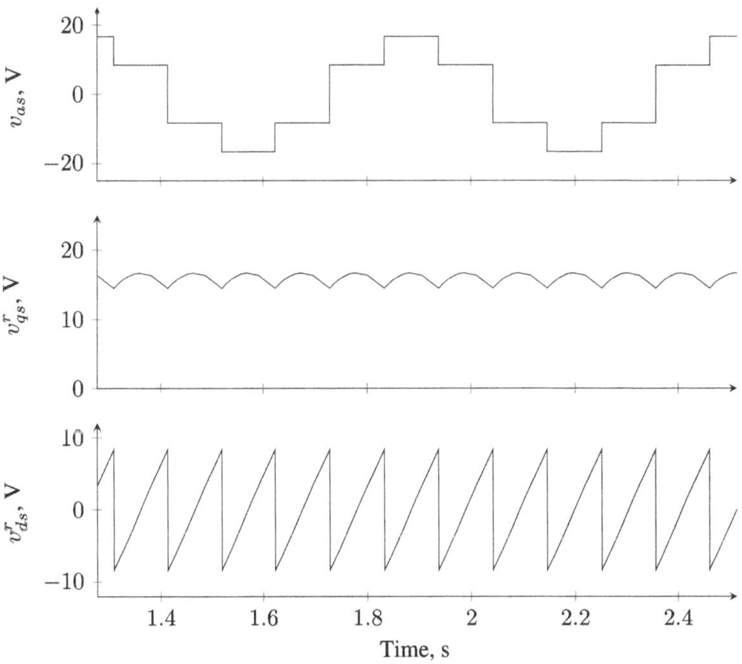


Figure 3.5-6 Plots of $v_{qs}^r$, $v_{ds}^r$, and $v_{as}$ of an $abc$ sequence with $V_{\text{in}} = 25$ V.

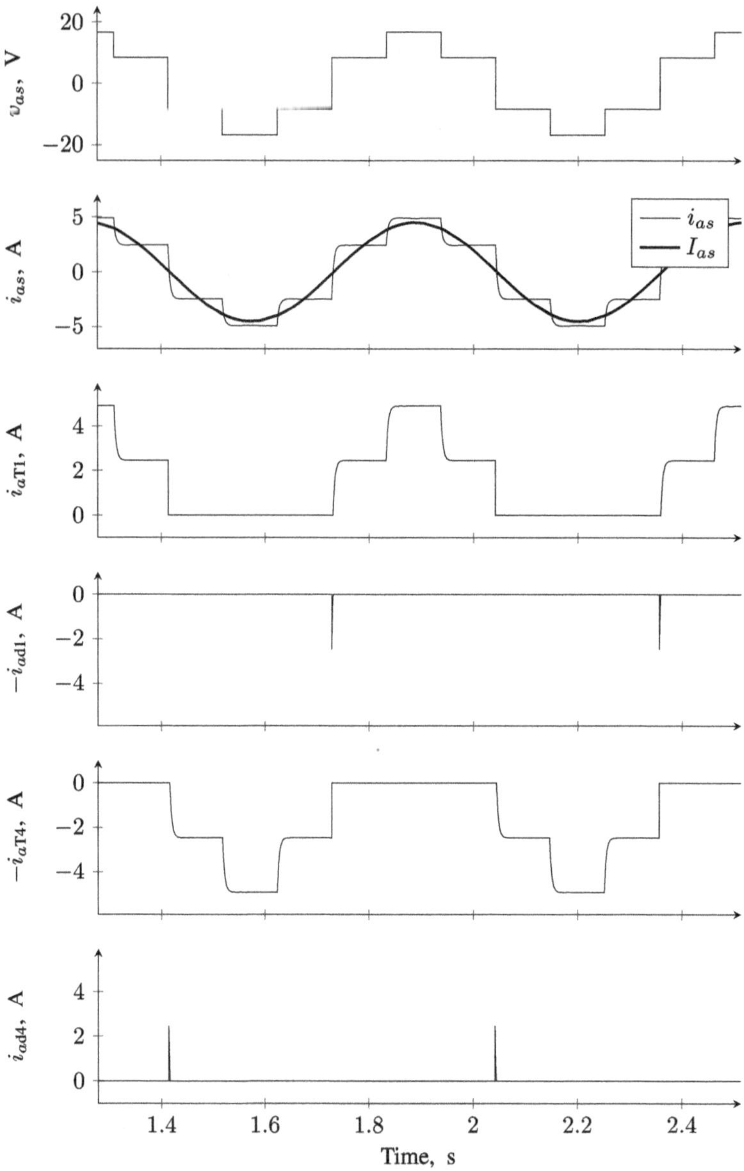


Figure 3.5-7 Waveforms from a three-phase inverter with $rL$ load, $V_{in} = 25$ V, $r_s = 3.4\,\Omega$, $L_s = 12.1$ mH and $f_{sw} = 1.58$ Hz.

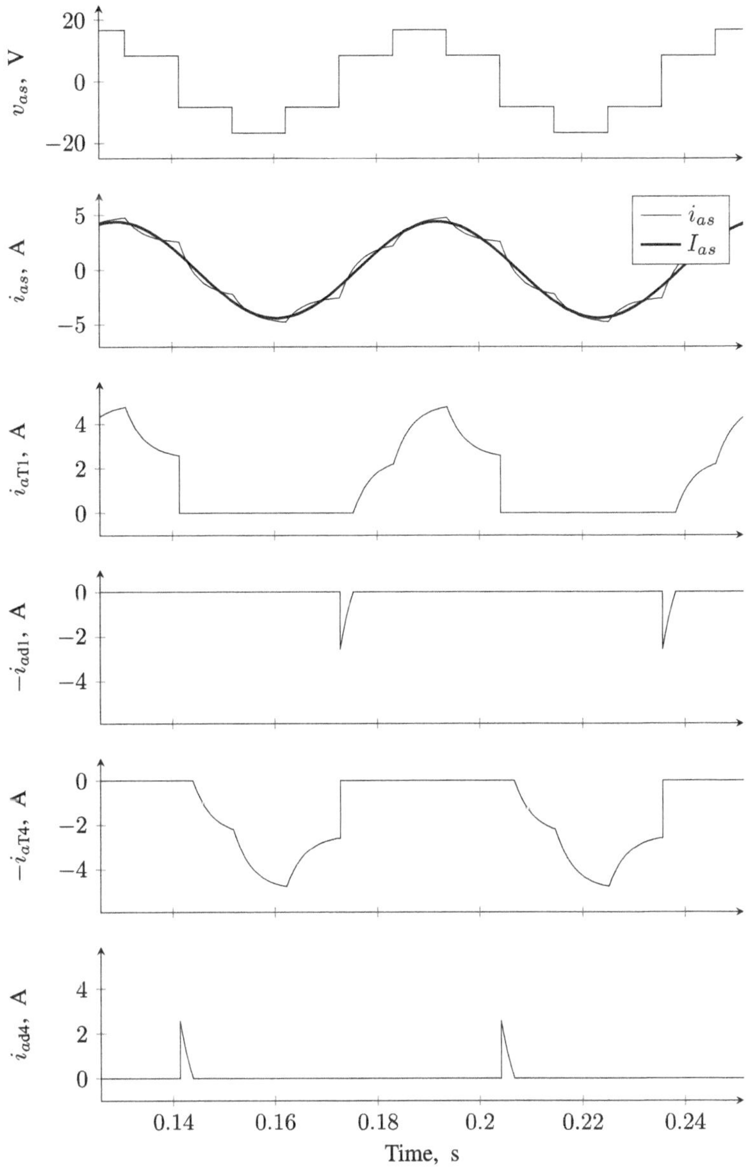


Figure 3.5-8 Waveforms from a three-phase inverter with $rL$ load, $V_{\text{in}} = 25$ V, $r_s = 3.4\,\Omega$, $L_s$ – 12.1 mH and $f_{\text{sw}} = 15.8$ Hz.

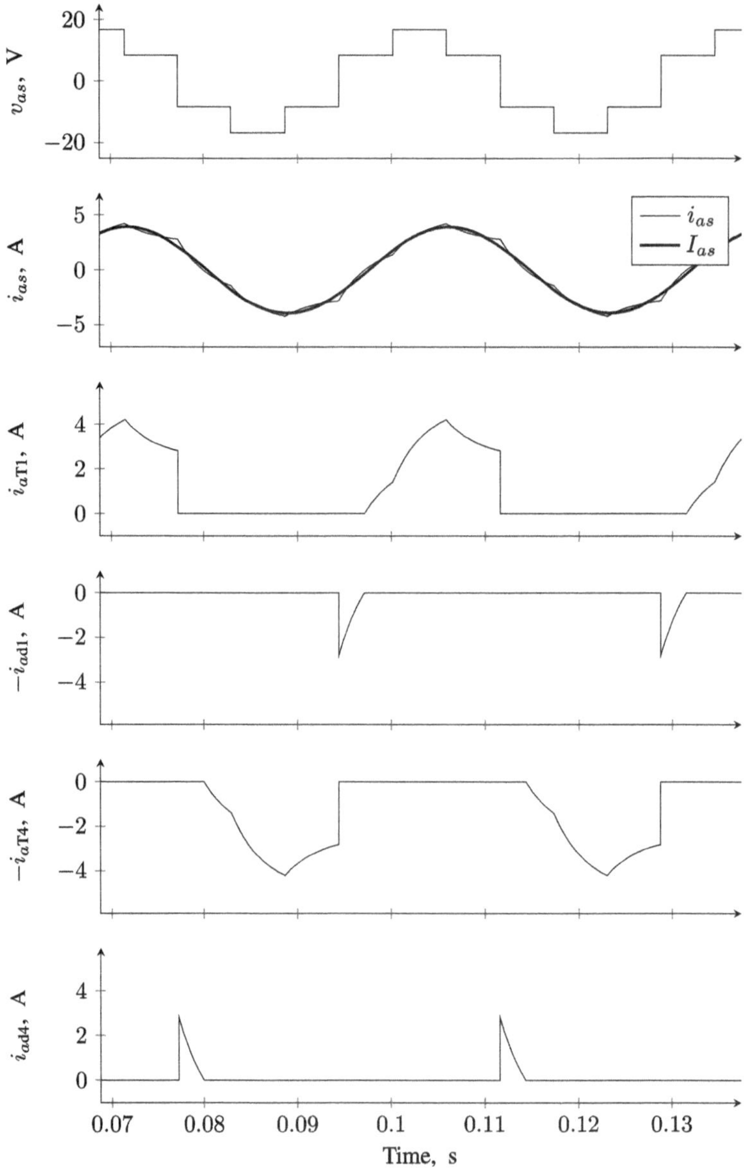


Figure 3.5-9 Waveforms from a three-phase inverter with $rL$ load, $V_{\text{in}} = 25$ V, $r_s = 3.4\ \Omega$, $L_s = 12.1$ mH and $f_{\text{sw}} = 28.9$ Hz.

that $-i_{aT4}$ is plotted to correspond with $i_{as}$ which is now negative. When T4 is turned off and T1 is turned on, the negative $i_{as}$ is diverted to d1 ($-i_{ad1}$ is plotted). Since $v_{as}$ has simultaneously been switched from negative to positive, $i_{aT1}$ ($i_{as}$) is driven to zero and T1 takes over. The cycle is complete.

For each switching frequency, we can express the fundamental component of $v_{as}$ from (3.5-11) and determine the fundamental component of $i_{as}$ shown in Figs. 3.5-7 through 3.5-9 by a Fourier analysis. For Fig. 3.5-7 where $f_{sw} = 1.58$ Hz ($\omega_r = 10$ rad/s); the fundamental components of $v_{as}$ and $i_{as}$ are

$$V_{as} = \frac{(2)(25)}{\pi} \cos 10t = 15.9 \cos 10t \tag{3.5-14}$$

$$I_{as} = 4.51 \cos(10t - 2.04°) \tag{3.5-15}$$

For Fig. 3.5-8 with $f_{sw} = 15.8$ Hz ($\omega_r = 100$ rad/s)

$$V_{as} = 15.9 \cos 100t \tag{3.5-16}$$

$$I_{as} = 4.4 \cos(100t - 19.7°) \tag{3.5-17}$$

For Fig. 3.5-9 with $f_{sw} = 28.9$ Hz ($\omega_r = 183$ rad/s)

$$V_{as} = 15.9 \cos 183t \tag{3.5-18}$$

$$I_{as} = 3.91 \cos(183t - 33.2°) \tag{3.5-19}$$

In the following example, we will calculate the fundamental component of each $i_{as}$ and compare with that obtained by a Fourier expansion. Although we would expect a good correlation since the fundamental component of the applied voltages is dominant, it is interesting, but perhaps not surprising, that the phase shift in $i_{as}$, due to the increase in switching frequency, is predicted by $\tan^{-1}(\omega_r L_s / r_s)$. Recall that we are using $\omega_r$ rather than $\omega_e$.

**Example 3H. Current calculation for three-phase inverter.** It is interesting, for comparison purposes, to calculate $I_{as}$ for each switching frequency using only the fundamental component of the voltage $V_{as}$. For $f_{sw} = 1.58$ Hz ($\omega_r = 10$ rad/s)

$$\tilde{I}_{as} = \frac{\tilde{V}_{as}}{Z}$$

$$= \frac{\dfrac{(2)(25)}{\pi\sqrt{2}}\underline{/0}}{3.4 + j(10)(12.1 \times 10^{-3})} = 3.31\underline{/-2.04°}\text{A} \tag{3H-1}$$

Thus,

$$I_{as} = 4.67 \cos(10t - 2.04°) \tag{3H-2}$$

This is to be compared to (3.5-15). For $f_{sw} = 15.8$ Hz ($\omega_r = 100$ rad/s)

$$\tilde{I}_{as} = \frac{11.25\underline{/0}}{3.4 + j(100)(12.1 \times 10^{-3})} = 3.09\underline{/-19.6°}\text{A} \tag{3H-3}$$

Thus

$$I_{as} = 4.37 \cos(100t - 19.6°) \tag{3H-4}$$

This is to be compared to (3.5-17). For $f_{sw} = 28.9$ Hz ($\omega_r = 183$ rad/s)

$$\tilde{I}_{as} = \frac{11.25\angle 0}{3.4 + j(183)(12.1 \times 10^{-3})} = 2.75\angle -33.7^\circ \text{A} \tag{3H-5}$$

Thus,

$$I_{as} = 3.89\cos(183t - 33.7^\circ) \tag{3H-6}$$

This is to be compared to (3.5-19).

---

We have now introduced fundamental power conversion concepts and applied them to dc-dc, ac-dc, and dc-ac conversion circuits. You might be asking, "What about ac-ac conversion?" As we have seen in Chapter 1, ac transformers work great for this, provided the transformation is at a fixed frequency. Since this is the case more often than not, power electronic converters are not as prevalent in this application. One exception to this is when the frequency must be changed. In that case, a transformer will not suffice and an ac-ac power electronic circuit – the cycloconverter is one example – is required. Another exception is when interconnecting two power systems that are at the same frequency. In this case, in order to control the power flow between the systems, often ac-dc-ac conversion is used. We will leave the cycloconverter and other ac-ac conversion methods as topics for advanced study. However, with full mastery of the techniques introduced in this chapter, ac-ac conversion circuit analysis is a direct extension.

---

**SP3.5-1.** Modify (3.2-22) to find the Fourier series of the single-phase inverter waveform shown in Fig. 3.5-2. [$v_{load}(t) = \frac{4V_{in}}{\pi} \sum_{n=1}^{\infty} \frac{\sin(n\pi/2)}{n} \cos(n\omega_{sw}t - n\pi/2)$]

**SP3.5-2.** It is noted in Fig. 3.5-7 that the peak value of the fundamental component of $i_{as}$ is less than the maximum value of the plot of $i_{as}$. Why? [$\frac{2}{\pi}$ vs $\frac{2}{3}$ and the $rL$ switching transient subsides early in switching period]

**SP3.5-3.** It is noted in Fig. 3.5-9 that $i_{as}$ is approaching a sinusoidal waveform. Why? [Switching period is completed early in $rL$ switching transient]

## 3.6 HARMONICS AND DISTORTION

The fact that the single and three-phase inverter waveforms introduced in Section 3.5 are not pure sinusoids indicates that they contain additional harmonics, which are a byproduct of the dc-ac inversion process. As seen in Figs. 3.5-3 and 3.5-4, respectively, filtering with inductive and resonant filters can attenuate, but not eliminate, many of these harmonics. Unfortunately, this is not an isolated example; a common characteristic of all power electronic converter circuits, not just inverters, is that they generate and inject harmonics into the power systems to which they are connected. Harmonics can have many undesirable effects, such as creating instability, increasing losses and heat, and causing electromagnetic interference (EMI). To characterize the extent of these potential issues, it is common to quantify the harmonic content of a power electronic circuit's waveforms. In the United States, the most common metric for this is the *total harmonic distortion* (THD) [1, 2].

In this text, we calculate THD for zero-average periodic (i.e., no dc offset) signals. THD can also be calculated for offset signals, but the derivations and definitions given below differ slightly from that case. This detail, however, should not complicate our discussion, as a mastery of the concepts and conclusions drawn below is sufficient to give the reader a fundamental understanding of the uses and importance of THD. Further, the material here can be applied to a dc-offset signal with no loss of generality if the offset is first removed.

If we write a zero-average periodic waveform $f(t)$ with its compact Fourier series (3.2-12):

$$f(t) = \sqrt{2}\sum_{n=1}^{\infty} F_n \cos(n\omega t + \theta_n) \tag{3.6-1}$$

where $F_n = c_n/\sqrt{2}$ and $\theta_n$ are the rms magnitude and phase, respectively, of the $n$th harmonic, the THD is defined as

$$\text{THD} \equiv \frac{\sqrt{\sum_{n=2}^{\infty} F_n^2}}{F_1} \tag{3.6 2}$$

The THD, often expressed as a percentage, is a measure of the ratio of the harmonic-to-fundamental content of a signal, and it is a convenient metric for quantifying the total harmonic content of a signal. If more detailed information about the specific harmonics contained in a signal is required, THD alone will not suffice. In those cases, the Fourier series is required.

The higher a signal's harmonic content, the higher is its THD. When no harmonics are present, as is the case in a sinusoid, the THD is zero. When harmonics are present, there is no mathematical upper limit on THD. For reference, the THD of a symmetric square wave such as $sq(t) = \text{sign}(\cos(\omega t))$ is 48.3% and the THD of a symmetric triangle wave is 12.1%. Between these two examples, the triangle wave more closely resembles a sinusoid and has lower harmonic content; thus, its THD is lower. It is left to the reader to confirm these numbers in the chapter problems.

Because the THD definition contains an infinite series, it is often cumbersome to work directly with (3.6-2) if a signal's Fourier series is complicated or unknown; however, if we note that the rms of $f(t)$, $F_{\text{rms}}$, is simply

$$F_{\text{rms}} = \sqrt{\sum_{n=1}^{\infty} F_n^2} \tag{3.6-3}$$

and substitute this into (3.6-2) we see that

$$\text{THD} = \frac{\sqrt{F_{\text{rms}}^2 - F_1^2}}{F_1} \tag{3.6-4}$$

Equation (3.6-4) is often easier to work with than (3.6-2), because a signal's total rms value and that of its fundamental can be found in several ways, including direct measurement with standard equipment. Because of this, it is a good idea to become familiar with both equations for calculating THD.

**Example 3I. THD of a switching function.** Calculate the THD of the switching function given in Example 3B and shown in Fig. 3B-1.

The THD can be calculated with (3.6-2); however, because the Fourier series of $q(t)$ is infinite (3B-5), this is not preferred. Rather, we will use (3.6-4). To calculate $F_{rms}$ we must first remove the dc offset from $q(t)$. By inspection,

$$\hat{q} = 1/4 \tag{3I-1}$$

After subtracting the dc offset the mean square of the shifted signal is calculated as

$$(q - \hat{q})_{ms} = \frac{1}{T_{sw}}\left[(3/4)^2 \times \frac{T_{sw}}{4} + (-1/4)^2 \times \frac{3}{4}T_{sw}\right] = \frac{3}{16} \tag{3I-2}$$

The rms is then

$$\begin{aligned} F_{rms} &= (q - \hat{q})_{rms} \\ &= \sqrt{(q - \hat{q})_{ms}} = \frac{\sqrt{3}}{4} \end{aligned} \tag{3I-3}$$

We can calculate $F_1$ directly from the switching function's compact Fourier series (3B-5):

$$\begin{aligned} F_1 &= \frac{1}{\sqrt{2}}\frac{2}{\pi}\sin(\pi/4) \\ &= \frac{1}{\pi} \end{aligned} \tag{3I-4}$$

The THD can now readily be calculated using (3.6-4):

$$\text{THD} = \frac{\sqrt{\frac{3}{16} - \frac{1}{\pi^2}}}{1/\pi} = 0.922\,(92.2\%) \tag{3I-5}$$

We conclude the Example by noting that the THD of the switching function with $D = 1/4$ (92.2%) is nearly twice the THD of a switching function with $D = 1/2$ (48.3%). This is because larger amplitude harmonics are required to create a signal with a smaller duty ratio. In fact, 48.3% is the minimum THD that can be achieved with a periodic square wave switching function defined by (3.2-22).

---

**SP3.6-1.** Calculate the THD of the periodic function $f(t)$ defined as the following: $f(t) = 10\sin(\omega t) + 8\cos(\omega t) - 2\sin(3\omega t) + \cos(8\omega t)$. [17.5%]

**SP3.6-2.** You wish to calculate the THD of $f(t) + 5$, where $f(t)$ is as defined in SP3.6-1, using (3.6-3) and (3.6-4). (a) What is the first step necessary? (b) What is the THD of $f(t) + 5$? [(a) Remove the dc component before calculating rms; (b) 17.5% (THD does not change with a dc offset)]

## REFERENCES

[1] P. T. Krein, *Elements of Power Electronics*, Oxford University Press, 1998.
[2] G. M. Masters, *Renewable and Efficient Electric Power Systems*, 2nd edition, Wiley-IEEE Press, 2013.
[3] R. W. Erickson and D. Maksimovic, *Fundamentals of Power Electronics*, 2nd edition, Springer, 2001.

## PROBLEMS

1. Show that the compact Fourier series of the periodic triangle wave, *tri*($t$), with period $T$ defined over the interval $[0, T]$ as

$$tri(t) = \begin{cases} 1 - \frac{4t}{T}, \forall t \in [0, T/2) \\ -1 + \frac{4(t - T/2)}{T}, \ \forall t \in [T/2, T) \end{cases}$$

is the following:

$$tri(t) = \frac{8}{\pi^2} \sum_{n=1,3,5\cdots}^{\infty} \frac{1}{n^2} \cos\left(\frac{2\pi n}{T} t\right)$$

It may be helpful to first calculate the Fourier series in the standard form, simplifying as appropriate using the rules in (3.2-16) through (3.2-19), and then to convert to the compact form.

2. Show that the compact Fourier series of the symmetric square wave switching function defined by $q(t) = 1/2[1 + \text{sign}(\sin(t))]$ is the following:

$$q(t) = \frac{1}{2} + \frac{2}{\pi} \sum_{n=1,3,5\cdots}^{\infty} \frac{(-1)^{\frac{n-1}{2}}}{n} \sin(nt)$$

It may help to sketch the function, compare it to Fig. 3.2-5, and use (3.2-22). Some simplification of the series may be necessary to match the given solution.

3. Repeat the buck converter design in Example 3D if instead the battery voltage is 24 V and the inductor peak-to-peak current ripple should be no more than 20% of the average current. All other specifications and assumptions are the same. Note that your results can vary from the solution, depending on your design choices.

4. For the buck converter you designed in Problem 3, sketch the inductor current waveform over three switching periods. Be sure to label key points on the $x$- and $y$-axes to explain your sketch.

5. You wish to interface a photovoltaic array, which produces dc current at 35 V, to a dc 140 V, 2.8 kW load. (a) Suggest dc-dc converter for this application. (b) Based on your choice of converter, find $D$, assuming no losses. (c) Calculate the load current. (d) Assuming no losses in the converter, calculate the photovoltaic array output current and power.

6. For the diode circuit and input voltage, $v_{\text{in}}(t)$, shown in Fig. 3.8-1, perform the diode trial method. (a) Create and fill in a trial method table similar to Table 3.1. (b) Based on your results in (a), sketch $v_{\text{load}}(t)$ over a few periods. (c) Calculate $\widehat{P}_{\text{load}}$.

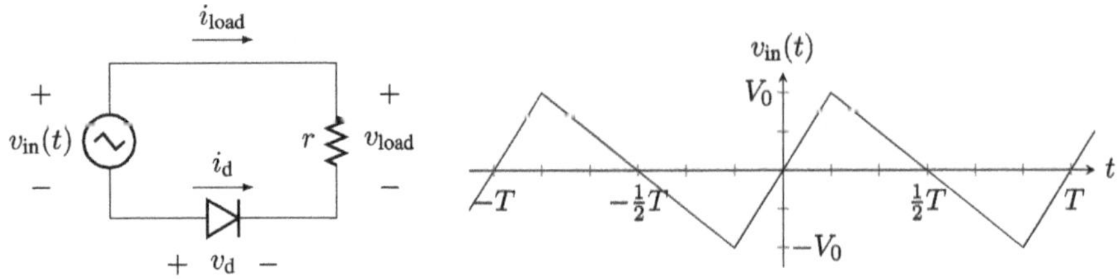


Figure 3.8-1 Diode circuit.

7. A single-phase inverter with $V_{in} = 10$ V is supplying a series $rL$ load with $r = 1\ \Omega$ and $L =$ 3 mH. The switching frequency, $f_{sw}$, is 60 Hz and relevant circuit waveforms are shown in Fig. 3.5-3. Find the load current phasor and average power for (a) the fundamental, and (b) the third harmonic. The series you found in SP3.5-1 will be useful.

8. For the three-phase inverter, with $rL$ load and $r_s = 3.4\ \Omega$, $L_s = 12.1$ mH, $V_{in} = 25$ V, and $\omega_r = 10$ rad/s, determine the peak amplitude of the 5th harmonic component of $i_{as}$ with $v_{as}$ given by (3.5-11).

9. Calculate the THD of the triangle wave, $tri(t)$, defined in Problem 1. The Fourier series calculated in Problem 1 may be useful.

10. Calculate the THD of the symmetric square wave switching function, $q(t)$, defined in Problem 2. The Fourier series calculated in Problem 2 may be useful.

CHAPTER 4

# PERFORMANCE AND SIMULATION OF AN ELECTRIC DRIVE

## 4.1 INTRODUCTION

For our purposes, an electric drive is broadly defined as a controlled inverter connected to an electric motor which drives a mechanical load. The electric drive of choice for low power applications has become the brushless dc drive which is a permanent-magnet ac motor supplied from a controlled three-phase six-step inverter. The permanent-magnet ac machine was analyzed in Chapter 2 where the analysis of it operating as a brushless dc motor was carried out by considering only the fundamental component of the six-step inverter voltages that are applied to the stator windings when supplied by the controlled inverter. In this chapter, a computer simulation of the brushless dc drive is used to illustrate its dynamic and steady-state performances with the sinusoidal approximation of the inverter voltages and a more realistic representation of the drive with six-step stator applied voltages. A direct comparison of the simulated performance with both sinusoidal applied voltages and with six-step voltages is made for three different values of phase shift, $\phi_v$, of $\tilde{V}_{as}$ relative to the $q$-axis; the normal mode of operation with $\phi_v = 0$, maximum torque per volt with $\phi_v = \phi_{vMT/V}$, and maximum torque per ampere with $\phi_v = \phi_{vMT/A}$. An idealized consideration of the controlled operating modes of a brushless dc drive is presented along with the implementation of the computer simulation of the inverter-brushless dc drive.

## 4.2 OPERATING MODES OF A BRUSHLESS DC MOTOR

In Section 2.5, equations were derived for the three-phase permanent-magnet ac machine. In Example 2F, steady-state operation was considered with $\phi_v$ set to zero. It was mentioned that this mode of operation ($\phi_v = 0$) "makes" the torque-speed characteristic of a permanent-magnet ac machine similar to the dc motor. Although this is a common mode of operation of the brushless dc motor, researchers in References

*Introduction to Electric Power and Drive Systems*, First Edition.
Paul Krause, Oleg Wasynczuk, Timothy O'Connell and Maher Hasan.
© 2017 by The Institute of Electrical and Electronics Engineers, Inc. Published 2017 by John Wiley & Sons, Inc.
Companion website: www.wiley.com/go/krause/electricpower

[1, 2] discovered experimentally that advancing $\phi_v$ could increase the torque at rotor speeds greater than zero. This was shown analytically in Reference [3] and illustrated by simulating the phase shifting (increasing $\phi_v$) of the applied voltages to obtain maximum torque per volt ($\phi_v = \phi_{vMT/V}$) at a given speed. These two modes of operation ($\phi_v = 0$ and $\phi_v = \phi_{vMT/V}$) along with operation with maximum torque per ampere ($\phi_v = \phi_{vMT/A}$) where $I_{ds}^r$ is made zero, are considered in this section using sinusoidal applied voltages.

## Brushless DC Motor Operation with $\phi_v = 0$

The parameters of the fractional-horsepower, four-pole, three-phase permanent-magnet ac machine considered in this section are those given in Example 2F; in particular, $r_s = 3.4\ \Omega$, $L_{ls} = 1.1$ mH, $L_{ms} = 11$ mH, and $\lambda'_m = 0.0826$ V $\cdot$ s/rad. Recall that $L_{ms}$ must be multiplied by $\frac{3}{2}$ for a three-phase machine. For brushless dc motor operation, $\omega_e$ is made equal to $\omega_r$ and $\phi_v = 0$, that is, the *as*-phase voltage is controlled to be

$$V_{as} = \sqrt{2}V_s \cos \omega_r t \tag{4.2-1}$$

with a balanced *abc* sequence and $V_s = 11.25$ V. With sinusoidal applied voltages and $\phi_v = 0$, the maximum positive value of $V_{as}$ coincides with the *q*-axis and this unison is fixed by controlling the frequency of the applied voltages to be equal to the electrical angular velocity of the rotor ($\omega_e = \omega_r$).

The free-acceleration characteristics (starting from stall with $T_L = 0$) are shown in Fig 4.2-1. The total inertia $J$ of the rotor and mechanical load is $5 \times 10^{-4}$ kg $\cdot$ m$^2$ and the damping coefficient $B_m$ is neglected. The torque versus the rotor electrical angular speed for Fig. 4.2-1 is shown in Fig. 4.2-2. The steady-state torque is also plotted in Fig. 4.2-2. The negative slope of the torque-speed characteristics ensures stable operation for motor operation. That is, for a given load torque, a slight slowing of the rotor from an operating point will cause $T_e$ to increase forcing the rotor back to the operating speed where $T_L = T_e$. A small increase in rotor speed causes $T_e$ to decrease whereupon the load torque slows the rotor back to the original operating point; a stable operating condition.

The voltage equation for a brushless dc motor during balanced steady-state operation may be obtained by setting $V_{ds}^r$ equal to zero, since $\phi_v = 0$ in (2.4-45), and solving (2.4-48) for $I_{ds}^r$ with $V_{ds}^r = 0$ yields

$$I_{ds}^r = \frac{\omega_r L_{ss}}{r_s} I_{qs}^r \tag{4.2-2}$$

Substituting (4.2-2) into (2.4-47) gives

$$V_{qs}^r = \frac{r_s^2 + \omega_r^2 L_{ss}^2}{r_s} I_{qs}^r + \omega_r \lambda_m'^r \tag{4.2-3}$$

During steady-state operation, all quantities in (4.2-2) and (4.2-3) are constants.

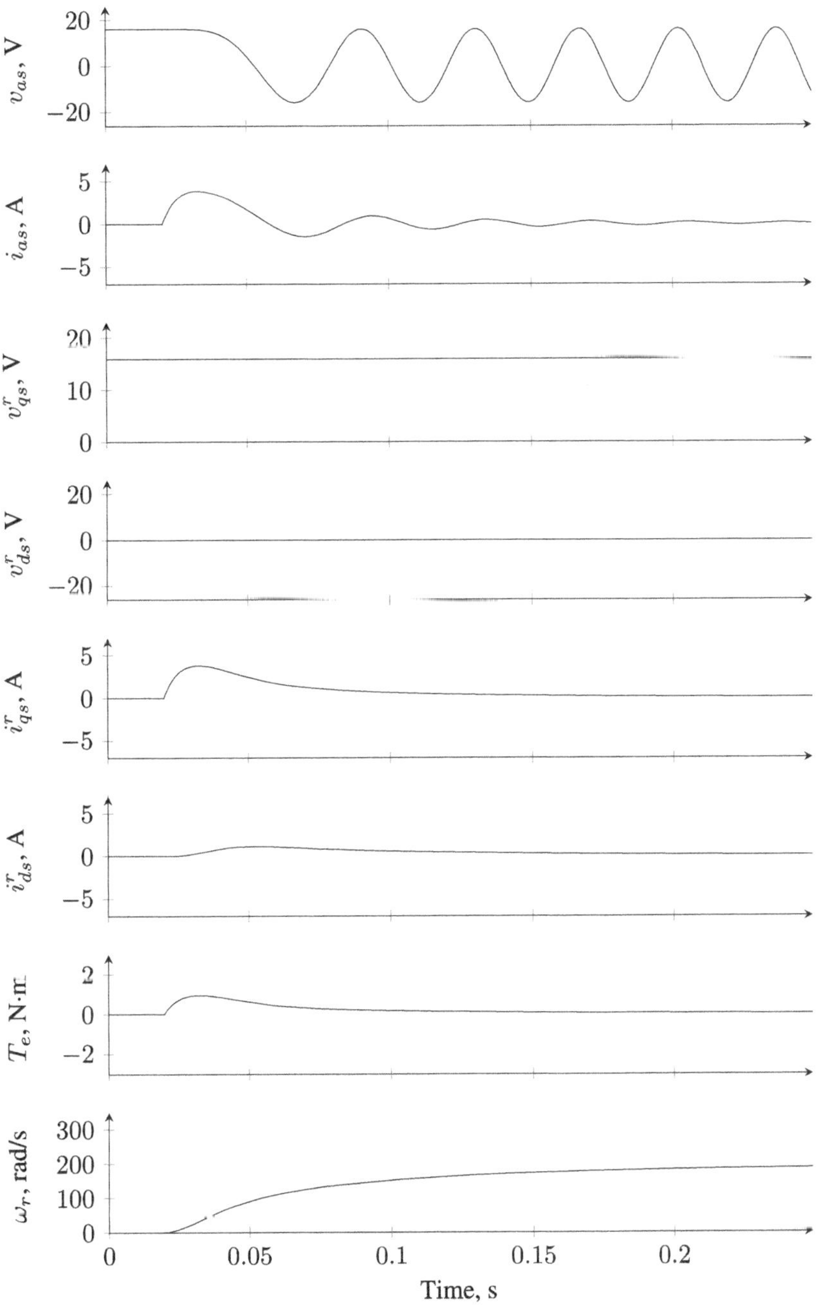


Figure 4.2-1 Free-acceleration characteristics of a brushless dc motor with $\phi_v = 0$ and a total inertia of $5 \times 10^{-4}$ kg · m$^2$.

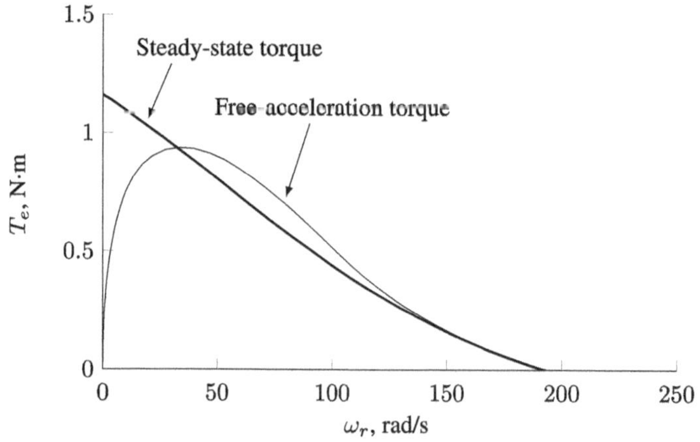


Figure 4.2-2 Torque-speed characteristics for the free acceleration shown in Fig. 4.2-1 with the steady-state torque shown for comparison purposes.

If we choose to work with the *abc* variables, the phasor voltage equation is given by (2.4-61), that is

$$\tilde{V}_{as} = (r_s + j\omega_r L_{ss})\tilde{I}_{as} + \tilde{E}_a \tag{4.2-4}$$

where from (2.4-60)

$$\tilde{E}_a = \frac{1}{\sqrt{2}}\omega_r \lambda_m'^r \angle 0° \tag{4.2-5}$$

We are dealing with a three-phase machine thus $\frac{3}{2}L_{ms}$ is used to evaluate $L_{ss}$. For the three-phase brushless dc machine, the torque is given by (2.5-35) which is

$$T_e = \frac{3}{2}\frac{P}{2}\lambda_m'^r i_{qs}^r \tag{4.2-6}$$

In summary, the brushless dc motor is so named because its torque versus speed characteristics resemble those of a dc motor. Therefore, it seems logical that there is a reference frame where the voltage and torque expressions for the brushless dc motor are dc rather than ac and we are aware of this from our work in Chapter 2. This fact is very apparent from the traces in Fig. 4.2-1 of $v_{qs}^r$, $i_{qs}^r$, and $T_e$. These variables are dc in the steady state and the torque-speed characteristics (Fig. 4.2-2) resemble those of a dc motor.

It is instructive to consider the operating characteristics of the brushless dc motor during load torque changes as shown in Fig 4.2-3. Initially, the machine is operating with $T_L = 0.1$ N · m. The load torque is suddenly stepped to 0.4 N · m. The motor slows and once steady-state operation is established the load torque is stepped back to 0.1 N · m. In this study, the inertia is $2 \times 10^{-4}$ kg · m$^2$. The decrease in the frequency of the electrical variables and the increase in currents as the load is increased are evident.

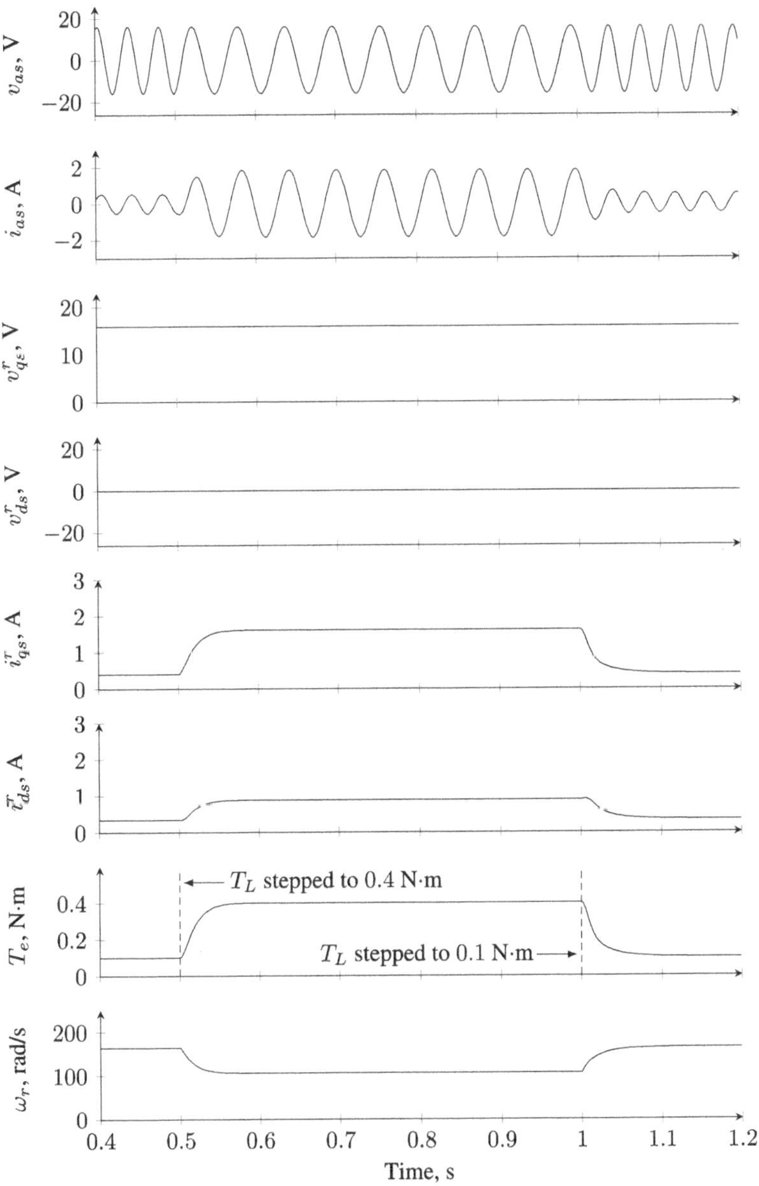


Figure 4.2-3 Dynamic performance of a brushless dc motor during step changes in load torque, total inertia of $2 \times 10^{-4}$ kg · m$^2$.

**Example 4A. Brushless dc motor operation with $\phi_v = 0$.** To facilitate a ready comparison, some of the information given in Example 2F will be repeated here. The example of three-phase four-pole permanent-magnet ac machine has the following parameters $r_s = 3.4\ \Omega$, $L_{ls} = 1.1$ mH, $L_{ms} = 11$ mH, and $\lambda'_m = 0.0826$ V · s/rad. It is wye-connected with balanced three-phase applied voltages. Since $\theta_r(0) = 0$ and $\phi_v = 0$, the *as*-phase voltage is $V_{as} = \sqrt{2}\,11.25 \cos \omega_e t$. The actual steady-state rotor speed is 750 r/min; therefore, the electrical angular velocity of the rotor, $\omega_r$, from (2F-1) is $50\pi$ rad/s which is the frequency of the applied voltages. We were asked to calculate $\tilde{I}_{as}$ and draw the phasor diagram and locate the stator and rotor poles. We will add power balance calculations in this example.

The following calculations were made:

$$I^r_{qs} = 0.521 \text{ A} \quad V^r_{qs} = \sqrt{2}(11.25) \text{ V} \quad \tilde{E}_a = 9.18\underline{/0^\circ} \text{ V}$$
$$I^r_{ds} = 0.424 \text{ A} \quad V^r_{ds} = 0 \quad \tilde{V}_{as} = 11.25\underline{/0^\circ} \text{ V}$$
$$\tilde{I}_{as} = 0.474\underline{/-39.1^\circ} \text{ A} \quad T_e = 0.129 \text{ N} \cdot \text{m}$$

The phasor diagram is shown in Fig. 4A-1.

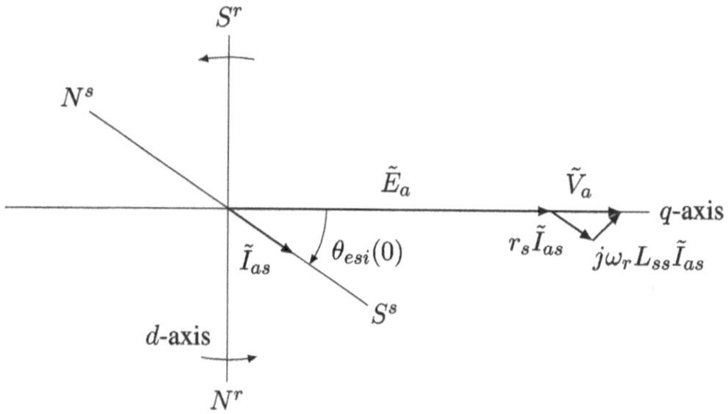


Figure 4A-1 Phasor diagram for brushless dc motor operation at $\omega_r = 50\pi$ rad/s with $\phi_v = 0$. Same as Fig. 2F-1.

We see from Fig. 4A-1 that the rotor poles can be considered as being "pulled" in the counterclockwise direction by the poles created by the stator currents; motor action.

The power balance calculations can be performed using the information given in Chapter 2. For balanced conditions, the power from the electrical system is

$$P_e = \frac{3}{2}\left(v^r_{qs} i^r_{qs} + v^r_{ds} i^r_{ds}\right) \tag{4A-1}$$

or

$$P_e = 3|\tilde{V}_{as}|\,|\tilde{I}_{as}| \cos[\theta_{esv}(0) - \theta_{esi}(0)] \tag{4A-2}$$

Using (4A-1),

$$P_e = \frac{3}{2}\left(\sqrt{2}11.25 \times 0.521\right) = 12.44 \text{ W} \tag{4A-3}$$

$$\begin{aligned} P_{\text{loss}} &= 3r_s|\tilde{I}_{as}|^2 \\ &= 3 \times 3.4(0.475)^2 = 2.30 \text{ W} \end{aligned} \tag{4A-4}$$

The power to the mechanical system is

$$\begin{aligned} P_m &= T_e\omega_{rm} \\ &= 0.129 \times \frac{2}{4} \times 50\pi = 10.13 \text{ W} \end{aligned} \tag{4A-5}$$

Note that $P_m = P_e - P_{\text{loss}}$ and the efficiency is

$$\text{eff} = \frac{P_m}{P_e} = \frac{10.13}{12.44} = 81.4\% \tag{4A-6}$$

## Maximum Torque Per Volt Operation of a Brushless DC Motor ($\phi_v = \phi_{vMT/V}$)

If the applied voltages are shifted relative to the magnetic field established by the permanent-magnet rotor, which is fixed in the $d$-axis, the torque versus speed characteristics can be changed over a wide range by shifting $\phi_v$ from zero to $2\pi$ [3, 4]. Here, we will limit our discussion to shifting $\phi_v$ for the purpose of maximizing torque. Torque is proportional to $i_{qs}^r$ or $I_{qs}^r$ and when $\phi_v$ is shifted from zero, $v_{ds}^r$ is nonzero, (2.4-45). For the purpose of deriving an expression for the maximum torque per volt at a given rotor speed ($\phi_{vMT/V}$) we will start with (2.4-47) and (2.4-48) for $V_{qs}^r$ and $V_{ds}^r$, respectively. In particular, for steady-state operation

$$V_{qs}^r = r_sI_{qs}^r + \omega_rL_{ss}I_{ds}^r + \omega_r\lambda_m'^r \tag{4.2-7}$$

$$V_{ds}^r = r_sI_{ds}^r - \omega_rL_{ss}I_{qs}^r \tag{4.2-8}$$

Please recall that these equations are valid for two- or three-phase devices for balanced steady-state operation if for the three-phase device, $\frac{3}{2}L_{ms}$ is used when calculating $L_{ss}$. We also need the expressions for $V_{qs}^r$ and $V_{ds}^r$ as functions of $\phi_v$ from (2.4-44) and (2.4-45), which are valid for transient and steady-state operation

$$V_{qs}^r = \sqrt{2}V_s\cos\phi_v \tag{4.2-9}$$

$$V_{ds}^r = -\sqrt{2}V_s\sin\phi_v \tag{4.2-10}$$

Solving (4.2-8) for $I_{ds}^r$ and substituting the result into (4.2-7) yields

$$V_{qs}^r = \frac{r_s^2 + \omega_r^2L_{ss}^2}{r_s}I_{qs}^r + \frac{\omega_rL_{ss}}{r_s}V_{ds}^r + \omega_r\lambda_m'^r \tag{4.2-11}$$

Now, solving (4.2-11) for $I_{qs}^r$ and substituting (4.2-9) and (4.2-10) for $V_{qs}^r$ and $V_{ds}^r$, respectively, with $\theta_r(0) = 0$, we have

$$I_{qs}^r = \frac{r_s}{r_s^2 + \omega_r^2 L_{ss}^2}\left(\sqrt{2}V_s \cos\phi_v + \frac{\omega_r L_{ss}}{r_s}\sqrt{2}V_s \sin\phi_v - \omega_r \lambda_m'^r\right) \quad (4.2\text{-}12)$$

It is interesting to note, from (4.2-12) that $V_{ds}^r$ aids $V_{qs}^r$ to increase $I_{qs}^r$ for a given rotor speed. Since this results in a negative $I_{ds}^r$, it is often referred to as field weakening even though $\lambda_m'$ is not decreased in magnitude.

Since $T_e$ is proportional to $I_{qs}^r$, (4.2-6), we can obtain the maximum torque for a given rotor speed by taking the derivative of $I_{qs}^r$ with respect to $\phi_v$ and setting the result equal to zero and then solving for $\phi_v$. Thus, from (4.2-12)

$$0 = -\sin\phi_v + \frac{\omega_r L_{ss}}{r_s}\cos\phi_v \quad (4.2\text{-}13)$$

whereupon

$$\frac{\sin\phi_v}{\cos\phi_v} = \frac{\omega_r L_{ss}}{r_s} \quad (4.2\text{-}14)$$

or

$$\phi_{vMT/V} = \tan^{-1}\frac{\omega_r L_{ss}}{r_s} \quad (4.2\text{-}15)$$

Equation (4.2-15) tells us that for a given positive rotor speed, $\phi_{vMT/V}$ will yield maximum possible torque per volt at that rotor speed.

The free-acceleration characteristics for $\phi_{vMT/V}$ are shown in Figs. 4.2-4 and 4.2-5. These characteristics may be compared to Figs. 4.2-1 and 4.2-2, respectively, where $\phi_v = 0$. Note the extended speed range with $\phi_{vMT/V}$ (Fig. 4.2-5) compared with $\phi_v = 0$ (Fig. 4.2-2). Also, note that $i_{ds}^r$ is small positive in Fig. 4.2-1 but a larger negative value in Fig. 4.2-4. In other words, an increase in torque and speed range occurs with an increase in the magnitude of $i_{ds}^r$.

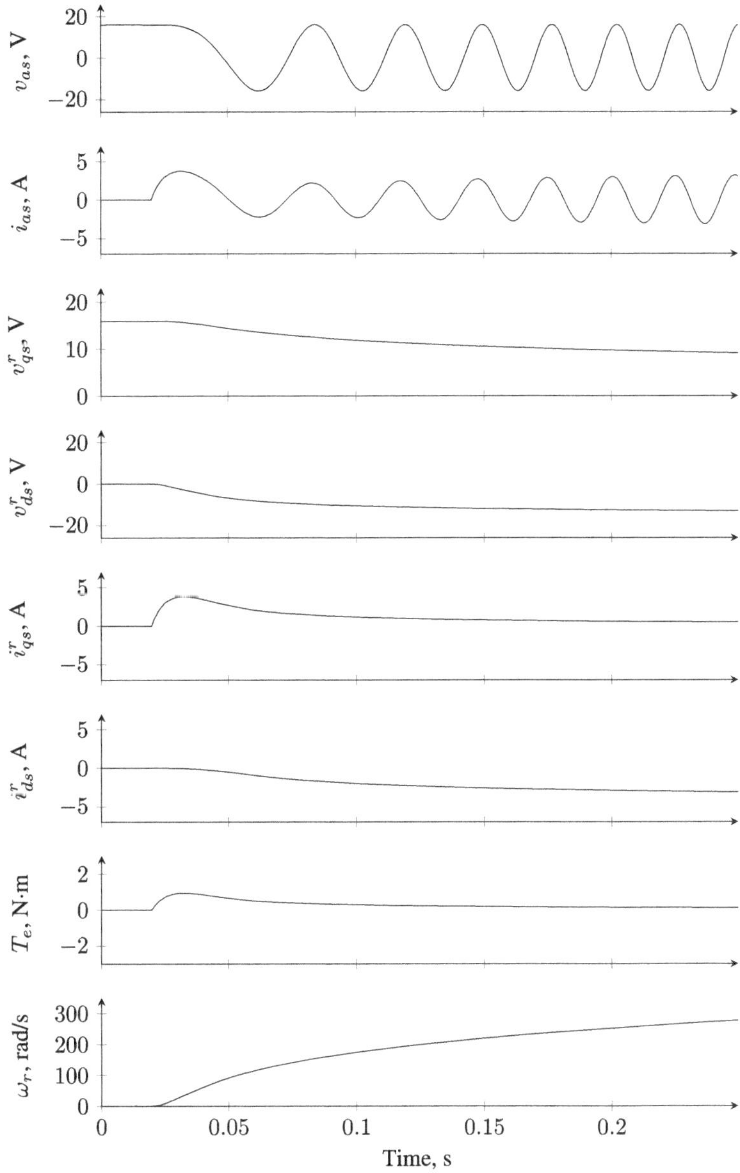


Figure 4.2-4 Free-acceleration characteristics of a brushless dc machine with $\phi_v = \phi_{vMT/V}$ and a total inertia of $5 \times 10^{-4}$ kg · m$^2$. Compare with Fig. 4.2-1 where $\phi_v = 0$.

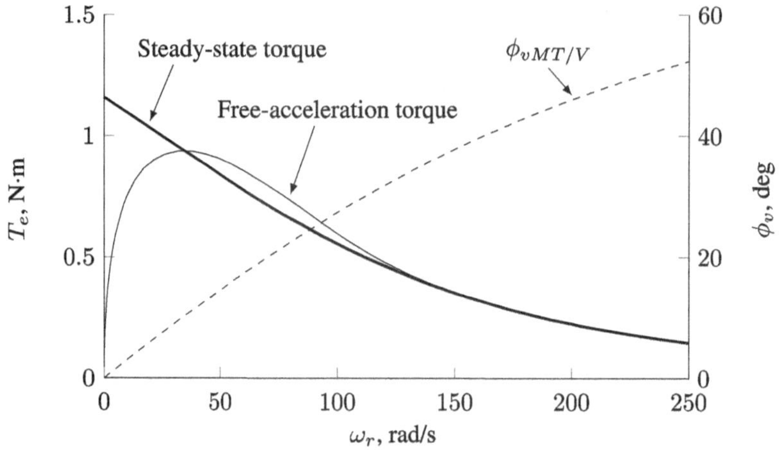


Figure 4.2-5 Torque-speed characteristics for free-acceleration shown in Fig. 4.2-4 with the steady-state torque also shown. Compare with Fig. 4.2-2 where $\phi_v = 0$.

**Example 4B. Brushless dc motor operation with $\phi_v = \phi_{vMT/V}$.** In Examples 2F and 4A, we determined the steady-state torque at 750 r/min ($\omega_r = 50\pi$ rad/s) with $\phi_v = 0$. In this example, we will calculate the torque at the same speed with $\phi_v = \phi_{vMT/V}$, draw the phasor diagram and locate the poles, and calculate the power balance. The parameters for the three-phase machine are repeated here for convenience; $r_s = 3.4\ \Omega$, $L_{ls} = 1.1$ mH, $L_{ms} = 11$ mH, $\lambda'_m = 0.0826$ V · s/rad. From (2F-1), $\omega_r = 50\pi$ rad/s. Now, for a three-phase machine,

$$\begin{aligned} L_{ss} &= L_{ls} + \frac{3}{2}L_{ms} \\ &= \left(1.1 + \frac{3}{2}11\right) \times 10^{-3} = 17.6 \text{ mH} \end{aligned} \tag{4B-1}$$

From (4.2-15),

$$\begin{aligned} \phi_{vMT/V} &= \tan^{-1}\frac{\omega_r L_{ss}}{r_s} \\ &= \tan^{-1}\frac{50\pi(17.6)\times 10^{-3}}{3.4} = 39.1^\circ \end{aligned} \tag{4B-2}$$

Therefore,

$$V_{as} = \sqrt{2}11.25\cos(\omega_r t + 39.1^\circ) \tag{4B-3}$$

From (2.4-61)

$$\tilde{V}_{as} = (r_s + j\omega_r L_{ss})\tilde{I}_{as} + \tilde{E}_a \tag{4B-4}$$

and from Example 4A, $\tilde{E}_a = 9.18\underline{/0^\circ}$. Solving for $\tilde{I}_{as}$,

$$\begin{aligned}\tilde{I}_{as} &= \frac{\tilde{V}_{as} - \tilde{E}_a}{r_s + j\omega_r L_{ss}} \\ &= \frac{11.25\underline{/39.1^\circ} - 9.18\underline{/0^\circ}}{3.4 + j50\pi(17.6)\times 10^{-3}} = 1.62\underline{/54.5^\circ}\ \text{A}\end{aligned} \tag{4B-5}$$

From (4.2-6), the steady-state torque is

$$\begin{aligned}T_e &= \frac{3}{2}\frac{P}{2}\lambda_m'^r I_{qs}^r \\ &= \left(\frac{3}{2}\right)\left(\frac{4}{2}\right)(0.0826)(\sqrt{2}\,1.62\cos 54.5^\circ) = 0.330\ \text{N}\cdot\text{m}\end{aligned} \tag{4B-6}$$

Since

$$\sqrt{2}\tilde{I}_{as} = I_{qs}^r - jI_{ds}^r \tag{4B-7}$$

we can solve for $I_{ds}^r$.

$$I_{ds}^r = -\sqrt{2}\,1.62\sin 54.5^\circ = -1.87\ \text{A} \tag{4B-8}$$

The phasor diagram is shown in Fig. 4B-1.

Calculating a power balance

$$\begin{aligned}P_{\text{in}} &= 3|\tilde{V}_{as}|\,|\tilde{I}_{as}|\cos[\theta_{esv}(0) - \theta_{esi}(0)] \\ &= 3\times 11.25\times 1.62\times\cos(39.1^\circ - 54.5^\circ) = 52.71\ \text{W}\end{aligned} \tag{4B-9}$$

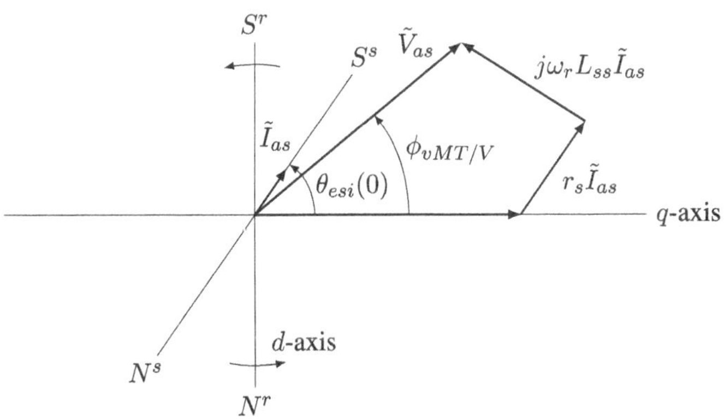


Figure 4B-1 Phasor diagram for brushless dc motor operation at $\omega_r = 50\pi$ rad/s with $\phi_v = \phi_{vMT/V}$.

$$P_{\text{loss}} = 3r_s|\tilde{I}_{as}|^2$$
$$= 3 \times 3.4 \times (1.62)^2 = 26.77 \text{ W} \tag{4B-10}$$

$$P_m = T_e\omega_{rm}$$
$$= 0.33 \times \frac{2}{4} \times 50\pi = 25.92 \text{ W} \tag{4B-11}$$

The efficiency is

$$\text{eff} = \frac{P_m}{P_e} = \frac{25.92}{57.71} = 49.2\% \tag{4B-12}$$

From Fig. 4B-1, we see that the poles created by the stator can be considered as "pushing" the rotor poles in the counterclockwise direction. In Example 2F, $T_e$ was calculated to be 0.129 N · m with $\phi_v = 0$. Here, with $\phi_{vMT/V}$, the torque is calculated to be 0.330 N · m; however, the magnitude of $\tilde{I}_{as}$ increased from 0.475 A to 1.62 A. The torque is increased by a factor of 2.6 while the current is increased by a factor of 3.4. This is the maximum torque per volt that this device can produce at $\omega_r = 50\pi$ rad/s (750 r/min) with $V_s = 11.25$ V, however, when shifting the phase of the terminal voltage to achieve maximum torque, one must not exceed rated conditions for an extended period.

## Maximum Torque Per Ampere Operation of a Brushless DC Motor ($\phi_v = \phi_{vMT/A}$)

Maximum torque per ampere operation occurs when $I^r_{ds}$ is made zero by controlling the position of $\tilde{V}_{as}$ relative to the permanent magnet of the rotor ($\phi_v = \phi_{vMT/A}$). The torque is directly related to the $q$-axis current (4.2-6). The $d$-axis current does not contribute to the torque and, in fact, it decreases the efficiency of the machine.

To derive an expression for $\phi_{vMT/A}$ for steady-state operation, we will substitute (4.2-9) and (4.2-10) into (4.2-7) and (4.2-8) for $V^r_{qs}$ and $V^r_{ds}$, respectively, and solve for $\cos\phi_v$ and $\sin\phi_v$. If we set $I^r_{ds} = 0$ and perform several mathematical manipulations we can express $\phi_{vMT/A}$, at a given rotor speed, as

$$\phi_{vMT/A} = \tan^{-1}\left[\omega_r\tau_s\left(\frac{-1 \pm \omega_r\tau_v\sqrt{1+\omega_r^2\tau_v^2(1-\omega_r^2\tau_v^2)}}{\omega_r^4\tau_s^2\tau_v^2 - 1}\right)\right] \tag{4.2-16}$$

where, for compactness

$$\tau_s = \frac{L_{ss}}{r_s} \tag{4.2-17}$$

$$\tau_v = \frac{\lambda'_m}{\sqrt{2}V_s} \tag{4.2-18}$$

The free-acceleration characteristics are shown in Fig. 4.2-6. The torque-speed characteristics are shown in Fig. 4.2-7 including a plot of $\phi_{vMT/A}$.

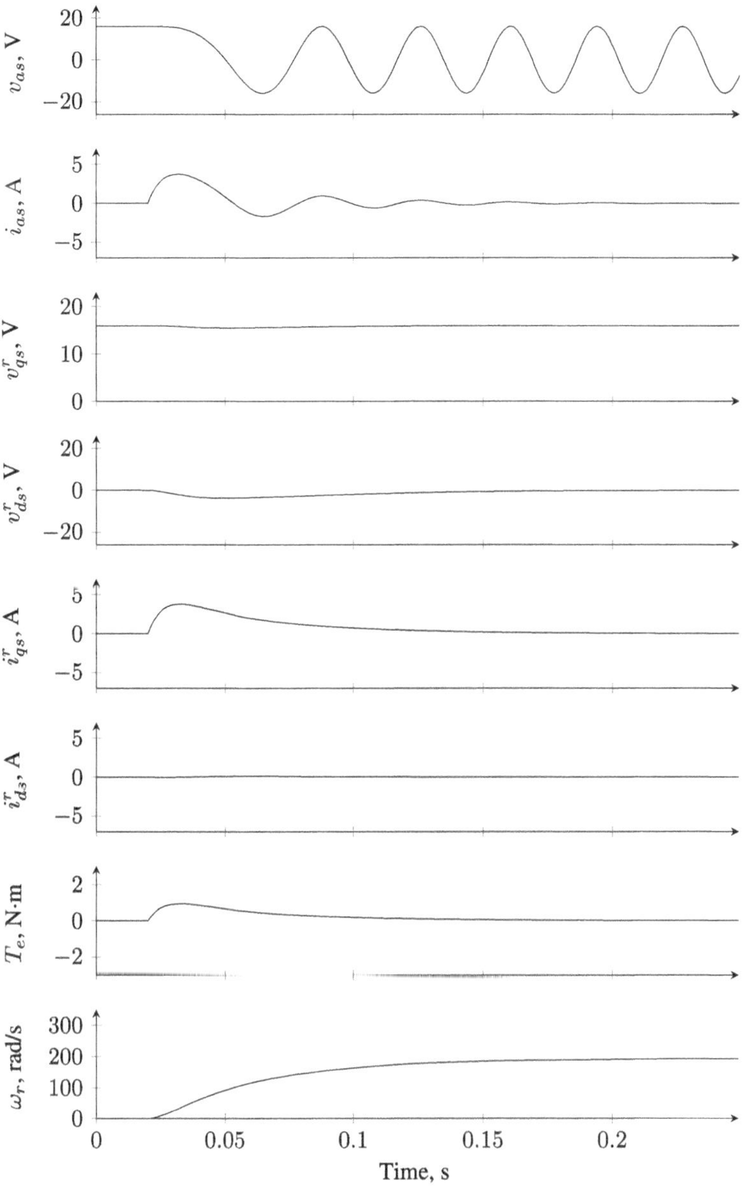


Figure 4.2-6 Free-acceleration characteristics of a brushless dc machine with $\phi_v = \phi_{vMT/A}$ ($I^r_{ds} = 0$) and a total inertia of $5 \times 10^{-4}$ kg · m$^2$. Compare with Figs. 4.2-1 and 4.2-4.

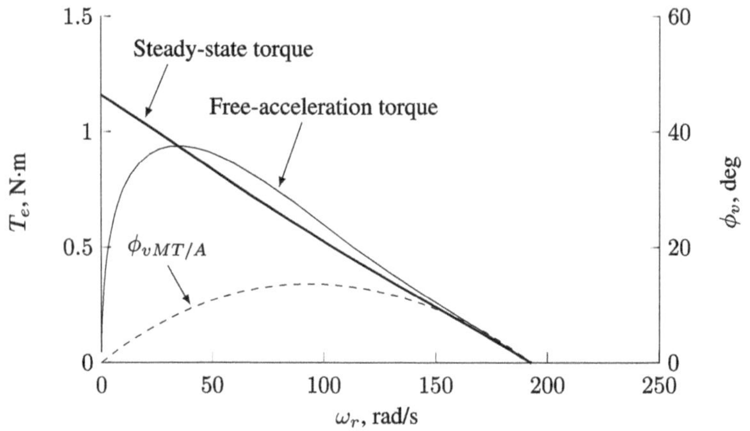


Figure 4.2-7 Torque-speed characteristics for free-acceleration shown in Fig. 4.2-6.

**Example 4C. Brushless dc motor operation with $\phi_v = \phi_{vMT/A}$.** In this example, we will calculate the torque with $\phi_v = \phi_{vMT/A}$ and draw the phasor diagram and locate the stator and rotor poles. The parameters are repeated again; $r_s = 3.4\ \Omega$, $L_{ls} = 1.1$ mH, $L_{ms} = 11$ mH, $\lambda'_m = 0.0826$ V · s/rad. In order to compare with Examples 4A and 4B, we will perform the calculations for $\omega_r = 50\pi$ rad/s. From (4.2-16) $\phi_{vMT/A}$ is calculated to be 8.15° and 17° at $\omega_r = 50\pi$. The larger angle (17°) from (4.2-16) is an extraneous root. Also, $V_s = 11.25$ V, $L_{ss} = 17.6$ mH, and

$$V_{as} = \sqrt{2}11.25\cos(\omega_r t + 8.15°) \tag{4C-1}$$

From (2.4-61)

$$\tilde{V}_{as} = (r_s + j\omega_r L_{ss})\tilde{I}_{as} + \tilde{E}_a \tag{4C-2}$$

and again $\tilde{E}_a = 9.18\underline{/0°}$. Solving for $\tilde{I}_{as}$,

$$\begin{aligned}\tilde{I}_{as} &= \frac{\tilde{V}_{as} - \tilde{E}_a}{r_s + j\omega_r L_{ss}} \\ &= \frac{11.25\underline{/8.15°} - 9.18\underline{/0°}}{3.4 + j50\pi(17.6)\times 10^{-3}} = 0.577\underline{/0°}\ \text{A}\end{aligned} \tag{4C-3}$$

From (4.2-6)

$$\begin{aligned}T_e &= \frac{3}{2}\frac{P}{2}\lambda'^r_m I^r_{qs} \\ &= \left(\frac{3}{2}\right)\left(\frac{4}{2}\right)\left(0.0826)\left(\sqrt{2}\right)(0.577\right) = 0.202\ \text{N}\cdot\text{m}\end{aligned} \tag{4C-4}$$

The phasor diagram is shown in Fig 4C-1. Note that the stator and rotor poles are orthogonal which yields the maximum torque per ampere for this device at $\omega_r = 50\pi$ rad/s.

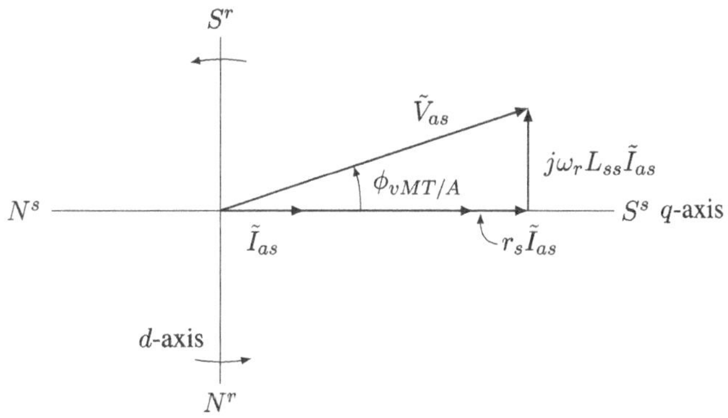


Figure 4C-1 Phasor diagram for brushless dc motor operation at $\omega_r = 50\pi$ rad/s with $\phi_v = \phi_{vMT/A}$.

The power balance becomes

$$P_e = 3|\tilde{V}_{as}|\,|\tilde{I}_{as}|\cos[\theta_{esv}(0) - \theta_{esi}(0)]$$
$$= 3 \times 11.25 \times 0.577 \cos(8.15^\circ - 0) = 19.3 \text{ W} \quad (4C\text{-}5)$$

$$P_{loss} = 3r_s|\tilde{I}_{as}|^2$$
$$= 3 \times 3.4(0.577)^2 = 3.4 \text{ W} \quad (4C\text{-}6)$$

$$P_m = T_e\omega_{rm}$$
$$= 0.202 \times \frac{2}{4} \times 50\pi = 15.9 \text{ W} \quad (4C\text{-}7)$$

The efficiency is

$$\text{eff} = \frac{P_m}{P_e} = \frac{15.9}{19.3} = 82.4\% \quad (4C\text{-}8)$$

---

It is instructive to summarize the results of Examples 4A through 4C. The three modes of operation $\phi_v = 0$, $\phi_v = \phi_{vMT/V}$, and $\phi_{vMT/A}$ were all performed at $\omega_r = 50\pi$ rad/s with the same parameters and $|\tilde{V}_{as}|$. The results of Examples 4A through 4C are summarized in Table 4.2-1. Recall that $\sqrt{2}\tilde{I}_{as} = I^r_{qs} - jI^r_{ds}$.

**TABLE 4.2–1 Summary of results of Examples 4A through 4C**

| $\phi_v$, degrees | $T_e$, N · m | $\tilde{I}_{as}$, A | $I^r_{qs}$, A | $I^r_{ds}$, A | eff, % |
|---|---|---|---|---|---|
| 0 | 0.129 | $0.475\angle{-39.1^\circ}$ | 0.521 | 0.424 | 81.4 |
| $\phi_{vMT/V}$ 39.1 | 0.330 | $1.62\angle{54.5^\circ}$ | 1.33 | −1.87 | 49.2 |
| $\phi_{vMT/A}$ 8.15 | 0.202 | $0.577\angle{0^\circ}$ | 0.816 | 0 | 82.4 |

**SP4.2-1.** In Fig 4.2-5 it is noted that $T_e$ at $\phi_v = 0$ and $\phi_{vMT/V}$ are equal at $\omega_r = 0$. Why? [$\phi_{vMT/V} = 0$]

**SP4.2-2.** Calculate the torque per volt, $T_{e/V}$, and torque per ampere, $T_{e/A}$, for the conditions given in Example 4A. Use rms values. [$T_{e/V} = 0.0147$ N · m/V; $T_{e/A} = 0.272$ N · m/A]

**SP4.2-3.** Repeat SP4.2-2 for Example 4B. [$T_{e/V} = 0.0293$ N · m/V; $T_{e/A} = 0.204$ N · m/A]

**SP4.2-4.** Repeat SP4.2-2 for Example 4C. [$T_{e/V} = 0.0179$ N · m/V; $T_{e/A} = 0.350$ N · m/A]

## 4.3 OPERATION OF A BRUSHLESS DC DRIVE

The brushless dc drive that we will consider consists of the three-phase, six-step inverter that was introduced in Chapter 3 and the permanent-magnet ac machine. The six-step inverter and the switching logic that were shown in Fig. 3.5-5 are repeated here in Fig. 4.3-1. The three-phase permanent-magnet ac machine is shown in Fig. 4.3-2. The three sensors shown in Fig. 4.3-2 are generally Hall-effect devices;

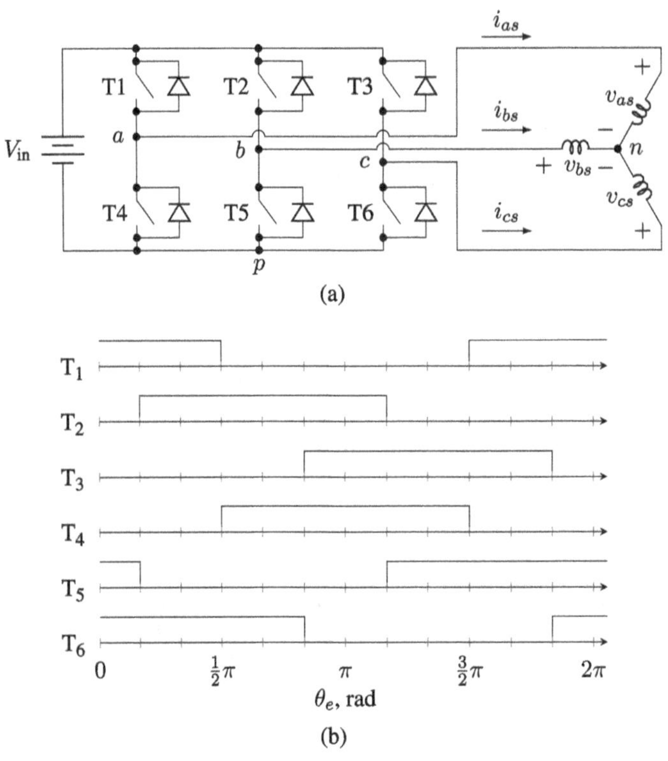


Figure 4.3-1 Inverter-machine drive. (*a*) Inverter configuration; (*b*) transistor switching logic.

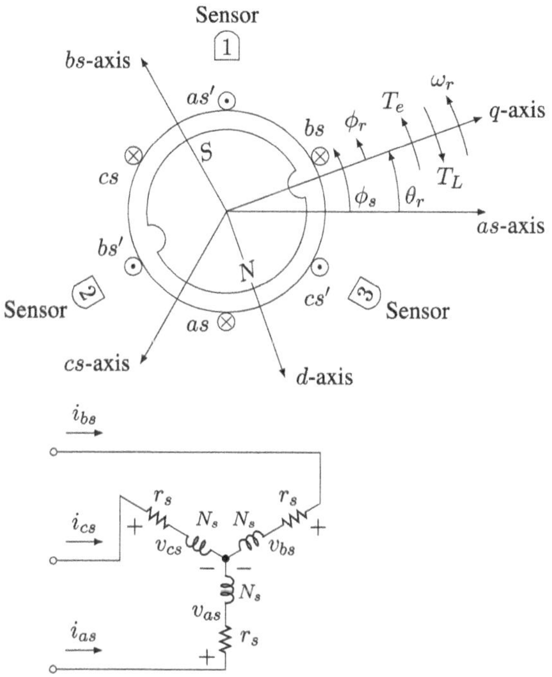


Figure 4.3-2 Two-pole three-phase permanent-magnet ac machine with sensors.

when the south pole of the permanent-magnet rotor is under a sensor, its output is nonzero; with the rotor north pole under the sensor, its output is zero. The states of the sensors are used to determine the switching logic for the inverter, which, in turn, determines the output frequency of the inverter. In the actual machine, the sensors are not positioned over the rotor as shown in Fig. 4.3-2. Instead, they are placed over a ring that is mounted on the shaft external to the stator and magnetized the same as the rotor.

Equations (3.5-8) through (3.5-10) are repeated here as

$$v_{as} = \frac{2}{3}v_{ap} - \frac{1}{3}(v_{bp} + v_{cp}) \tag{4.3-1}$$

$$v_{bs} = \frac{2}{3}v_{bp} - \frac{1}{3}(v_{cp} + v_{ap}) \tag{4.3-2}$$

$$v_{cs} = \frac{2}{3}v_{cp} - \frac{1}{3}(v_{ap} + v_{bp}) \tag{4.3-3}$$

In (4.3-1) through (4.3-3), $v_{ap}$, $v_{bp}$, and $v_{cp}$ are either $V_{in}$ or zero depending upon the state of transistors $T_1$ through $T_6$ (see Fig. 4.3-1a).

Recall, from Fig. 3.5-5c, that the maximum value of the switched phase voltages is $\frac{3}{2}V_{in}$; however, from (3.5-11), the amplitude of the fundamental component of the phase voltage is $\frac{2}{\pi}V_{in}$. Now, from (2.4-46)

$$\phi_v = \theta_{esv} - \theta_r \tag{4.3-4}$$

where, in the steady state,

$$\theta_{esv} = \omega_e t + \theta_{esv}(0) \tag{4.3-5}$$

$$\theta_r = \omega_r t + \theta_r(0) \tag{4.3-6}$$

The voltages $v_{qs}^r$ and $v_{ds}^r$ may be changed in magnitude by changing $V_{\text{in}}$ or by changing the phase of the applied stator voltages by changing $\phi_v$. We will select the time-zero position of the $q$-axis to be at $\theta_r = 0$, thus, $\theta_r(0) = 0$. The six-step $as$-phase voltage may be written

$$v_{as} = \frac{2}{\pi} V_{\text{in}} \left[\cos \theta_{esv} + \frac{1}{5} \cos 5\theta_{esv} - \frac{1}{7} \cos 7\theta_{esv} + \ldots \right] \tag{4.3-7}$$

where $\omega_e$ is controlled to be $\omega_r$. Now, from (4.3-4), we could replace $\theta_{esv}$ with $\phi_v + \theta_r$ and since $\theta_r(0) = 0$, $\theta_r = \omega_r t$ and $\phi_v$ is zero for conventional brushless dc motor operation. That is, with $\theta_{esv}(0) = 0$, $\phi_v = 0$, (4.3-7) becomes (3.5-11) and $v_{qs}^r$ and $v_{ds}^r$ become (3.5-12) and (3.5-13), respectively, which are repeated below for convenience where we have set $\omega_e = \omega_r$. Since it is rather involved to go through the trigonometry to express $v_{qs}^r$ and $v_{ds}^r$ with $\theta_{esv}(0)$ nonzero, we will not do it; however, it is carried out in Reference [2].

$$v_{qs}^r = \frac{2V_{\text{in}}}{\pi} \left(1 + \frac{2}{35} \cos 6\omega_r t - \frac{2}{143} \cos 12\omega_r t + \cdots \right) \tag{4.3-8}$$

$$v_{ds}^r = \frac{2V_{\text{in}}}{\pi} \left(\frac{12}{35} \sin 6\omega_r t - \frac{24}{143} \sin 12\omega_r t + \cdots \right) \tag{4.3-9}$$

## Operation of Brushless DC Drive with $\phi_v = 0$

The free-acceleration characteristic of this brushless dc motor-inverter drive with $\phi_v = 0$ is shown in Fig. 4.3-3. The machine parameters are those given previously. For purposes of comparison with Fig. 4.2-1, the inverter voltage $V_{\text{in}}$ is 25 V, where the constant component of $v_{qs}^r$ from (4.3-8) with $\phi_v = 0$ is equal to the value used in the case of sinusoidal applied stator voltages. The torque versus speed characteristics for this free acceleration are shown in Fig. 4.3-4. The dynamic performance of this system during load-torque switching is shown in Fig. 4.3-5. As in Fig 4.2-4, the inertia is $2 \times 10^{-4}$ kg · m$^2$, and the load torque is switched from 0.1 to 0.4 N · m and then back to 0.1 N · m.

Let us take a little closer look at the voltage and current waveforms. For this purpose, it is sufficient to consider only $v_{as}$ and $i_{as}$. The voltage $v_{as}$ is given by (4.3-1), wherein $v_{ap}$ is equal to $V_{\text{in}}$ when $T_1$ is on and $T_4$ is off and is zero when $T_4$ is on and $T_1$ is off. A similar situation exists for $v_{bp}$ with regard to the states of $T_2$ and $T_5$ and for $v_{cp}$ with regard to $T_3$ and $T_6$.

The harmonic content of the machine variables due to switching of the inverter is apparent. Actually, from our work in Chapter 3, we knew that harmonics would be present; however, it is generally not a serious drawback for most applications except perhaps for low inertia loads. The sinusoidal approximation that we have used previously should be sufficient to analyze most modes of operation. If, however, it is desirable to calculate the steady-state harmonics for all values of $\phi_v$, the method set forth in Reference [3] may be used.

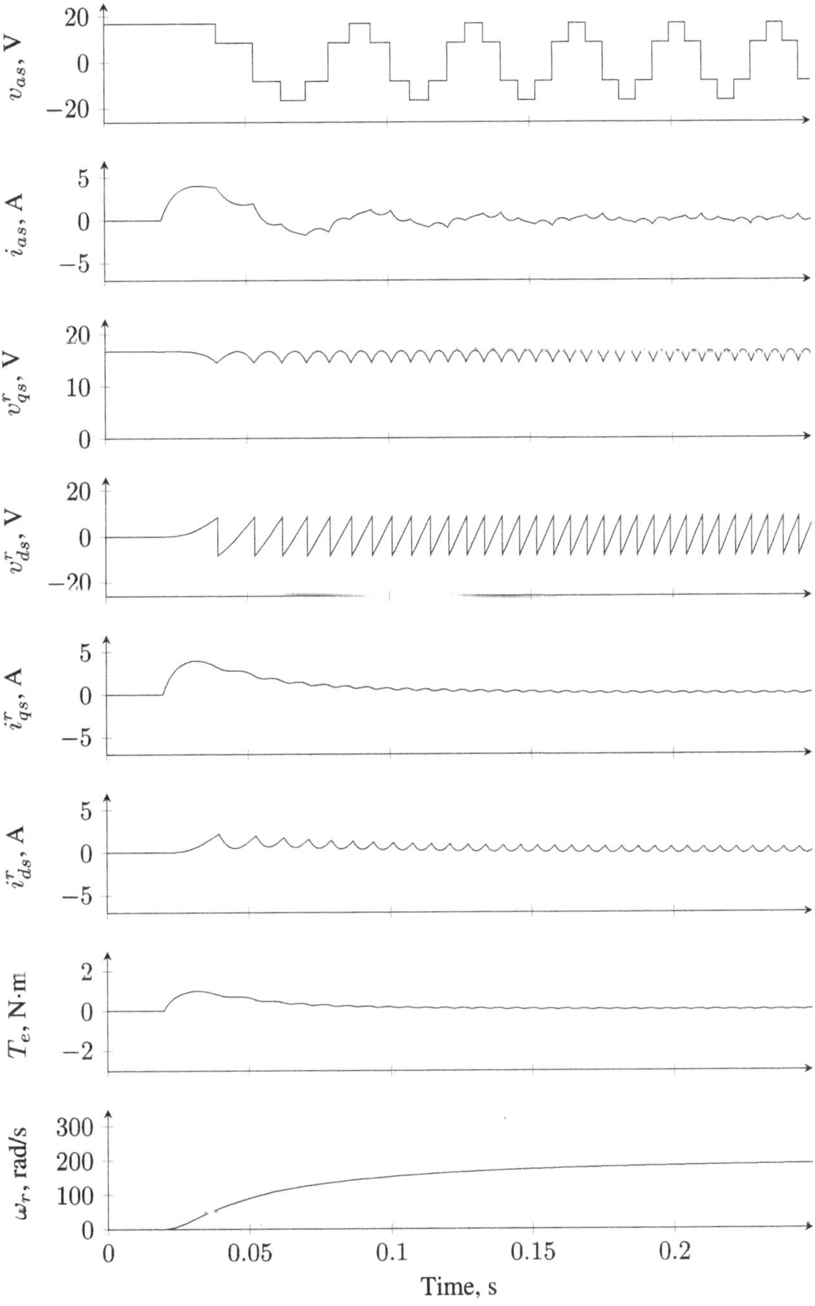


Figure 4.3-3 Free-acceleration characteristics of a three-phase brushless dc motor supplied by a six-step inverter with $\phi_v = 0$; compare with Fig 4.2-1.

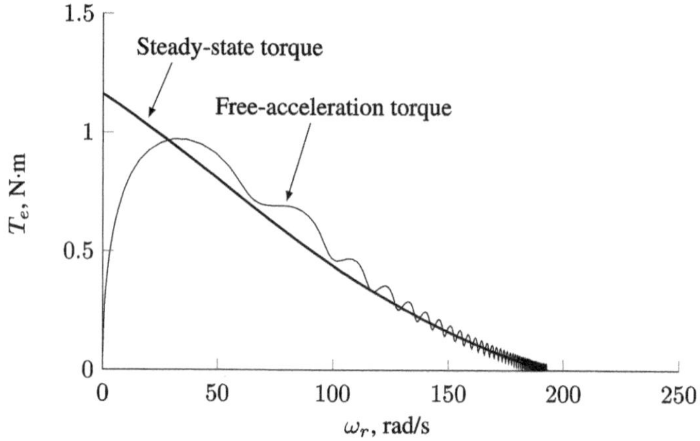


Figure 4.3-4 Torque-speed characteristics for free acceleration shown in Fig 4.3-3; compare with Fig. 4.2-2.

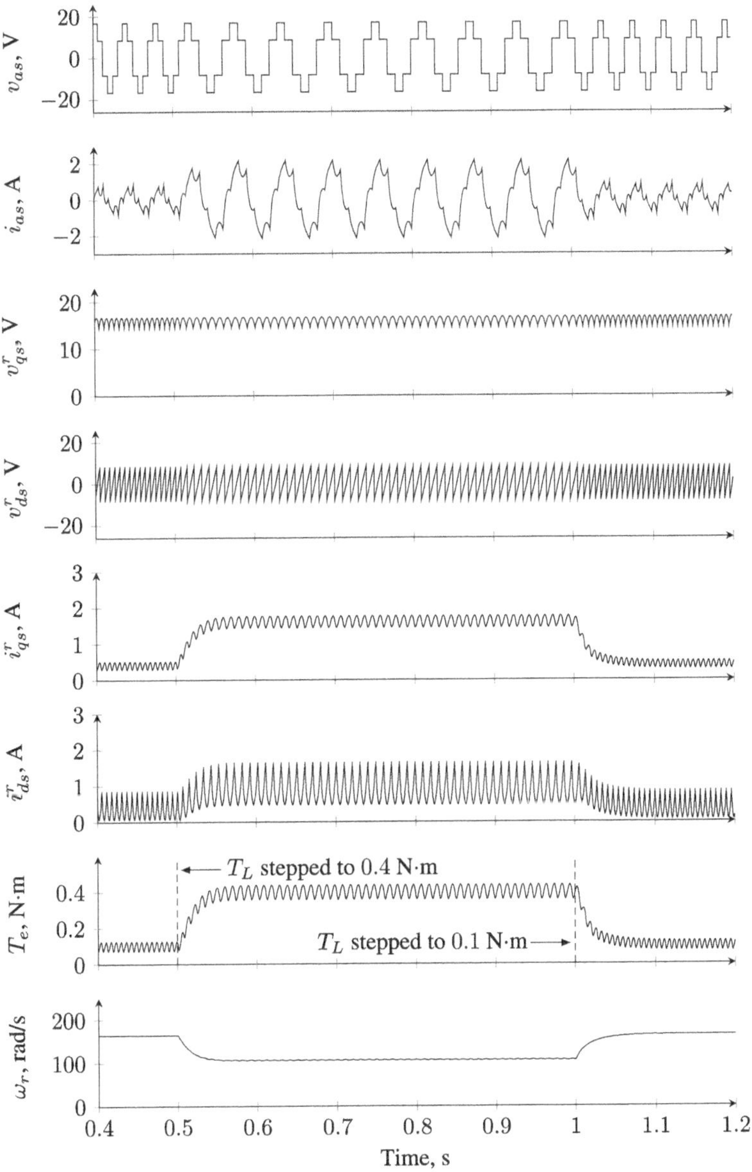


Figure 4.3-5 Dynamic performance during step changes in load torque of a three-phase, brushless dc motor supplied by an inverter with $\phi_v = 0$, compare with Fig. 4.2 3.

**Example 4D. Consideration of the sixth harmonic torque.** Set forth the procedure for calculating the 6th harmonic torque in the brushless dc drive with $\phi_v = 0$. From (4.3-8) and (4.3-9) the 6th harmonic voltages are

$$V_{qs}^r = \left(\frac{2V_{\text{in}}}{\pi}\right)\left(\frac{2}{35}\right)\cos 6\omega_r t \tag{4D-1}$$

$$V_{ds}^r = \left(\frac{2V_{\text{in}}}{\pi}\right)\left(\frac{12}{35}\right)\sin 6\omega_r t \tag{4D-2}$$

From (2.5-31) and (2.5-32), the steady-state voltage equations become

$$V_{qs}^r = (r_s + pL_{ss})I_{qs}^r + \omega_r L_{ss} I_{ds}^r + \omega_r \lambda_m'^r \tag{4D-3}$$

$$V_{ds}^r = (r_s + pL_{ss})I_{ds}^r - \omega_r L_{ss} I_{qs}^r \tag{4D-4}$$

At stall, $\omega_r = 0$, $I_{qs}^r$ is constant, and $I_{ds}^r$ is zero. Clearly, a 6th harmonic torque is not present at stall.

Now, we need to calculate $I_{qs}^r$ from (4D-3) and (4D-4) in order to determine the 6th harmonic torque for $\omega_r > 0$. This can be accomplished by solving (4D-3) and (4D-4) for $I_{qs}^r$ or phasors could be used. That is,

$$\tilde{V}_{qs}^r = (r_s + j6\omega_r L_{ss})\tilde{I}_{qs}^r + \omega_r L_{ss}\tilde{I}_{ds}^r + \omega_r \lambda_m'^r \tag{4D-5}$$

$$\tilde{V}_{ds}^r = (r_s + j6\omega_r L_{ss})\tilde{I}_{ds}^r - \omega_r L_{ss}\tilde{I}_{qs}^r \tag{4D-6}$$

For any given rotor speed, we can solve (4D-6) for $\tilde{I}_{ds}^r$ and substitute it into (4D-5). We can then solve for $\tilde{I}_{qs}^r$ from which we can get the instantaneous expression for $I_{qs}^r$. Substituting this into the expression for torque given by (4.2-6) yields the expression for the steady-state instantaneous 6th harmonic torque.

## Operation of Brushless DC Drive with $\phi_v = \phi_{vMT/V}$

Figures 4.3-6 and 4.3-7 show the free-acceleration characteristics with $\phi_v = \phi_{vMT/V}$. These figures may be compared to Figs. 4.2-4 and 4.2-5, respectively where the applied voltages were sinusoidal and also to Figs. 4.3-3 and 4.3-4 where $\phi_v = 0$. When we compare Figs. 4.3-4 and 4.3-7, we see that the speed range is extended when $\phi_v = \phi_{vMT/V}$ compared to $\phi_v = 0$; however, the peak-to-peak 6th harmonic torque increases.

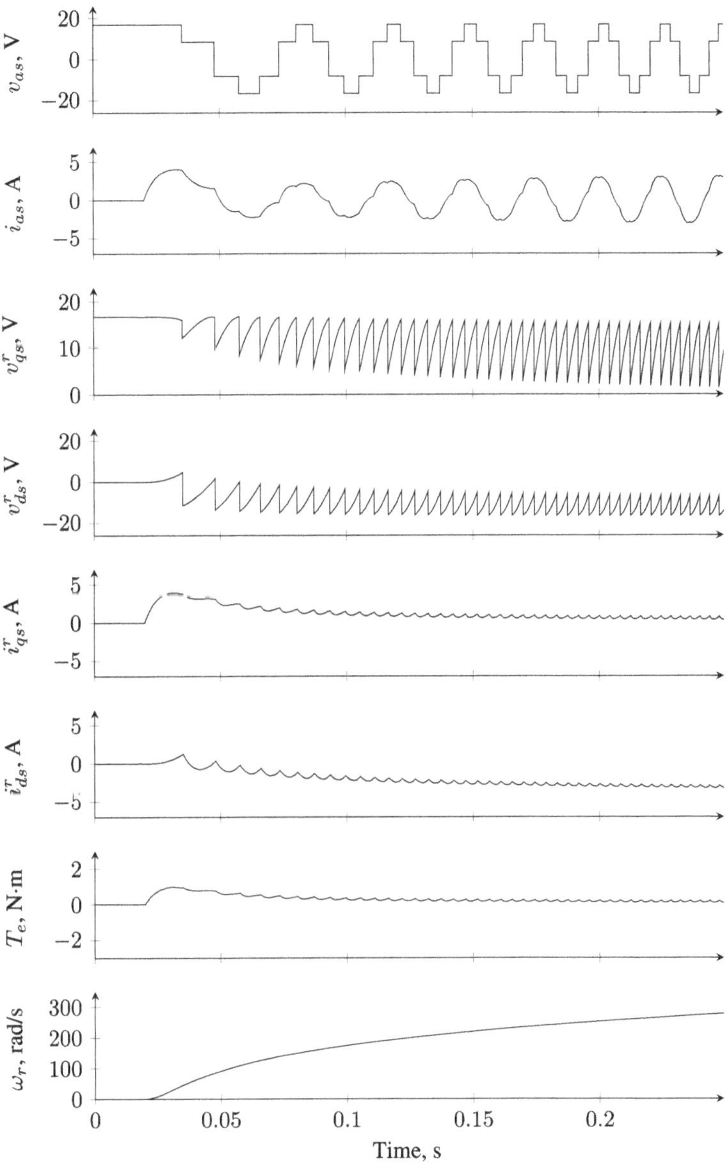


Figure 4.3-6 Free-acceleration characteristics of a brushless dc machine with $\phi_v = \phi_{vMT/V}$.

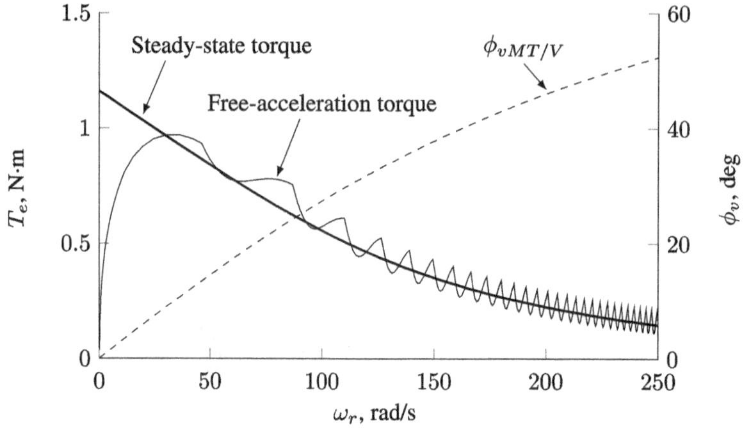


Figure 4.3-7 Torque-speed characteristics for acceleration shown in Fig. 4.3-6 with $\phi_v = \phi_{vMT/V}$.

## Operation of Brushless DC Drive with $\phi_v = \phi_{vMT/A}$

Figures 4.3-8 and 4.3-9 show the free-acceleration characteristics with $\phi_v = \phi_{vMT/A}$. These figures may be compared to the free-acceleration characteristics for $\phi_v = 0$ and $\phi_v = \phi_{vMT/V}$.

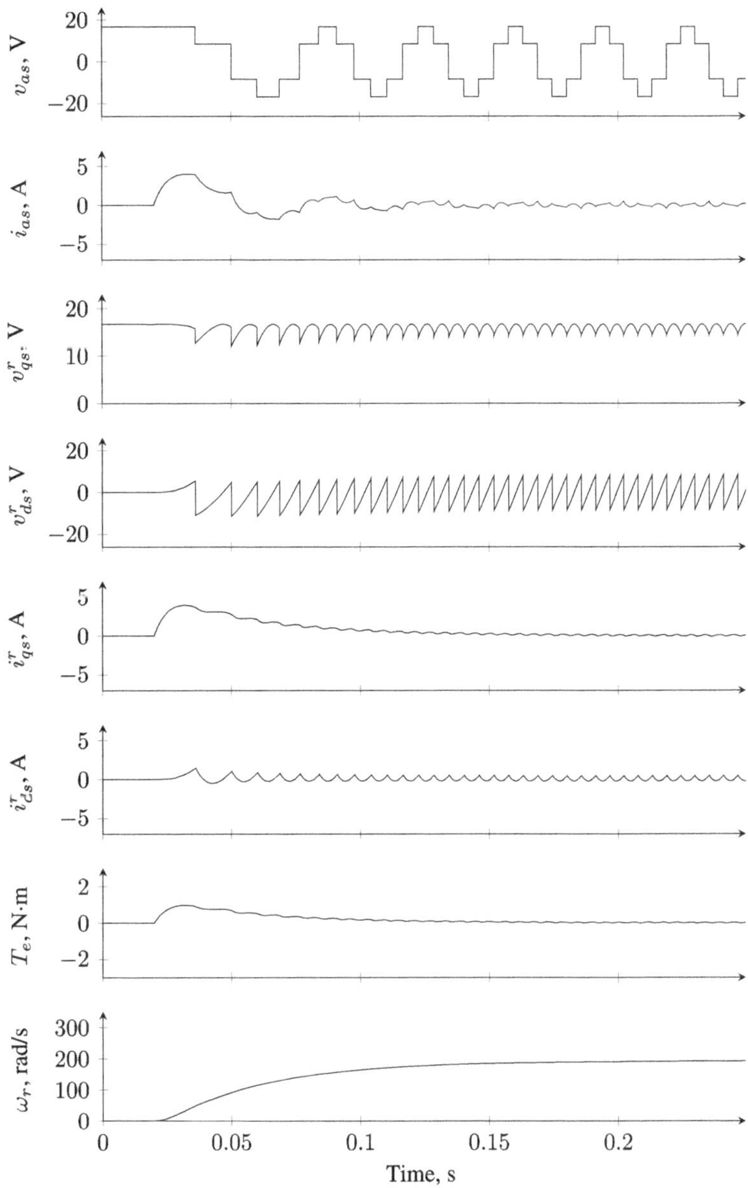


Figure 4.3-8 Free-acceleration characteristics of a brushless dc machine with $\phi_{vMT/A}$.

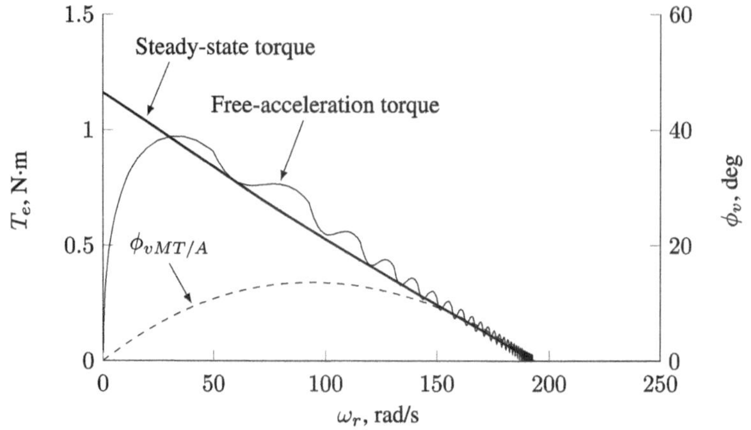


Figure 4.3-9 Torque-speed characteristics shown in Fig. 4.3-8 with $\phi_{vMT/A}$.

## Modes of Control of a Brushless DC Drive

Although we are not going to become involved in the actual control of the brushless dc drive; we can give some insight to the common controlled modes through an idealized approach. We have considered three modes of operation: $\phi_v = 0$, $\phi_v = \phi_{vMT/V}$, and $\phi_v = \phi_{vMT/A}$ with rated applied voltage. Zero phase shift of $\tilde{V}_{as}$ ($\phi_v = 0$) is generally considered as the common mode of operation; however, the torque can be increased at a given speed and voltage by advancing the phase of $\tilde{V}_{as}$ and maximum torque per volt is achieved when $\phi_v = \phi_{vMT/V}$ (4.2-15). Advancing the phase to where $I^r_{ds} = 0$ yields maximum torque per ampere which is very desirable since it is the most efficient mode of operation as illustrated in Examples 4A through 4C.

In addition to advancing the phase of $\tilde{V}_{as}$, varying the magnitude of $\tilde{V}_{as}$ provides a means of speed control. Clearly, the brushless dc drive can provide a wide array of controlled operating characteristics by appropriate control of the phase shift and the magnitude of $\tilde{V}_{as}$. Depending on the sophistication of the drive system, the control can become quite involved. We will leave the actual control of $\phi_v$ and $|\tilde{V}_{as}|$ to an advanced course; instead, we will consider the resulting drive capabilities when ideally controlled.

The steady-state torque versus rotor speed characteristics for the example permanent-magnet machine with $\phi_v = 0$ and $|\tilde{V}_{as}|$ equal to 11.25 V, 7.5 V, 4 V, and 2 V are shown in Fig. 4.3-10. The same characteristics for $\phi_v = \phi_{vMT/V}$ are shown in Fig. 4.3-11. Therein $\phi_{vMT/V}$ is calculated by (4.2-15) and is the same for all voltages depending on $r_s$, $L_{ss}$, and $\omega_r$. The torque-speed characteristics for $\phi_{vMT/A}$ are shown in Fig. 4.3-12 and since (4.2-16) and (4.2-18) tell us that $\phi_{vMT/A}$ is a function of the magnitude of $\tilde{V}_{as}$, Fig. 4.3-13 shows $\phi_{vMT/A}$ versus $\omega_r$ for each $|\tilde{V}_{as}|$.

We see that the phase shifting of the applied voltage, $\tilde{V}_{as}$, relative to the permanent-magnet rotor and/or the change in the magnitude of the applied voltages provide a wide range of control options.

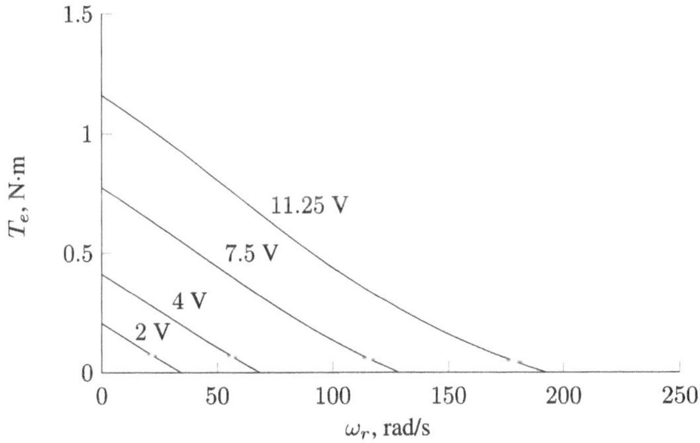


Figure 4.3-10 Torque-speed characteristics for $\phi_v = 0$ for different values of $|\tilde{V}_{as}|$.

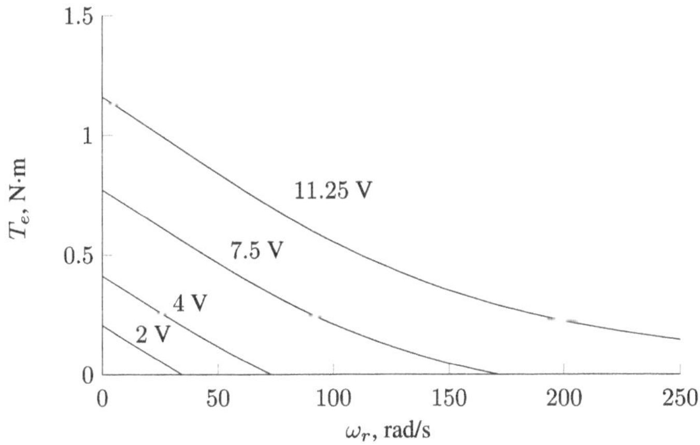


Figure 4.3-11 Torque-speed characteristics for $\phi_v = \phi_{vMT/V}$ for different values of $|\tilde{V}_{as}|$.

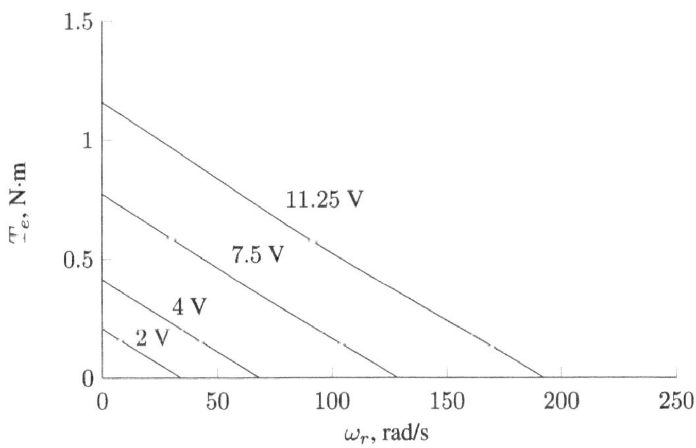


Figure 4.3-12 Torque-speed characteristics for $\phi_v = \phi_{vMT/A}$ for different values of $|\tilde{V}_{as}|$.

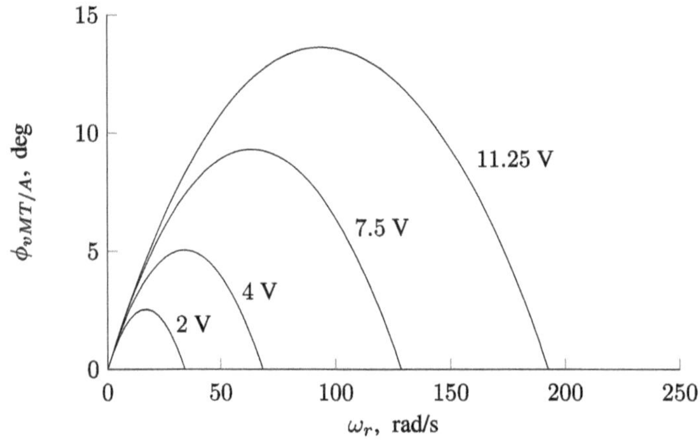


Figure 4.3-13 $\phi_{vMT/A}$ ($I_{ds}^r = 0$) versus $\omega_r$ for different value of $|\tilde{V}_{as}|$.

We are aware that phase shifting of the applied voltages relative to the position of the permanent-magnet rotor is possible by controlling the firing of the inverter based on the information retrieved from the Hall effect devices or some other means of determining the position of the rotor. We are, however, not aware how the magnitude of $\tilde{V}_{as}$, which is $V_s$, can be changed. We will assume that the maximum voltage of the example machine is 11.25 V. We have not discussed how this voltage can be decreased. We know that the output voltage of the inverter can be changed by changing the input voltage $V_{in}$. In most cases, however, the inverter input voltage is supplied from a constant source such as a battery. In this case, voltage control of the inverter output voltage is achieved by pulse width modulation (PWM). This was mentioned in passing in Section 3.5. We will add to that discussion. For this purpose let us consider Fig. 4.3-14 wherein one of the output phase voltages of a six-step inverter with a PWM capability is shown. The "notching" of the six-step phase voltages is accomplished by instantaneously connecting all three phases to the same side of the dc source by appropriate switching of the transistors. Recall from Chapter 3 that during normal operation of the inverter, two phases are connected to one terminal of the dc source while the third is connected to the other terminal. During the time when all three phases of the machine are connected to the same side of the inverter, the stator windings are short-circuited; thus, the machine is effectively disconnected from the inverter.

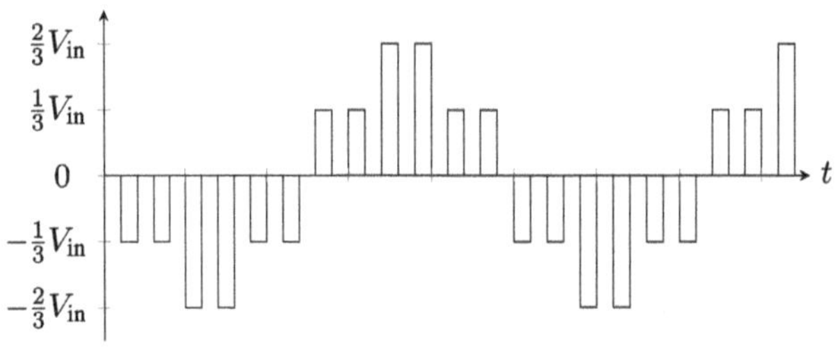


Figure 4.3-14 Pulse width modulated six-step inverter phase voltage.

There are different PWM procedures, the type shown in Fig. 4.3-14 is one where the on and off times of the transistors are equal. In this case, a change in the effective value of the stator phase voltages is achieved by controlling the time that the phase voltages are held at zero during one cycle of the voltages. Regardless of the variation of PWM used, the purpose is to change the magnitude of the effective sinusoidal approximation of the six-step phase voltages.

We now have some understanding of how the shifting of the phase and the change of the magnitude of the applied voltages can be achieved. Please be aware that (4.2-15) and (4.2-16) for $\phi_{vMT/V}$ and $\phi_{vMT/A}$, respectively, have been derived assuming an "idealized" machine, in that the assumptions include sinusoidal distribution of the windings, sinusoidal voltages, constant parameters at all operating temperatures, and that $L_{ms}$ is the same in the $q$- and $d$-axes. Unfortunately, these assumptions are idealizations of the real world. Nevertheless, making these approximations for analysis purposes allows the salient operating characteristics to be readily portrayed. When controlling a "real world" drive, the ideal machine offers only guidelines. Nevertheless, an understanding of the operating characteristics with these simplifying assumptions is very helpful; however, in reality, it is necessary to implement a closed-loop control. For example, an online determination of $I_{ds}^r$ implemented with a means of correcting any variation in $I_{ds}^r$ from zero is necessary for $\phi_{vMT/A}$ operation. Regardless of all this, it is our purpose to establish an insight to the controlled potential of the brushless dc drive as simple and straightforward as possible and to lay the ground work for an advanced treatment of electric drives.

We can see that numerous controlled operating characteristics are possible. For example, a load may require a constant torque at a given speed. This requirement can be satisfied by voltage magnitude and phase control with $\phi_v$ equal to zero, $\phi_{vMT/V}$, or $\phi_{vMT/A}$. Hopefully the following example will demonstrate this capability of a permanent-magnet brushless dc motor drive. It is understood, however, that the drive should not be designed to operate above rated conditions for any extended period of time.

**Example 4E. Constant torque and constant speed operation of a brushless dc drive.** We are to calculate the magnitude and phase shift of $\tilde{V}_{as}$ for a constant torque of $T_e = 0.4\ \text{N}\cdot\text{m}$ at a speed of $\omega_r = 50$ rad/s. That is, the load requires 0.4 N · m and is to operate at 50 rad/s. For (a) $\phi_v = 0$, (b) $\phi_v = \phi_{vMT/V}$, and (c) $\phi_v = \phi_{vMT/A}$. The parameters of the example four-pole, three-phase machine are $r_s = 3.4\ \Omega$, $L_{ls} = 1.1$ mH, $L_{ms} = 11$ mH, and $\lambda_m' = 0.0826\ \text{V}\cdot\text{s/rad}$.

For part (a), there is no phase shift and thus $V_{ds}^r = 0$. But first, from (4.2-6)

$$T_e = \frac{3}{2}\frac{P}{2}\lambda_m' I_{qs}^r \tag{4E-1}$$

thus

$$I_{qs}^r = \frac{0.4}{\left(\frac{3}{2}\right)\left(\frac{4}{2}\right)(0.0826)} = 1.61\ \text{A} \tag{4E-2}$$

The current will be the same for parts (a) through (c). From (4.2-3)

$$\begin{aligned} V_{qs}^r &= \frac{r_s^2 + \omega_r^2 L_{ss}^2}{r_s} I_{qs}^r + \omega_r \lambda_m'^r \\ &= \left[\frac{(3.4)^2 + (50)^2(0.0176)^2(1.61)}{3.4}\right] 1.61 + (50)(0.0826) \\ &= 9.97 \text{ V} \end{aligned} \tag{4E-3}$$

Since $V_{ds}^r = 0$,

$$\tilde{V}_{as} = \frac{9.97}{\sqrt{2}} \underline{/0^\circ} = 7.05 \underline{/0^\circ} \text{ V} \tag{4E-4}$$

and

$$V_{as} = 9.97 \cos 50t \text{ V} \tag{4E-5}$$

We can use the information given in Fig. 4.3-10 as a check of (4E-5).

For part (b), $V_{ds}^r$ is not zero; however, from (4.2-15)

$$\begin{aligned} \phi_{vMT/V} &= \tan^{-1} \frac{\omega_r L_{ss}}{r_s} \\ &= \tan^{-1} \frac{(50)(0.0176)}{3.4} = 14.5^\circ \end{aligned} \tag{4E-6}$$

From (4.2-11),

$$V_{qs}^r - \frac{\omega_r L_{ss}}{r_s} V_{ds}^r = \frac{r_s^2 + \omega_r^2 L_{ss}^2}{r_s} I_{qs}^r + \omega_r \lambda_m'^r \tag{4E-7}$$

It will save us some work to note that the right-hand side of (4E-3) and (4E-7) are the same. Now, from (4.2-9) and (4.2-10), respectively,

$$V_{qs}^r = \sqrt{2} |\tilde{V}_{as}| \cos \phi_{vMT/V} \tag{4E-8}$$

$$V_{ds}^r = -\sqrt{2} |\tilde{V}_{as}| \sin \phi_{vMT/V} \tag{4E-9}$$

Wait a minute, should not (4E-8) and (4E-9) be in terms of $V_s$ rather than $|\tilde{V}_{as}|$? Well yes; however, we now have a PWM control (notched six step) which changes the effective value of the equivalent sinusoidal approximation of the applied voltage. Using $|\tilde{V}_{as}|$ rather than $V_s$ helps to call attention to this.

Substituting (4E-6) into (4E-8) and (4E-9) and since the left-hand side of (4E-7) is equal to 9.97 V as calculated in (4E-3), we have

$$\sqrt{2} |\tilde{V}_{as}| \left( \cos 14.5^\circ + \frac{\omega_r L_{ss}}{r_s} \sin 14.5^\circ \right) = 9.97 \text{ V} \tag{4E-10}$$

There are a couple of things we should take a minute to point out. Remember back following (4.2-12), we mentioned that $V_{ds}^r$ helps $V_{qs}^r$ to increase $I_{qs}^r$. We now see this "demagnetizing" effect very clearly from (4E-7) through (4E-10). Thus from (4E-10),

$$\sqrt{2}|\tilde{V}_{as}| = \frac{9.97}{\cos 14.5^\circ + \dfrac{(50)(0.0176)\sin 14.5^\circ}{3.4}}$$

$$= \frac{9.97}{1.03} = 9.68 \text{ V} \tag{4E-11}$$

Therefore,

$$\tilde{V}_{as} = \frac{9.68}{\sqrt{2}}\underline{/14.5^\circ}$$

$$= 6.85\underline{/14.5^\circ} \text{ V} \tag{4E-12}$$

and

$$V_{as} = 9.68\cos(50t + 14.5^\circ) \text{ V} \tag{4E-13}$$

Here, Fig. 4.3-11 gives a check of (4E-13).

For part (c), $I^r_{ds} = 0$, thus from (2.4-47) and (2.4-48) with $I^r_{ds} = 0$

$$V^r_{qs} = r_s I^r_{qs} + \omega_r \lambda'^r_m \tag{4E-14}$$

$$V^r_{ds} = -\omega_r L_{ss} I^r_{qs} \tag{4E-15}$$

Now, since $I^r_{qs} = 1.61$ A, from (4E-14) and (4E-15),

$$V^r_{qs} = (3.4)(1.61) + (50)(0.0826) = 9.6 \text{ V} \tag{4E-16}$$

$$V^r_{ds} = -(50)(0.0176)(1.61) = -1.42 \text{ V} \tag{4E-17}$$

Thus,

$$\tilde{V}_{as} = \frac{1}{\sqrt{2}}(9.6 + j1.42)$$

$$= 6.86\underline{/8.4^\circ} \text{ V} \tag{4E-18}$$

and

$$V_{as} = 9.7\cos(50t + 8.4^\circ) \text{ V} \tag{4E-19}$$

Since we were given torque and $\omega_r$, we were able to determine $\phi_{vMT/A}$ without solving (4.2-16). We can check (4E-19) from Figs. 4.3-12 and 4.3-13.

---

**SP4.3-1.** The stall torque of an inverter-driven brushless dc machine is slightly larger than the sinusoidal approximation. Why? $\left[\frac{2V_{in}}{3} \text{ vs. } \frac{2V_{in}}{\pi}\right]$

**SP4.3-2.** Why is $\omega_r$ in the last two terms of (4D-5) and in the last term of (4D-6) rather than $6\omega_r$? [$\frac{d}{dt}$ of the 6th harmonic yields $6\omega_r$; the rotor speed is still $\omega_r$]

**SP4.3-3.** Calculate the output mechanical power $P_m$ for all parts of Example 4E. [10 W]

## 4.4 SIMULATION OF A BRUSHLESS DC DRIVE

In recent years, computer simulation has been used increasingly in the analysis and the early stages of design of electric drives. Although a computer-aided design must be followed by hardware prototyping, computer simulation is becoming a viable means to reduce design costs. Since computer simulation will continue to become more engrained in electric drive design, there is ample reason for us to touch on it by setting forth the simulation of the brushless dc drive presented in the previous section.

### Simulation of Coupled Circuits

It is instructive to first develop the simulation for coupled circuits since this allows the method used to simulate electric machines to be readily established. We will follow the work of C. H. Thomas who set forth a method of simulating electric machines that is still used widely [5]. He formulated the voltage equations with flux linkages rather than currents as state variables.

The voltage equations for the transformer equivalent *T* circuit given in Fig. 1.4-2 are (1.4-20) and (1.4-22)

$$v_1 = r_1 i_1 + p\lambda_1 \tag{4.4-1}$$

$$v_2' = r_2' i_2' + p\lambda_2' \tag{4.4-2}$$

From (1.4-23) and (1.4-24),

$$\lambda_1 = L_{l1} i_1 + \lambda_{m1} \tag{4.4-3}$$

$$\lambda_2' = L_{l2}' i_2' + \lambda_{m1} \tag{4.4-4}$$

where

$$\lambda_{m1} = L_{m1}(i_1 + i_2') \tag{4.4-5}$$

Since we will use flux linkages as state variables, let us solve (4.4-3) and (4.4-4) for $i_1$ and $i_2'$, respectively.

$$i_1 = \frac{1}{L_{l1}}(\lambda_1 - \lambda_{m1}) \tag{4.4-6}$$

$$i_2' = \frac{1}{L_{l2}'}(\lambda_2' - \lambda_{m1}) \tag{4.4-7}$$

Substituting (4.4-6) and (4.4-7) into (4.4-5) and after a little elementary algebra,

$$\lambda_{m1} = L_M \left( \frac{\lambda_1}{L_{l1}} + \frac{\lambda_2'}{L_{l2}'} \right) \tag{4.4-8}$$

where

$$L_M = \left( \frac{1}{L_{m1}} + \frac{1}{L_{l1}} + \frac{1}{L_{l2}'} \right)^{-1} \tag{4.4-9}$$

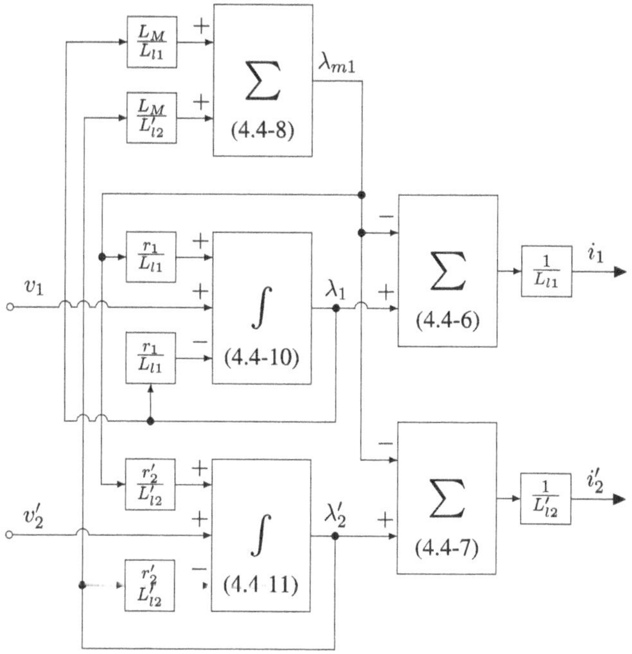


Figure 4.4-1 Block diagram of computer simulation of coupled circuits.

Substituting (4.4-6) into (4.4-1) and (4.4-7) into (4.4-2) and solving for $\lambda_1$ and $\lambda'_2$, respectively, we obtain

$$\lambda_1 = \frac{1}{p}\left[v_1 - \frac{r_1}{L_{l1}}(\lambda_1 - \lambda_{m1})\right] \tag{4.4-10}$$

$$\lambda'_2 = \frac{1}{p}\left[v'_2 - \frac{r'_2}{L'_{l2}}(\lambda'_2 - \lambda_{m1})\right] \tag{4.4-11}$$

where $\frac{1}{p}$ indicates integration and (4.4-6) through (4.4-11) make up the computer equations. A block diagram of the computer simulation is shown in Fig. 4.4-1.

## Simulation of Drive Inverter and Transformation

We will refer to Figs. 4.3-1 and 4.3-2 and many of the equations that we have derived previously. In particular, the inverter voltages applied to the machine are (4.3-1) through (4.3-3), that is

$$v_{as} = \frac{2}{3}v_{ap} - \frac{1}{3}(v_{bp} + v_{cp}) \tag{4.4-12}$$

$$v_{bs} = \frac{2}{3}v_{bp} - \frac{1}{3}(v_{cp} + v_{ap}) \tag{4.4-13}$$

$$v_{cs} = \frac{2}{3}v_{cp} - \frac{1}{3}(v_{ap} + v_{bp}) \tag{4.4-14}$$

where

$$v_{ap} = v_{as} + v_{np} \tag{4.4-15}$$

$$v_{bp} = v_{bs} + v_{np} \tag{4.4-16}$$

$$v_{cp} = v_{cs} + v_{np} \tag{4.4-17}$$

and since we are dealing with a wye-connected symmetrical system

$$v_{as} + v_{bs} + v_{cs} = 0 \tag{4.4-18}$$

whereupon

$$v_{np} = \frac{1}{3}(v_{ap} + v_{bp} + v_{cp}) \tag{4.4-19}$$

From Fig. 3.5-5, recall that $v_{ap}$ is $V_{\text{in}}$ if $T_1$ is high and $T_4$ is low, and zero when $T_1$ is low and $T_4$ is high, Similarly for $v_{bp}$ with $T_2$ and $T_5$ and $v_{cp}$ with $T_3$ and $T_6$.

From (2.5-17) and (2.5-18), the transformation is

$$\mathbf{K}_s^r = \frac{2}{3}\begin{bmatrix} \cos\theta_r & \cos\left(\theta_r - \frac{2}{3}\pi\right) & \cos\left(\theta_r + \frac{2}{3}\pi\right) \\ \sin\theta_r & \sin\left(\theta_r - \frac{2}{3}\pi\right) & \sin\left(\theta_r + \frac{2}{3}\pi\right) \\ \frac{1}{2} & \frac{1}{2} & \frac{1}{2} \end{bmatrix} \tag{4.4-20}$$

$$(\mathbf{K}_s^r)^{-1} = \begin{bmatrix} \cos\theta_r & \sin\theta_r & 1 \\ \cos\left(\theta_r - \frac{2}{3}\pi\right) & \sin\left(\theta_r - \frac{2}{3}\pi\right) & 1 \\ \cos\left(\theta_r + \frac{2}{3}\pi\right) & \sin\left(\theta_r + \frac{2}{3}\pi\right) & 1 \end{bmatrix} \tag{4.4-21}$$

where

$$\omega_r = \frac{d\theta_r}{dt} \tag{4.4-22}$$

The block diagram of the computer simulation of the drive inverter and the transformation is given in Fig. 4.4-2a. The zero-variables are zero in a three-wire symmetrical system.

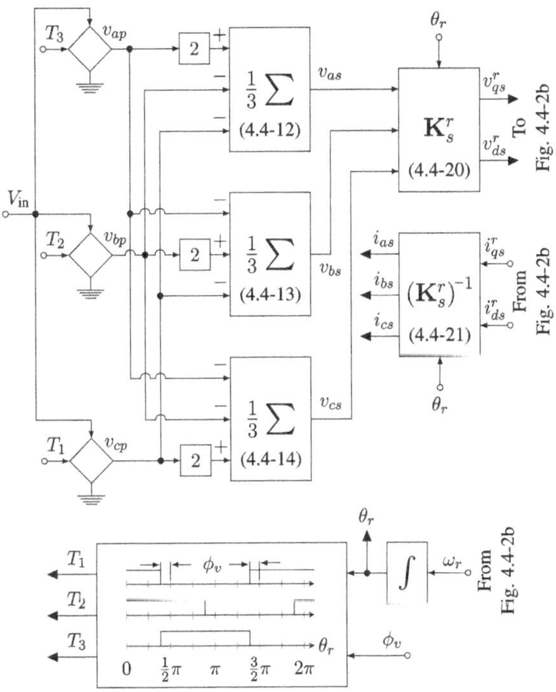


Figure 4.4-2a Block diagram of computer simulation of brushless dc drive – inverter and transformation.

## Simulation of Permanent-Magnet AC Machine

The $qs^r$- and $ds^r$-voltages and flux linkage equations are repeated from (2.5-24) through (2.5-29)

$$v_{qs}^r = r_s i_{qs}^r + \omega_r \lambda_{ds}^r + p\lambda_{qs}^r \tag{4.4-23}$$

$$v_{ds}^r = r_s i_{ds}^r - \omega_r \lambda_{qs}^r + p\lambda_{ds}^r \tag{4.4-24}$$

$$v_{0s} = r_s i_{0s} + p\lambda_{0s} \tag{4.4-25}$$

where

$$\lambda_{qs}^r = L_{ss} i_{qs}^r \tag{4.4-26}$$

$$\lambda_{ds}^r = L_{ss} i_{ds}^r + \lambda_m'^r \tag{4.4-27}$$

$$\lambda_{0s} = L_{ls} i_{0s} \tag{4.4-28}$$

From (2.5-35),

$$T_e = \frac{3}{2}\frac{P}{2}\lambda_m'^r i_{qs}^r \tag{4.4-29}$$

From (2.3-16),

$$T_e = J\left(\frac{2}{P}\right)\frac{d\omega_r}{dt} + B_m\left(\frac{2}{P}\right)\omega_r + T_L \tag{4.4-30}$$

Since $v_{as} + v_{bs} + v_{cs} = 0$, we need only concern ourselves with the $qs^r$- and $ds^r$-variables. From (4.4-26) and (4.4-27),

$$i^r_{qs} = \frac{1}{L_{ss}}\lambda^r_{qs} \tag{4.4-31}$$

$$i^r_{ds} = \frac{1}{L_{ss}}\left(\lambda^r_{ds} - \lambda'^r_m\right) \tag{4.4-32}$$

If we substitute (4.4-31) into (4.4-23) and solve for $\lambda^r_{qs}$; and substitute (4.4-32) into (4.4-24) and solve for $\lambda^r_{ds}$, we obtain the following integral equations

$$\lambda^r_{qs} = \frac{1}{p}\left(v^r_{qs} - \omega_r\lambda^r_{ds} - \frac{r_s}{L_{ss}}\lambda^r_{qs}\right) \tag{4.4-33}$$

$$\lambda^r_{ds} = \frac{1}{p}\left[v^r_{ds} + \omega_r\lambda^r_{qs} - \frac{r_s}{L_{ss}}\left(\lambda^r_{ds} - \lambda'^r_m\right)\right] \tag{4.4-34}$$

Now, from (4.4-30)

$$\omega_r = \left(\frac{1}{p}\right)\left[\left(\frac{P}{2J}\right)\left(T_e - B_m\frac{2}{P}\omega_r - T_L\right)\right] \tag{4.4-35}$$

The block diagram given in Fig. 4.4-2b depicts the implementation of the computer simulation of the permanent-magnet ac machine (brushless dc machine).

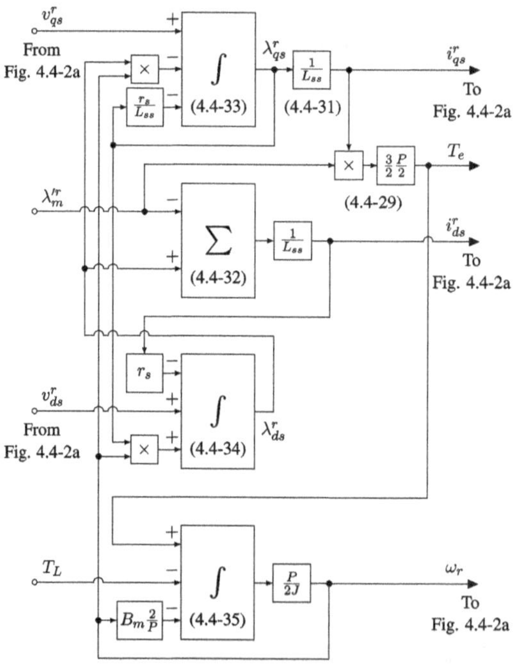


Figure 4.4-2b Block diagram of computer simulation of brushless dc drive permanent-magnetic ac machine (brushless dc machine).

**Example 4F.** It is instructive to consider the modifications necessary to the simulation of two coupled circuits shown in Fig. 4.4-1 to model a short-circuited and an open-circuited winding. The modification to simulate a short circuit across the 2-winding, for example, is simply to set $v_2' = 0$. Simulating an open circuit of the 2-winding is a bit more involved. With $i_2' = 0$, $v_2'$ becomes

$$v_2' = p\lambda_{m1} \tag{4F-1}$$

It may be helpful to refer to Fig. 1.4-2. Now since $i_2' = 0$, (4.4-5) becomes

$$\lambda_{m1} = L_{m1} i_1 \tag{4F-2}$$

Substituting (4.4-6) into (4F-2) yields

$$\begin{aligned}\lambda_{m1} &= \frac{L_{m1}}{L_{l1}}(\lambda_1 - \lambda_{m1}) \\ &= \frac{L_{m1}}{L_{l1} + L_{m1}}\lambda_1 \end{aligned} \tag{4F-3}$$

Substituting (4F-3) into (4F-1) yields

$$v_2' = \frac{L_{m1}}{L_{l1} + L_{m1}} p\lambda_1 \tag{4F-4}$$

Equation (4F-4) may be obtained without taking the derivative of $\lambda_1$ since $p\lambda_1$ is available at the input of the integration block for (4.4-10) in Fig. 4.4-1. Therefore, (4F-4) may be made up by summing the integrand of (4.4-10), which is $p\lambda_1$, multiplying it by $L_{m1}(L_{l1} + L_{m1})^{-1}$ and applying it as $v_2'$ in (4.4-11). It is important that this should be applied as $v_2'$ only when $i_2' = 0$ whereupon $i_2'$ will remain at zero. If, for example, $i_2'$ is sinusoidal, $v_2'$ for $i_2 = 0$ must be applied at the instant $i_2'$ passes through zero. If, however, $i_2'$ is zero at $t = 0$ and it is to remain at zero, then $v_2'$ for $i_2' = 0$ would be applied at $t = 0$.

---

**SP4.4-1.** We are to simulate a brushless dc drive with three-phase balanced sinusoidal voltages applied with $\phi_v = 0$ and $v_{as} = \sqrt{2}\, V_s \cos \omega_r t$. Modify the simulation given in Fig. 4.4-2 to accommodate this simulation. [Disregard Fig. 4.4-2a and set $v_{qs}^r = \sqrt{2}\, V_s$ and $v_{ds}^r = 0$]

**SP4.4-2.** What is the advantage of using flux linkages as state variables rather than currents in the case of coupled circuits? [$pi_1$ and $pi_2'$ appear in both voltage equations]

**SP4.4-3.** Why does not the flux linkage formulation offer the same advantage in the case of the brushless dc motor? [$\lambda_m$ is constant and transformer coupling does not exist with the $ds^r$ winding; however, the formulation is still convenient]

**SP4.4-4.** Is the simulation given in Fig. 4.4-2 valid for unbalanced stator voltages? Why? [Yes; three-wire symmetrical stator]

**SP4.4-5.** Modify the simulation in Fig. 4.4-2 to accommodate $v_{as}$, $v_{bs}$, and $v_{cs}$ of any form. [Replace $v_{as}$, $v_{bs}$, and $v_{cs}$ in Fig. 4.4-2a with the given $v_{as}$, $v_{bs}$, and $v_{cs}$; eliminate the inverter simulation]

## REFERENCES

[1] T. W. Nehl, F. A. Fouad, and N. A. Demerdash, "Digital simulation of power conditioner–Machine interaction for electronically commutated DC permanent magnet machines," *IEEE Transactions on Magnetics*, vol. 17, November 1981, pp. 3284–3286.

[2] T. M. Jahns, "Torque production in permanent-magnet synchronous motor drives with rectangular current excitations," *IAS Conference Record*, October 1983, pp. 476–487.

[3] P. C. Krause, *Analysis of Electric Machinery*, McGraw-Hill Book Company, New York, 1986.

[4] P. C. Krause, O. Wasynczuk, and S. D. Pekarek, *Electromechanical Motion Devices*, 2nd edition, Wiley-IEEE Press, New York, 2012.

[5] C. H. Thomas, "Discussion of 'analogue computer representation of synchronous generators in voltage-regulation studies'," *Transactions of AIEE (Power Apparatus and Systems)*, vol. 75, December 1956, pp. 1182–1184.

## PROBLEMS

1. The permanent-magnet ac machine considered in Example 4A is stalled with the permanent magnet ($d$-axis) aligned with the positive $as$-axis. The parameters are $r_s = 3.4\ \Omega$, $L_{ls} = 1.1$ mH, $L_{ms} = 11$ mH, and $\lambda'_m = 0.0826$ V · s/rad. Calculate the steady-state stall torque for balanced steady-state conditions with $\tilde{V}_{as} = 11.25\angle 0°$. Justify your answer from the pole positions. Next, calculate $P_e$ and $P_{loss}$ which should be the same.

2. Assume $\phi_v = \frac{\pi}{2}$ and the parameters of the three-phase machine are the same as in Problem 1. Determine the rotor speed where the steady-state torque is zero if $\frac{L_{ss}}{r_s} = \frac{\lambda'_m}{\sqrt{2}V_s}$.

3. Consider Fig. 4.2-7. Justify that $\phi_{vMT/A}$ is zero at stall and at the rotor speed where $T_e$ is zero.

4. Draw the phasor diagram for the mode of operation considered in Example 4E part (a) where $\phi_v = 0$ with $T_e$ constant at 0.4 N · m at a speed of 50 rad/s. The parameters of the three-phase permanent-magnet ac machine are; $r_s = 3.4\ \Omega$, $L_{ls} = 1.1$ mH, $L_{ms} = 11$ mH, and $\lambda'_m = 0.0826$ V · s/rad. Show values and angles of $\tilde{V}_{as}$, $\tilde{I}_{as}$, $r_s\tilde{I}_{as}$, $jX_{ss}\tilde{I}_{as}$, $\tilde{E}_a$, and the positions of $N^r$, $S^r$, $N^s$, and $S^s$.

5. Repeat Problem 4 for part (b) of Example 4E where $\phi_v = \phi_{vMT/V}$.

6. Repeat Problem 4 for part (c) of Example 4E where $\phi_v = \phi_{vMT/A}$.

7. Using information from Fig. 3.5-1 and Fig. 4.4.2, develop a computer simulation of a two-phase permanent-magnet ac motor supplied from a two-phase square wave voltage source.

# CHAPTER 5

# POWER SYSTEMS

## 5.1 INTRODUCTION

Although power system engineering dominated the power area in the first half of the twentieth century, electric drives, power electronics, and machine design have emerged as important areas of power engineering. It is understandable that power system analysis having evolved first, established conventions and analysis techniques appropriate for power systems and are not necessarily convenient for the areas of power engineering that came later. For example, the power system engineer prefers to consider positive current out of the machine since they work primarily with power generation while the electric drive's engineer deals primarily with motors and the power system's engineering convention is not the most convenient.

In this chapter, several of the analytical aspects of power system engineering are set forth. Although the presentation may be a bit disjointed, the power system engineer is faced with a host of challenges and our purpose is to make the reader aware of some of these challenges without becoming too involved. For example, the presentation of the synchronous machine is arranged so that the derivation of the dynamic voltage equations can be bypassed, allowing the reader to focus on the steady-state performance. Likewise, the material on the per-unit system is arranged so that per unitizing the synchronous generator may be omitted to focus on per unitizing the transmission system.

## 5.2 THREE-PHASE TRANSFORMER CONNECTIONS

Single-phase transmission exists in urban distribution systems and sparsely populated rural areas. Three-phase systems have traditionally been associated with electric power generation and high voltage transmission. Therefore, three-phase transformer connections warrant a brief consideration in an introductory treatment of power system engineering.

In a two-phase system, transformers are generally arranged as two separate single-phase transformers without interconnection; however, in a three-phase system, the phase transformers are generally interconnected and, although there are numerous variations, there are essentially only three basic transformer configurations. We will

*Introduction to Electric Power and Drive Systems*, First Edition.
Paul Krause, Oleg Wasynczuk, Timothy O'Connell and Maher Hasan.
© 2017 by The Institute of Electrical and Electronics Engineers, Inc. Published 2017 by John Wiley & Sons, Inc.
Companion website: www.wiley.com/go/krause/electricpower

touch on each of these. Three-phase transformers may be three separately constructed single-phase transformers or three single-phase transformers wound on a common iron core. We will deal only with three separate single-phase transformers.

## Wye-Wye Connection

The so-called wye-wye (Y-Y) connection is illustrated in Fig. 5.2-1. The transformers are identical; hence $r_{a1} = r_{b1} = r_{c1}$ and $r_{a2} = r_{b2} = r_{c2}$. Here, we have changed notation to accommodate the three-phase variables. Rather than denoting the windings as winding 1 and winding 2, as in previous sections, the notation of $a1$ and $a2$ winding, respectively, is used for the $a$-phase transformer, $b1$ and $b2$ winding for the $b$-phase transformer, and $c1$ and $c2$ winding for the $c$-phase transformer. In Fig. 5.2-1$a$, the three single-phase transformers are arranged to show the transformer coupling in each phase; however, the circuit diagram shown in Fig. 5.2-1$b$ is generally used since it most readily depicts the type of connection being used. Although the Y-Y connection is not used too often in power systems, it is a good place to start.

Note that only three conductors enter and three leave the transformers. This does not seem correct. For a single-phase system, we would need two conductors and for a two-phase system, we would need four. Why are only three required for a three-phase system? Well, we know that a balanced three-phase set of currents consists of, by definition, equal-amplitude sinusoidal currents displaced by 120°. The sum of these currents is zero; hence, if the system is balanced, $i_{a1} + i_{b1} + i_{c1} = 0$, that is, the

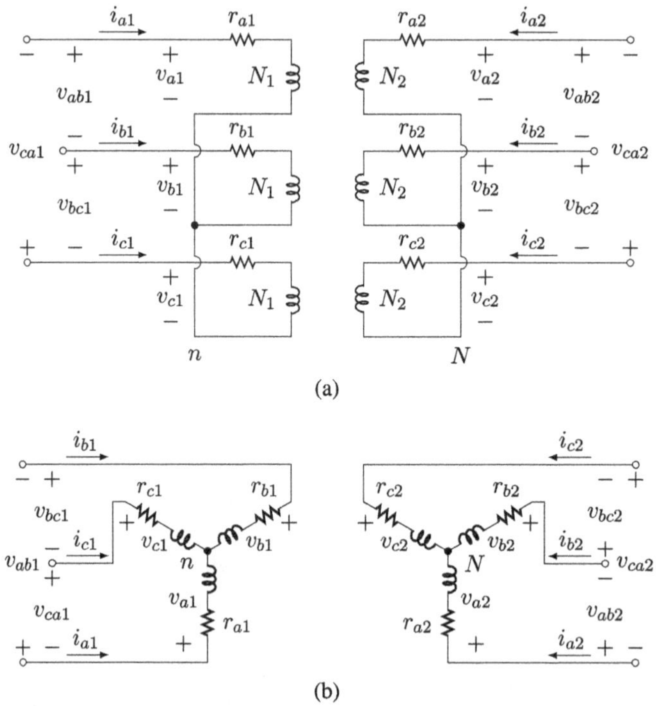


Figure 5.2-1 A Y-Y three-phase transformer connection.

current entering one phase of the 1-winding returns through the other phases of the 1-windings. Actually, a fourth wire is not needed regardless if the system is balanced or unbalanced. We are now starting to see the economic advantage of transmitting three-phase electric power. There is also another advantage of two- or three-phase electromechanical devices over their single-phase little sister. As we mentioned earlier, the instantaneous power or torque in a single-phase system pulsates about an average value; however, in a multiphase system, the instantaneous power or torque is constant for balanced steady-state conditions and the power system operator strives to keep the system balanced.

In a Y-Y connection, the assigned negative-potential sides of the windings are all connected to form what is called the *neutral point*; *n* for 1-windings in Fig. 5.2-1 and *N* for the 2-windings. Either or both neutral points may be grounded or left to "float." The voltages across each transformer winding, $v_{a1}, v_{a2}, v_{b1}$, etc., are referred to as line-to-neutral voltages while the voltages between two of the three terminals of the 1-windings (2-windings) are the line-to-line voltages. The individual transformer currents $i_{a1}, i_{a2}, i_{b1}$, etc. are the line currents. The line-to-neutral and the line-to-line voltages are related as

$$v_{ab1} = v_{a1} - v_{b1} \tag{5.2-1}$$

$$v_{bc1} = v_{b1} - v_{c1} \tag{5.2-2}$$

$$v_{ca1} = v_{c1} - v_{a1} \tag{5.2-3}$$

Similar expressions may be written for the 2-windings.

If the system is balanced, then we can express the steady-state voltages of the 1-windings for an *abc* sequence as

$$V_{a1} = \sqrt{2}V_{s1} \cos \omega_e t \tag{5.2-4}$$

$$V_{b1} = \sqrt{2}V_{s1} \cos\left(\omega_e t - \frac{2\pi}{3}\right) \tag{5.2-5}$$

$$V_{c1} = \sqrt{2}V_{s1} \cos\left(\omega_e t + \frac{2\pi}{3}\right) \tag{5.2-6}$$

where the uppercase letters are used to denote steady-state conditions. In phasor form,

$$\tilde{V}_{a1} = V_{s1}\underline{/0^\circ} \tag{5.2-7}$$

$$\tilde{V}_{b1} = V_{s1}\underline{/-120^\circ} \tag{5.2-8}$$

$$\tilde{V}_{c1} = V_{s1}\underline{/120^\circ} \tag{5.2-9}$$

From (5.2-1) through (5.2-3), the line-to-line voltage may be expressed as

$$\begin{aligned}\tilde{V}_{ab} &= V_{s1}\underline{/0^\circ} - V_{s1}\underline{/-120^\circ} \\ &= \sqrt{3}V_{s1}\underline{/30^\circ}\end{aligned} \tag{5.2-10}$$

$$\begin{aligned}\tilde{V}_{bc} &= V_{s1}\underline{/-120^\circ} - V_{s1}\underline{/120^\circ} \\ &= \sqrt{3}V_{s1}\underline{/-90^\circ}\end{aligned} \tag{5.2-11}$$

$$\begin{aligned}\tilde{V}_{ca} &= V_{s1}\underline{/120^\circ} - V_{s1}\underline{/0^\circ}\\ &= \sqrt{3}V_{s1}\underline{/150^\circ}\end{aligned} \tag{5.2-12}$$

Hence, the line-to-line voltages form a balanced three-phase set which are $\sqrt{3}$ times the magnitude of the line-to-neutral voltages and shifted in phase by 30°.

For balanced steady-state conditions, the operation of each phase may be considered separately. In fact, we need to consider only one phase since, once we have determined the variables associated with one phase of the transformers, we can express the other phase variables by shifting the phase by either 120° or −120°.

## Delta-Delta Connection

The delta-delta (Δ-Δ) connection is illustrated in Fig. 5.2-2. In this type of connection, the line-to-line voltages are the voltages across the individual transformers, that is, $v_{a1} = v_{ab1}$, $v_{a2} = v_{ab2}$, $v_{b1} = v_{bc1}$, etc. There is no neutral connection. The line currents are made up of the currents from two of the transformers. In particular,

$$i_{ac1} = i_{a1} - i_{c1} \tag{5.2-13}$$

$$i_{ba1} = i_{b1} - i_{a1} \tag{5.2-14}$$

$$i_{cb1} = i_{c1} - i_{b1} \tag{5.2-15}$$

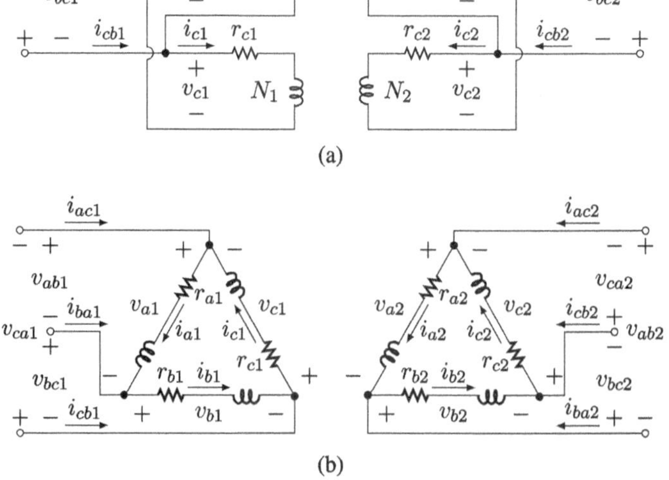


Figure 5.2-2 A Δ-Δ three-phase transformer connection.

If, for example, the transformer currents form a balanced *abc* sequence, then

$$\tilde{I}_{a1} = I_{s1}\angle 0° \tag{5.2-16}$$
$$\tilde{I}_{b1} = I_{s1}\angle -120° \tag{5.2-17}$$
$$\tilde{I}_{c1} = I_{s1}\angle 120° \tag{5.2-18}$$

The line currents become

$$\begin{aligned}\tilde{I}_{ac1} &= I_{s1}\angle 0° - I_{s1}\angle 120° \\ &= \sqrt{3}I_{s1}\angle -30°\end{aligned} \tag{5.2-19}$$

Thus, $\tilde{I}_{ac1}$ is $\sqrt{3}$ times the amplitude of $\tilde{I}_{a1}$ and shifted 30° back in phase from it. Similarly, $\tilde{I}_{ba1}$ is shifted back 30° from $\tilde{I}_{b1}$, and $\tilde{I}_{cb1}$ back 30° from $\tilde{I}_{c1}$. For balanced conditions, a phase current in a Y-Y connection is $\sqrt{3}$ times the phase current in a Δ-Δ connection for the same line current.

## Wye-Delta or Delta-Wye Connection

The final three-phase transformer connection that we will consider is the wye-delta (Y-Δ) or delta-wye (Δ-Y) connection. The Δ-Y is generally used to "step up" the voltage, thus the Y-Δ is used for "stepping down" the voltage. The Y-Δ connection is illustrated in Fig. 5.2-3.

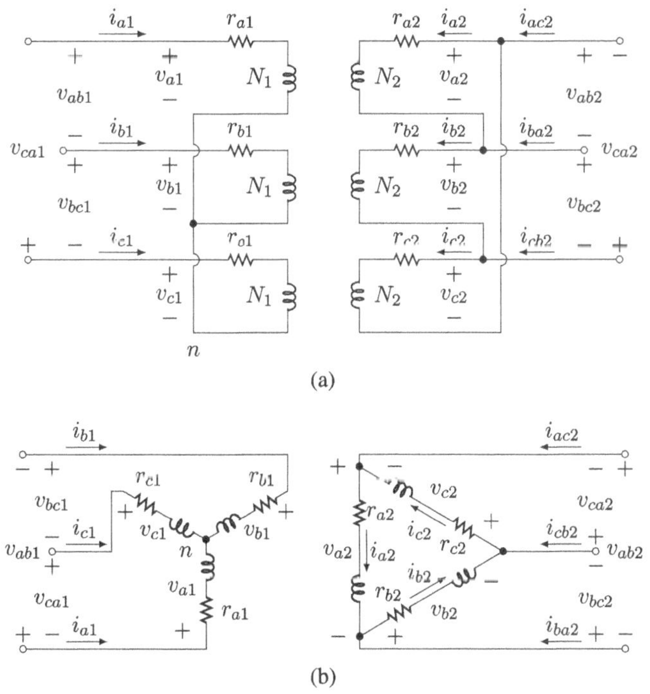


Figure 5.2-3 A Y-Δ three-phase transformer connection.

## Ideal Transformers

Since in electromechanical motion devices the magnetizing flux must traverse the air gap, the leakage inductances and the magnetizing current must be considered when analyzing these devices. However, in the analysis of a power system, the system transformers are often assumed to be ideal. In this case, the winding resistance, leakage inductance, and magnetizing current are all neglected and the voltages (currents) of the transformer windings differ only by the turns ratio. The relations between 1- and 2-winding variables for the Y-Y, Δ-Δ, and Y-Δ connections of ideal transformers are shown in Fig. 5.2-4.

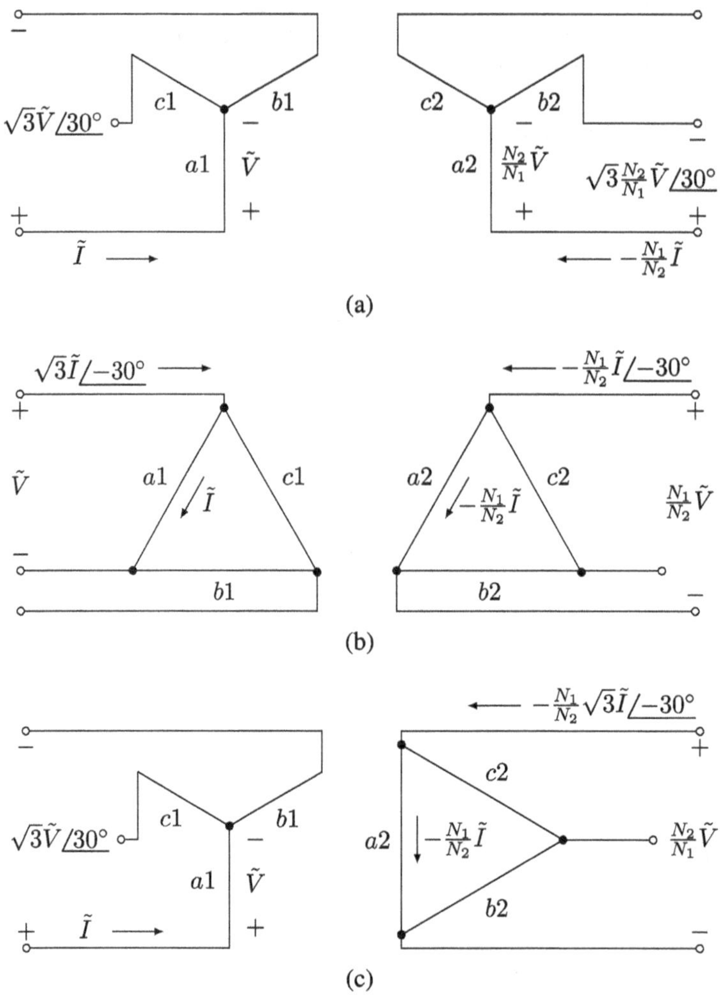


Figure 5.2-4 Three-phase ideal transformers for *abc* sequence: (*a*) Y-Y connection; (*b*) Δ-Δ connection; (*c*) Y-Δ connection.

**Example 5A. Equivalent wye- and delta-connected impedances.** Determine the delta and wye equivalent impedances.

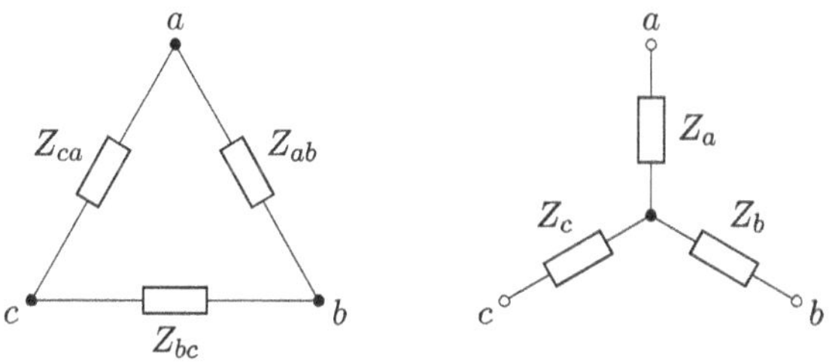


Figure 5A-1 Delta- and wye-connected impedances.

From Fig. 5A-1

$$Z_a + Z_b = \frac{Z_{ab}(Z_{ca} + Z_{bc})}{Z_{ab} + Z_{bc} + Z_{ca}} \tag{5A-1}$$

$$Z_b + Z_c = \frac{Z_{bc}(Z_{ab} + Z_{ca})}{Z_{ab} + Z_{bc} + Z_{ca}} \tag{5A-2}$$

$$Z_c + Z_a = \frac{Z_{ca}(Z_{bc} + Z_{ab})}{Z_{ab} + Z_{bc} + Z_{ca}} \tag{5A-3}$$

Show that if $Z_{ab}, Z_{bc}$, and $Z_{ca}$ are each equal to $Z_\Delta$; and $Z_a, Z_b$, and $Z_c$ equal to $Z_Y$ then

$$Z_\Delta = 3Z_Y \tag{5A-4}$$

---

**SP5.2-1.** Write the expression for the steady-state line-to-line voltage $V_{ab}$ of a Y-Y-connected three-phase system for a balanced set of (a) *abc* sequence voltages and (b) *acb* sequence voltages. Assume that the steady-state line-to-neutral voltage of phase a is $V_{a1} = \sqrt{2}V_s \cos(\omega_e t - 30°)$. [(a) $V_{ab} = \sqrt{6}V_s \cos \omega_e t$; (b) $V_{ab} = \sqrt{6}V_s \cos(\omega_e t - 60°)$]

**SP5.2-2.** Refer to Fig. 5.2-2. Calculate $I_{ac1}$ for an *acb* sequence. Assume $I_{a1} = I_{s1}\angle 0°$. [$\tilde{I}_{ac1} = \sqrt{3}I_{s1}\angle 30°$]

**SP5.2-3.** In Fig. 5.2-1, $V_{ab1} = \sqrt{6}\cos \omega_e t$ and $I_{a1} = \sqrt{2}\cos(\omega_e t - 30°)$. Assume balanced operation with an *abc* sequence. Calculate the total power flow into the Y-Y transformer. [3 W]

**SP5.2-4.** Show that the power supplied to a balanced wye and an equivalent balanced delta-connected resistive load are equal.

## 5.3 SYNCHRONOUS GENERATOR

Although the synchronous machine can be operated as either a generator or a motor, it is generally operated as a generator. In the steady state, the rotor rotates in synchronism with the electrical angular velocity corresponding to that of the system to which it is connected; 60 Hz in the United States, 50 Hz in Europe. In Chapter 4, we used the permanent-magnet ac machine, which is also a synchronous machine, as the motor in an electric drive. In that case, the stator windings of the permanent-magnet ac machine were supplied from an inverter that was controlled so that the frequency of the applied stator voltages was made equal to the electrical angular velocity of the rotor. In other words, the machine is always operating in synchronism with the supply voltages. That is, the motor slows when it is loaded and the frequency of the stator voltages is decreased in unison with the decrease in rotor speed.

The synchronous machine operating in a power grid is a much different situation. In this case, the power grid operates at essentially a constant frequency and the synchronous machine must operate in synchronism with the frequency of the power grid. The synchronous generator "falling out" of synchronism with the power system during a disturbance is a huge concern for power system operators. Therein lies an area of system stability that presents many engineering challenges. Our purpose here is to introduce the synchronous generator and in doing so we will draw on material in Section 2.5.

In the case of isolated operation of a small electrical power system with one or two generators, the frequency (electrical angular velocity) of the system may vary somewhat depending on the load and the power generated. Some aircraft, ships, or microgrids are examples; however, we will not consider this mode of operation.

A three-phase round-rotor synchronous machine is shown in Fig. 5.3-1. We have replaced the permanent-magnet rotor of the permanent-magnet ac machine with a field winding; the *fd*-winding. The voltage equations are

$$v_{as} = r_s i_{as} + \frac{d\lambda_{as}}{dt} \tag{5.3-1}$$

$$v_{bs} = r_s i_{bs} + \frac{d\lambda_{bs}}{dt} \tag{5.3-2}$$

$$v_{cs} = r_s i_{cs} + \frac{d\lambda_{cs}}{dt} \tag{5.3-3}$$

$$v_{fd} = r_f i_{fd} + \frac{d\lambda_{fd}}{dt} \tag{5.3-4}$$

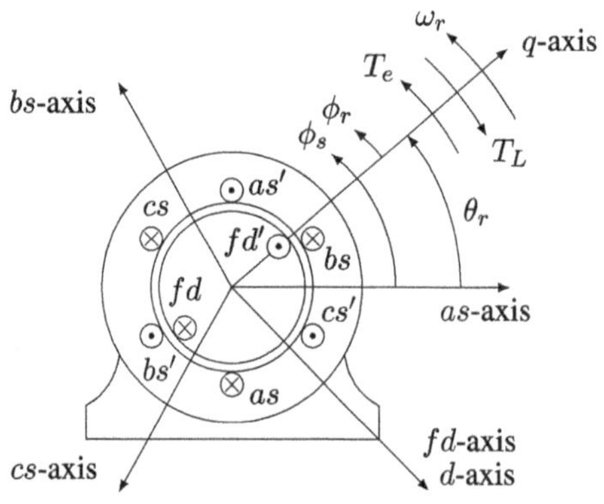


Figure 5.3-1 Three-phase round-rotor synchronous machine.

Equations (5.3-1) through (5.3-3) are identical in form to (2.5-1) through (2.5-3) for the three-phase permanent-magnet ac machine. Therefore, $v_{qs}^r$, $v_{ds}^r$, and $v_{0s}$ will be identical in form as (2.5-24) through (2.5-26). Moreover, the *fd*-winding is physically in the rotor frame of reference and therefore will not require a change of variables. In fact, the steady-state voltage equations are similar to those of the permanent-magnet ac machine, differing only in that the synchronous machine (generator) has a field winding rather than a permanent magnet. If the focus is on the analysis of steady-state operation, the reader may wish to skip to the subsection following (5.3-34) and follow the derivation leading to (5.3-48).

In order to derive the voltage equations valid for dynamic operation, it is necessary to consider the flux linkage equations; in particular

$$\begin{bmatrix} \lambda_{as} \\ \lambda_{bs} \\ \lambda_{cs} \\ \lambda_{fd} \end{bmatrix} = \begin{bmatrix} L_{ls}+L_{ms} & -\frac{1}{2}L_{ms} & -\frac{1}{2}L_{ms} & L_{sfd}\sin\theta_r \\ -\frac{1}{2}L_{ms} & L_{ls}+L_{ms} & -\frac{1}{2}L_{ms} & L_{sfd}\sin(\theta_r-120^\circ) \\ -\frac{1}{2}L_{ms} & -\frac{1}{2}L_{ms} & L_{ls}+L_{ms} & L_{sfd}\sin(\theta_r+120^\circ) \\ L_{sfd}\sin\theta_r & L_{sfd}\sin(\theta_r-120^\circ) & L_{sfd}\sin(\theta_r+120^\circ) & L_{lfd}+L_{mfd} \end{bmatrix} \begin{bmatrix} i_{as} \\ i_{bs} \\ i_{cs} \\ i_{fd} \end{bmatrix} \tag{5.3-5}$$

This may be written as

$$\begin{bmatrix} \boldsymbol{\lambda}_{abcs} \\ \lambda_{fd} \end{bmatrix} = \begin{bmatrix} \mathbf{L}_s & \mathbf{L}_{sr} \\ (\mathbf{L}_{sr})^T & L_{fdfd} \end{bmatrix} \begin{bmatrix} \mathbf{i}_{abcs} \\ i_{fd} \end{bmatrix} \tag{5.3-6}$$

where

$$\mathbf{L}_s = \begin{bmatrix} L_{ls}+L_{ms} & -\frac{1}{2}L_{ms} & -\frac{1}{2}L_{ms} \\ -\frac{1}{2}L_{ms} & L_{ls}+L_{ms} & -\frac{1}{2}L_{ms} \\ -\frac{1}{2}L_{ms} & -\frac{1}{2}L_{ms} & L_{ls}+L_{ms} \end{bmatrix} \tag{5.3-7}$$

$$\mathbf{L}_{sr} = \begin{bmatrix} L_{sfd}\sin\theta_r \\ L_{sfd}\sin(\theta_r-120^\circ) \\ L_{sfd}\sin(\theta_r+120^\circ) \end{bmatrix} \tag{5.3-8}$$

$$L_{fdfd} = L_{lfd} + L_{mfd} \tag{5.3-9}$$

Since we are assuming a magnetically linear system, the inductances may be expressed as

$$L_{ms} = \frac{N_s^2}{\mathcal{R}_m} \tag{5.3-10}$$

$$L_{mfd} = \frac{N_f^2}{\mathcal{R}_m} = \frac{N_f^2}{N_s^2}L_{ms} \tag{5.3-11}$$

$$L_{sfd} = \frac{N_s N_f}{\mathcal{R}_m} = \frac{N_f}{N_s} L_{ms} \tag{5.3-12}$$

where $N_s(N_f)$ is the number of turns of the stator windings (*fd*-winding). Now, from (2.5-15), (5.3-6), and (2.5-23)

$$\begin{aligned}\boldsymbol{\lambda}_{qd0s}^r &= \mathbf{K}_s^r \boldsymbol{\lambda}_{abcs} + \mathbf{K}_s^r \mathbf{L}_{sr} i_{fd}^r \\ &= \mathbf{K}_s^r \mathbf{L}_s (\mathbf{K}_s^r)^{-1} \mathbf{i}_{qd0s}^r + \mathbf{K}_s^r \mathbf{L}_{sr} i_{fd}^r\end{aligned} \tag{5.3-13}$$

Equation (5.3-13) may be written as

$$\begin{aligned}\begin{bmatrix} \lambda_{qs}^r \\ \lambda_{ds}^r \\ \lambda_{0s} \end{bmatrix} &= \begin{bmatrix} L_{ls} + \frac{3}{2}L_{ms} & 0 & 0 \\ 0 & L_{ls} + \frac{3}{2}L_{ms} & 0 \\ 0 & 0 & L_{ls} \end{bmatrix} \begin{bmatrix} i_{qs}^r \\ i_{ds}^r \\ i_{0s} \end{bmatrix} \\ &\quad + L_{sfd} \begin{bmatrix} 0 \\ 1 \\ 0 \end{bmatrix} i_{fd}^r\end{aligned} \tag{5.3-14}$$

In (5.3-13) and (5.3-14), we have added an $r$ superscript to $i_{fd}$ to emphasize that it is in the rotor reference frame. Now, in preparation for an equivalent circuit with $\frac{3}{2}L_{ms}$ as the mutual inductance between stator and rotor circuits, let us express (5.3-14) as

$$\lambda_{qs}^r = L_{ls} i_{qs}^r + \frac{3}{2} L_{ms} i_{qs}^r \tag{5.3-15}$$

$$\lambda_{ds}^r = L_{ls} i_{ds}^r + \frac{3}{2} L_{ms} \left( i_{ds}^r + i_{fd}'^r \right) \tag{5.3-16}$$

$$\lambda_{0s} = L_{ls} i_{0s} \tag{5.3-17}$$

where

$$i_{fd}'^r = \frac{2}{3} \frac{N_f}{N_s} i_{fd}^r \tag{5.3-18}$$

Comparing the last term of (5.3-14) and the last term of (5.3-16) we see that

$$\begin{aligned}L_{sfd} i_{fd}^r &= \frac{N_f}{N_s} L_{ms} \frac{N_s}{N_f} \frac{3}{2} i_{fd}'^r \\ &= \frac{3}{2} L_{ms} i_{fd}'^r\end{aligned} \tag{5.3-19}$$

If we let

$$v_{fd}'^r = \frac{N_s}{N_f} v_{fd}^r \tag{5.3-20}$$

$$\lambda_{fd}'^r = \frac{N_s}{N_f} \lambda_{fd}^r \tag{5.3-21}$$

and after considerable algebraic manipulation, the turns-ratio referred field-voltage equation becomes

$$v_{fd}'^{r} = r_{fd}' i_{fd}'^{r} + p\lambda_{fd}'^{r} \tag{5.3-22}$$

where

$$\lambda_{fd}'^{r} = L_{lfd}' i_{fd}'^{r} + \frac{3}{2}L_{ms}\left(i_{ds}^{r} + i_{fd}'^{r}\right) \tag{5.3-23}$$

$$r_{fd}' = \frac{3}{2}\left(\frac{N_s}{N_f}\right)^2 r_{fd} \tag{5.3-24}$$

$$L_{lfd}' = \frac{3}{2}\left(\frac{N_s}{N_f}\right)^2 L_{lfd} \tag{5.3-25}$$

Care must be taken since

$$v_{fd}^{r} i_{fd}^{r} \neq v_{fd}'^{r} i_{fd}'^{r} \tag{5.3-26}$$

However, (5.3-18) allows a convenient equivalent circuit that results by combining (5.3-22) with (2.5-24) and (2.5-25) which are repeated here

$$v_{qs}^{r} = r_s i_{qs}^{r} + \omega_r \lambda_{ds}^{r} + p\lambda_{qs}^{r} \tag{5.3-27}$$

$$v_{ds}^{r} = r_s i_{ds}^{r} - \omega_r \lambda_{qs}^{r} + p\lambda_{ds}^{r} \tag{5.3-28}$$

where, in the case of the three-phase machine

$$\lambda_{qs}^{r} = L_{ls} i_{qs}^{r} + \frac{3}{2}L_{ms} i_{qs}^{r} \tag{5.3-29}$$

$$\lambda_{ds}^{r} = L_{ls} i_{ds}^{r} + \frac{3}{2}L_{ms}\left(i_{ds}^{r} + i_{fd}'^{r}\right) \tag{5.3-30}$$

The equivalent circuit suggested by (5.3-22), (5.3-23), and (5.3-27) through (5.3-30) is shown in Fig. 5.3-2. The zero variables are not shown since we will not consider unbalanced conditions.

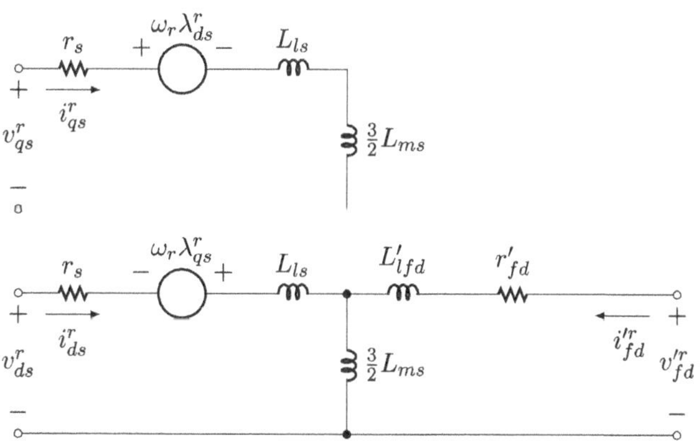


Figure 5.3-2 The $qs^r$ and $ds^r$ equivalent circuit for a three-phase synchronous machine.

## Damper Windings

We need to take a moment to mention a characteristic of the synchronous machine that, in the past, caused some concern and to indicate how it was remedied. The device shown in Fig. 5.3-1 provides very little damping to rotor oscillations following system disturbances that can cause undesirable fluctuations in power flow. Damping of these rotor oscillations is achieved by currents flowing in short-circuited rotor bars which provide a damping torque by induction motor action. That is, currents are introduced in these "windings" only when the rotor is not revolving in unison with the rotating air-gap mmf established by the stator currents. Since these windings do not play a role during balanced steady-state operation, we will not consider these windings further.

## Torque

Recall in Section 1.6 we expressed the torque by relating the energy in the coupling field, the energy entering the coupling field from the electrical system, and the energy entering from the mechanical system. There are other methods of deriving an expression for torque. For example, once the variables are expressed in the rotor reference frame, torque can be conveniently expressed using a "power balance" approach. For balanced conditions, the electrical power of the generator can be expressed as

$$P_e = \frac{3}{2}\left(v_{qs}^r i_{qs}^r + v_{ds}^r i_{ds}^r\right) \tag{5.3-31}$$

The $\frac{3}{2}$ factor comes about due to the transformation of a three-phase system to a two-phase system. The mechanical power can be expressed as

$$P_m = T_e \left(\frac{2}{P}\right) \omega_r \tag{5.3-32}$$

where $T_e$ is positive for motor action. Equating (5.3-31) and (5.3-32) and substituting (5.3-27) and (5.3-28) for $v_{qs}^r$ and $v_{ds}^r$, respectively, and then equating coefficients of $\omega_r$ yields the electromagnetic torque which is positive for motor action.

$$T_e = \left(\frac{3}{2}\right)\left(\frac{P}{2}\right)\left(\lambda_{ds}^r i_{qs}^r - \lambda_{qs}^r i_{ds}^r\right) \tag{5.3-33}$$

The torque and speed may be related as given by (2.3-16) which is repeated here

$$T_e = J\left(\frac{2}{P}\right)\frac{d\omega_r}{dt} + B_m\left(\frac{2}{P}\right)\omega_r + T_L \tag{5.3-34}$$

where $P$ is the number of poles, $J$ is the inertia in $\text{kg} \cdot \text{m}^2$, $B_m$ is the damping coefficient in $\text{N} \cdot \text{m} \cdot \text{s/rad}$ due to friction and windage losses, and $T_L$ is positive for a load torque.

## Steady-State Operation and Rotor Angle

The analysis of steady-state operation of the synchronous generator is very similar to that carried out in the case of the permanent-magnet ac machine in Sections 2.4 and 2.5. Therein, it was convenient to fix the time-zero position of the $q$-axis of the rotor on the real axis and for conventional brushless dc motor operation, the firing of the inverter is controlled to fix the time-zero position of the *as*-phase voltage (phasor $\tilde{V}_{as}$) on the real axis, coinciding with the $q$-axis as shown in Fig. 2F-1. However, as discussed in Chapter 4, we can shift the applied voltages by controlling the firing of the inverter to maximize torque. The important difference in the operation of the permanent-magnet ac machine in an electric drive system and that of a synchronous generator connected to a large power system is that the power system is generally not influenced by the mode of operation of the generator. That is, the power system voltages remain fixed in frequency, phase, and magnitude. In this case, the system is referred to as an "infinite bus" since its mechanical and electrical momentums are not influenced by the operation of one synchronous generator. Therefore, in the analysis of the synchronous generator the phase of the system voltages is generally assumed to remain fixed and the time-zero position is selected so that, the system voltage, $\tilde{V}_{as}$ is fixed on the real axis. The $q$- and $d$-axes, which are fixed in the rotor, shift in position relative to the synchronous speed of the electric power system (phase of $\tilde{V}_{as}$) as a result of a change in shaft torque. Therefore, we are not surprised that there could be a problem keeping the generator operating in synchronism with the power grid. This is of particular concern during a disturbance such as a fault near the generator resulting in a disruption of power flow from the generator to the system. In this case, the torque input to the generator would accelerate the generator above synchronous speed. If the fault is not cleared and generator operation rapidly restored, its power delivery capability to the system is lost. Clearly, this was not of concern in the case of the permanent-magnet ac machine in an inverter drive since the source frequency is controlled to be the same as the electrical angular velocity of the rotor speed.

During steady-state operation of the synchronous generator, the $q$- and $d$-axis voltage equations, (5.3-27), (5.3-28), and (5.3-22), respectively become

$$V_{qs}^r = r_s I_{qs} + \omega_r \lambda_{ds}^r \tag{5.3-35}$$

$$V_{ds}^r = r_s I_{ds}^r - \omega_r \lambda_{qs}^r \tag{5.3-36}$$

$$V_{fd}'^r = r_{fd}' I_{fd}'^r \tag{5.3-37}$$

The rotor angle, which is the displacement of the $q$-axis relative to $\tilde{V}_{as}$, is defined as

$$\begin{aligned} \delta &= \theta_r - \theta_{esv} \\ &= \int_0^t [\omega_r(\xi) - \omega_e(\xi)]d\xi + \theta_r(0) - \theta_{esv}(0) \end{aligned} \tag{5.3-38}$$

In the steady state, $\omega_r = \omega_e$ and

$$\delta - \theta_r(0) - \theta_{esv}(0) \tag{5.3-39}$$

Let us now consider a balanced three-phase system with

$$F_{as} = \sqrt{2} F_s \cos \theta_{esf} \tag{5.3-40}$$

For steady-state conditions, $\theta_{esf}$ is $\omega_e t + \theta_{esf}(0)$ and the $qs^r$- and $ds^r$-variables are

$$F_{qs}^r = \sqrt{2}F_s \cos[\theta_{esf}(0) - \theta_r(0)] \tag{5.3-41}$$

$$F_{ds}^r = -\sqrt{2}F_s \sin[\theta_{esf}(0) - \theta_r(0)] \tag{5.3-42}$$

If we solve (5.3-39) for $\theta_r(0)$ and substitute the result into (5.3-41) and (5.3-42), we will obtain

$$F_{qs}^r = \sqrt{2}F_s \cos[\theta_{esf}(0) - \theta_{esv}(0) - \delta] \tag{5.3-43}$$

$$F_{ds}^r = -\sqrt{2}F_s \sin[\theta_{esf}(0) - \theta_{esv}(0) - \delta] \tag{5.3-44}$$

As we have mentioned, when the synchronous generator connected to a power system, the system voltage is unchanged in magnitude and phase by the synchronous generator. We generally select time-zero when the system voltage at the point of connection is maximum, thus $\theta_{esv}(0) = 0$. Therefore, $\theta_{esv}(0)$ is zero in (5.3-39) and if (5.3-40) through (5.3-42) are used to express voltages, $\theta_{esf}(0)$ in these equations is zero since in this case it is $\theta_{esv}(0)$.

Now, from (5.3-40)

$$\tilde{F}_{as} = F_s e^{j\theta_{esf}(0)} \tag{5.3-45}$$

Multiplying both sides of (5.3-45) by $e^{-j\delta}$ yields

$$\tilde{F}_{as} e^{-j\delta} = F_s \cos[\theta_{esf}(0) - \delta] + jF_s \sin[\theta_{esf}(0) - \delta] \tag{5.3-46}$$

where $\theta_{esv}(0)$ in (5.3-43) and (5.3-44) has been set equal to zero. Comparing (5.3-46) with (5.3-43) and (5.3-44) we see that, since $\theta_{esv}(0)$ is zero, $\tilde{F}_{as}$ may be expressed as

$$\sqrt{2}\tilde{F}_{as} = \left(F_{qs}^r - jF_{ds}^r\right)e^{j\delta} \tag{5.3-47}$$

Following the same procedure as in the case of a round-rotor permanent-magnet ac machine we obtain

$$\tilde{V}_{as} = (r_s + j\omega_e L_{ss})\tilde{I}_{as} + \frac{1}{\sqrt{2}}\omega_e \left(\frac{3}{2}\right) L_{ms} I_{fd}'^r e^{j\delta} \tag{5.3-48}$$

If we compare the last term of (2.4-59) and (5.3-48), we see that $\lambda_m'^r$ is analogous to $\frac{3}{2}L_{ms}I_{fd}'^r$. Also in (2.4-59), the last term, which we called $\tilde{E}_a$, was on the $q$-axis, fixed at zero degrees whereas the last term in (5.3-48), which we are also going to call $\tilde{E}_a$, is also along the $q$-axis but positioned at the angle $\delta$. That is

$$\tilde{E}_a = \frac{1}{\sqrt{2}}\omega_e \left(\frac{3}{2}\right) L_{ms} I_{fd}'^r e^{j\delta} \tag{5.3-49}$$

In the case of the brushless dc drive, we have the capability to change instantaneously the phase relation between the permanent magnet of the rotor and the applied voltages by controlling the phase (shifting) of the applied voltages. The permanent magnet is fixed in the $d$-axis and the $q$-axis of the rotor, and thus $\tilde{E}_a$ is fixed 90° ahead (counterclockwise) of the $d$-axis. In the case of the synchronous generator connected to an electric power grid, however, the stator voltages are fixed in phase and magnitude by the power grid. In this case, the system voltages are the fixed reference, $\tilde{V}_{as}$ is selected at zero degrees, and the $q$-axis and thus $\tilde{E}_a$ shifts relative to $\tilde{V}_{as}$ depending on the operating conditions (loading) of the generator.

The steady-state torque may be expressed from (5.3-29), (5.3-30), and (5.3-33) as

$$T_e = \left(\frac{P}{2}\right)\left(\frac{3}{2}\right)^2 L_{ms} I_{qs}^r I_{fd}^{\prime r} \tag{5.3-50}$$

This relationship can also be obtained by realizing from (5.3-48) that $\lambda_m^{\prime r}$ is analogous to $\frac{3}{2}L_{ms}I_{fd}^{\prime r}$; thus, (5.3-50) can be justified for steady-state operation from (2.5-35).

It is common to neglect $r_s$ in (5.3-35) and (5.3-36), which is a good approximation in the case of a 60 Hz power system since the inductive reactance dominates the machine impedance. If we then solve (5.3-35) and (5.3-36) for $I_{qs}^r$ and substitute it into (5.3-50) we can express the steady-state torque, with $r_s$ neglected, as

$$T_e = -\left(\frac{P}{2}\right)\left(\frac{3}{2}\right)\left(\frac{1}{\omega_e}\right)\left(\frac{1}{X_{ss}}\right)E_x^{\prime}\sqrt{2}V_s \sin\delta \tag{5.3-51}$$

where $\omega_r = \omega_e$, and $E_x^{\prime}$, the excitation voltage, is

$$E_x^{\prime} = \omega_e\left(\frac{3}{2}\right)L_{ms}I_{fd}^{\prime r} = \sqrt{2}\,\left|\tilde{E}_a\right| \tag{5.3-52}$$

From (5.3-51), it is clear that stable steady-state operation occurs for $-\frac{\pi}{2} < \delta < \frac{\pi}{2}$, where motor operation occurs for $-\frac{\pi}{2} < \delta < 0$ and generator operation for $0 < \delta < \frac{\pi}{2}$.

Let us take a few minutes to discuss the operation of a synchronous generator connected to a power system. We will assume that the generator is synchronized with the system and the generator is "floating" on the line. Meaning that there is no input torque to the generator from the prime mover (e.g., steam turbine). The power system is supplying a small amount of power to the generator to satisfy the rotational and ohmic losses; thus, the rotor angle $\delta$ is essentially zero.

In order to supply electrical power to the power system, an input torque must be supplied from the steam turbine by appropriate steam valving. This is done in increments whereupon each incremental increase in input torque causes the rotor speed to increase temporarily, thus increasing the rotor angle $\delta$. This incremental loading continues until the generator is supplying the prescribed amount of power to the power system without exceeding the stable operating range of the generator. Example 5B should help to illustrate typical rated, steady-state operation of a synchronous generator.

It is common to express the torque and rotor speed relationship given by (5.3-34) in terms of rotor angle. From (5.3-38) we can write

$$\begin{aligned}\frac{d\delta}{dt} &= \frac{d\theta_r}{dt} - \frac{d\theta_{esv}}{dt} \\ &= \omega_r - \omega_e\end{aligned} \tag{5.3-53}$$

where $\omega_e$ is the electrical angular velocity of the power system which is generally assumed to be constant. Thus,

$$\frac{d\omega_r}{dt} = \frac{d^2\delta}{dt^2} \tag{5.3-54}$$

Therefore, (5.3-34) can be written as

$$T_e - T_L = J\left(\frac{2}{P}\right)\frac{d^2\delta}{dt^2} \tag{5.3-55}$$

The friction and windage term containing $B_m$ is generally neglected. We see that when the torque output, $T_e$, is greater than the load torque, $T_L$, the machine accelerates. When $T_L > T_e$ it slows.

We have mentioned that the power system engineers assume positive current out of the synchronous generator. With this reversing of the sense of positive current, power and torque are positive for generator action. Thus, the negative sign is deleted from (5.3-51). Also, the first term on the right-hand side of (5.3-48) would be negative.

There is one last thing before considering Example 5B; location of the stator poles. For this purpose we need (2.2-15) which is repeated here as

$$\text{mmf}_s = \frac{N_s}{2}\sqrt{2}I_s\cos[(\omega_e - \omega)t + \theta_{esi}(0) - \theta(0) - \phi] \tag{5.3-56}$$

where $\omega_e$ is the electrical angular velocity of the stator variables, $\omega$ is the electrical angular velocity of the $q$-axis, $\theta(0)$ is the time-zero position of the $q$-axis, and $\phi$ is the displacement from the $q$-axis to the south pole of the stator. Now, two things; first (5.3-56) is for a two-phase stator, but we know that the three-phase differs only by the $\frac{3}{2}$ factor, so we are not going to worry about that. Secondly, in the case of the permanent-magnet ac machine $\theta(0)$ was zero; here $\theta(0)$ is $\delta$. If now we let $\omega = \omega_e$ and $\theta(0) = \delta$ (5.3-56) becomes

$$\text{mmf}_s^e = \frac{N_s}{2}\sqrt{2}I_s\cos(\theta_{esi}(0) - \delta - \phi_e) \tag{5.3-57}$$

We are now moving at $\omega_e$, the location of the south pole due to the stator currents is located at

$$\phi_e = \theta_{esi}(0) - \delta \tag{5.3-58}$$

relative to the $q$-axis or in phase with $\tilde{I}_{as}$ which is rotating at $\omega_e$. We would have probably predicted this from our work in Chapter 2.

**Example 5B. Phasor diagram of steam turbine generator operation.** Consider a steam turbine generator with the following ratings and parameters

Power rating: 835 MVA, line-to-line voltage: 26 kV
Power factor: 0.85, Poles: 2, Speed: 3600 r/min
Inertia: $J = 0.0658 \times 10^6$ J · s$^2$ (turbine and generator)

Resistances and reactances in ohms:

$$r_s = 2.43 \times 10^{-3} \qquad X_{ls} = 0.1538$$

$$\frac{3}{2}X_{ms} = 1.3032 \qquad X_{ss} = 1.457$$

$$r'_{fd} = 7.5 \times 10^{-4} \quad X'_{lfd} = 0.1145$$

For balanced, steady-state, rated operation, calculate $\tilde{E}_a$, $E'_x$, and $T_e$, draw the closed phasor diagram and locate the stator and rotor poles.

From (5.3-48) and since the last term is $\tilde{E}_a$

$$\tilde{E}_a = \tilde{V}_{as} - (r_s + j\omega_e L_{ss})\tilde{I}_{as} \tag{5B-1}$$

The magnitude of the phase current is the MVA per phase divided by the rms value of the phase voltage. According to (5.2-10), the line-to-line voltage divided by $\sqrt{3}$ yields the phase voltage. Therefore,

$$|\tilde{I}_{as}| = \frac{MVA}{3}\left(\frac{1}{|\tilde{V}_{as}|}\right)$$
$$= \frac{835}{3}\left(\frac{\sqrt{3}}{26 \times 10^3}\right) = 18.54\,\text{kA} \tag{5B-2}$$

The power factor is given as 0.85. We must be careful here. If we were dealing with motor operation and positive current into the machine, then $\tilde{I}_{as}$ would be positioned at $\theta_{esi}(0) = \pm 31.8°$ with the time-zero position of $\tilde{V}_{as}$ at zero degrees. The reactive power is said to be absorbed if $\sin[\theta_{esv}(0) - \theta_{esi}(0)]$ is positive with positive current in the direction of a voltage drop. Since a generator will generally deliver reactive power and since we are assuming positive current into the machine, $\theta_{esi}(0) = 180° - 31.8°$ or 148.2°. Whereupon, power is negative (generator action) and reactive power is also negative (producing reactive power).

In calculating $\tilde{E}_a$ using (5B-1), we are neglecting $r_s$ since it is small relative to the inductive reactances. Thus, from (5B-1)

$$\tilde{E}_a = \frac{26 \times 10^3}{\sqrt{3}}\underline{/0°} - (j1.457)18.54 \times 10^3 \underline{/180° - 31.8°}$$
$$= 15.03 \times 10^3 + 26.9 \times 10^3 \underline{/90 - 31.8} = 37.1 \times 10^3 \underline{/38.1°} \tag{5B-3}$$

We can calculate $E'_x$ from (5.3-52). The torque may be calculated from (5.3-51)

$$T_e = -\left(\frac{P}{2}\right)\left(\frac{3}{2}\right)\left(\frac{1}{\omega_e}\right)\left(\frac{1}{X_{ss}}\right)E'_x\sqrt{2}V_s \sin\delta$$
$$= -\left(\frac{2}{2}\right)\left(\frac{3}{2}\right)\left(\frac{1}{377}\right)\left(\frac{1}{1.457}\right)\left(\sqrt{2}37.1 \times 10^3\right)\left(\sqrt{2}15.03 \times 10^3\right)\sin 38.1°$$
$$= -1.88 \times 10^6\ \text{N}\cdot\text{m} \tag{5B-4}$$

Power is being supplied to the electric system by the generator.

The phasor diagram becomes that shown in Fig. 5B-1.

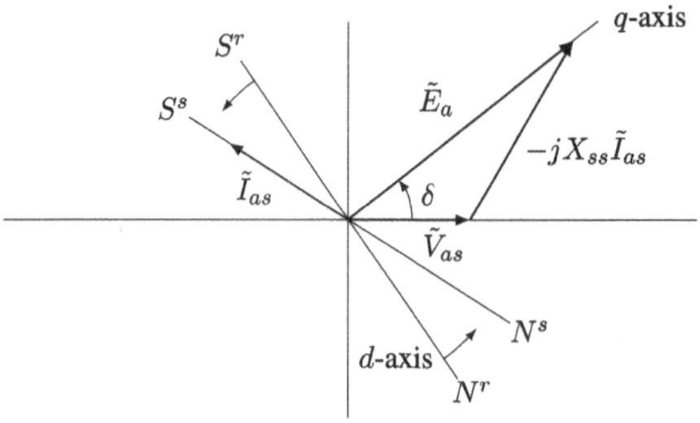


Figure 5B-1 Phasor diagram with $r_s$ neglected for generator action and with the direction of positive current selected to be into the machine.

From Fig. 5B-1 we see that the rotor poles are "pushing" the stator poles in the counterclockwise direction; generator action.

---

**SP5.3-1.** Consider the steam turbine generator given in Example 5B. Determine $\theta_{esi}(0)$ for operation at rated power output with rated power factor and absorbing reactive power. [$\theta_{esi}(0) = 211.8°$]

**SP5.3-2.** Calculate $\tilde{E}_a$ for SP5.3-1 and sketch the phasor diagram. [$\tilde{E}_a = 22.8 \times 10^3 \underline{/87.8°}$ V]

**SP5.3-3.** Calculate $T_e$ using (5.3-51). [$T_e = -1.88 \times 10^6$ N · m]

**SP5.3-4.** Why is this mode of operation undesirable? [$\delta = 87.8°$, near the steady-state power limit]

**SP5.3-5.** Suggest how this undesirable mode of operation can be corrected. [Increase $\tilde{E}_a$ by increasing $I_{fd}$]

## 5.4 REACTIVE POWER AND POWER FACTOR CORRECTION

In Chapter 1, we introduced reactive power which really is not power but an interchange of energy between the capacitances and inductances of the system. That is, energy is stored in the electric field for one-half of the cycle of the sinusoidal current and in the electromagnetic field for the other half. A measure of this energy interchange is calculated using (1.2-32); that is

$$Q = |\tilde{V}||\tilde{I}| \sin[\theta_{ev}(0) - \theta_{ei}(0)] \tag{5.4-1}$$

where positive current is assumed in the direction of voltage drop. Reactive power has the units of volt-ampere reactive (var) and it is said to be "absorbed" if $Q$ is positive and "generated" if $Q$ is negative. In other words, $Q$ is absorbed by an inductance and supplied by a capacitance.

The vast majority of electric power system loads are inductive and the reactive power absorbed by the inductive loads must be generated or supplied by capacitive system elements. How is this "supplying" done? Well, in Example 5B, the steam turbine generator was operating at rated conditions. The phasor diagram depicting this mode of operation is shown in Fig. 5B-1. Let us calculate $Q$ for this operation; here

$$\begin{aligned} Q &= 3(15\text{ kV})(18.5\text{ kA})\sin[0 - 148^\circ] \\ &= -441\text{ Mvar} \end{aligned} \tag{5.4-2}$$

The generator is supplying both real and reactive power; it is "acting" as a capacitor. How does this happen? Well, the per-phase steady-state voltage equation for a synchronous machine is given by (5.3-48), which can be written as

$$\tilde{V}_{as} = jX_{ss}\tilde{I}_{as} + \tilde{E}_a \tag{5.4-3}$$

where the stator resistance, $r_s$, is neglected,

$$X_{ss} = \omega_e L_{ss} \tag{5.4-4}$$

and

$$\tilde{E}_a = \frac{3}{2\sqrt{2}}\omega_e L_{ms} I_{fd}'^r e^{j\delta} \tag{5.4-5}$$

For this explanation, let us assume that the machine is connected to the power system. If there is no external load or input torque to the shaft; the machine is said to be "floating on the line." Actually, the small amount of power being supplied to the machine from the power system is the $|\tilde{I}_s|^2 r_s$ and the small friction and windage losses. We will allow this small violation of a floating machine and assume that the rotor angle to be zero. The phasor diagram for this mode of operation is shown in Fig. 5.4-1. In Fig. 5.4-1a, $\tilde{E}_a(I_{fd}'^r)$ is zero; $\tilde{I}_{as}$ is lagging $\tilde{V}_{as}$ by 90° and $jX_{ss}\tilde{I}_{as}$ is equal

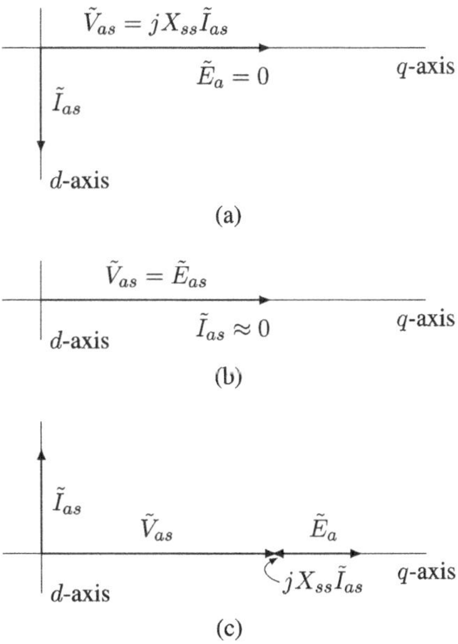


Figure 5.4-1 (a) $I_{fd}'^r(\tilde{E}_a)$ = zero; (b) $\tilde{E}_a = \tilde{V}_{as}$; (c) $\tilde{E}_a > \tilde{V}_{as}$.

to $\tilde{V}_{as}$. The synchronous machine "looks like" an inductance. In Fig. 5.4-1b, $I'^r_{fd}$ has been increased to where $\tilde{I}_{as}$ is negligibly small, sufficient to satisfy the small rotational and ohmic losses. In this case, $\tilde{E}_a$ is nearly equal to $\tilde{V}_{as}$. In Fig. 5.4-1c, $\tilde{E}_a(I'^r_{fd})$ is increased and $\tilde{E}_a > \tilde{V}_{as}$ and, in order to satisfy (5.4-3), $\tilde{I}_{as}$ must lead $\tilde{V}_{as}$ by 90° since we are neglecting $r_s$. Here the machine "looks like" a capacitor, delivering reactive power. The mode of operation of the synchronous generator illustrated in Fig. 5.4-1a is said to be under-excited while the mode of operation shown in Fig. 5.4-1c is over-excited. If $|\tilde{I}_{as}|$ is plotted on the $y$-axis and $I'^r_{fd}$ or $|\tilde{E}_a|$ on the $x$-axis we would obtain the so-called "$V$" curve. Therefore, the synchronous generators connected to the power system are generally over-excited in order to supply reactive power to the inductive power system loads. A synchronous generator operated without a torque input or load, but instead operated solely to supply reactive power is referred to as a *synchronous condenser*.

There is a good reason why synchronous generators are not used to supply all of the reactive power to the inductive loads. We will use Fig. 5.4-2 to help explain this. The elementary transmission system shown in Fig. 5.4-2 could represent single-phase transmission as in rural areas or one phase of a three-phase system. Therein, $\tilde{V}_G$ and $\tilde{I}_G$ can be thought of as phase variables of a three-phase generator. The transmission line resistance and reactance are $r_l$ and $x_l$, respectively. The load is assumed to be a series $RL$; $R_s$ and $X_s$. It is convenient to consider the series load as the parallel equivalent whereupon Fig. 5.4-2 becomes Fig. 5.4-3. In Fig. 5.4-3,

$$R_p = \frac{R_s^2 - (jX_s)^2}{R_s} = \frac{R_s^2 + X_s^2}{R_s} \tag{5.4-6}$$

$$jX_p = -\frac{R_s^2 - (jX_s)^2}{jX_s} = j\frac{(R_s^2 + X_s^2)}{X_s} \tag{5.4-7}$$

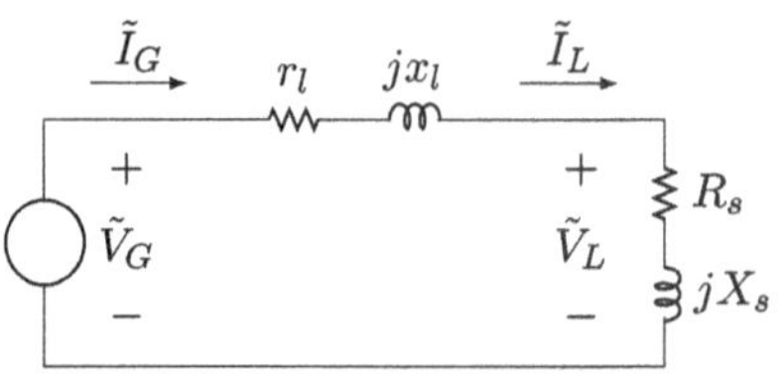


Figure 5.4-2 One-line diagram of an elementary power system with a series $R_s$ and $X_s$ load.

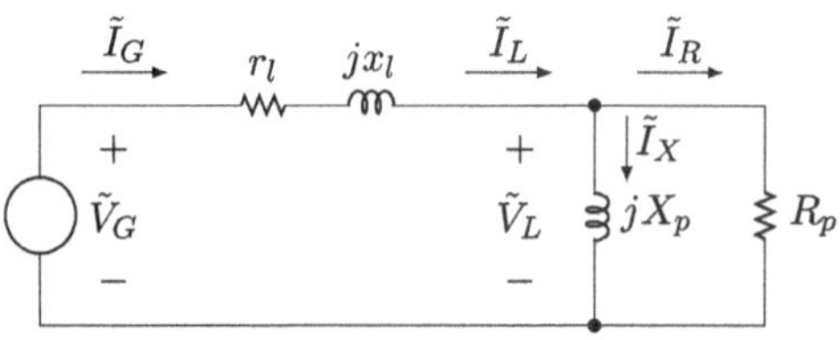


Figure 5.4-3 One-line diagram of an elementary power system with equivalent parallel-connected load.

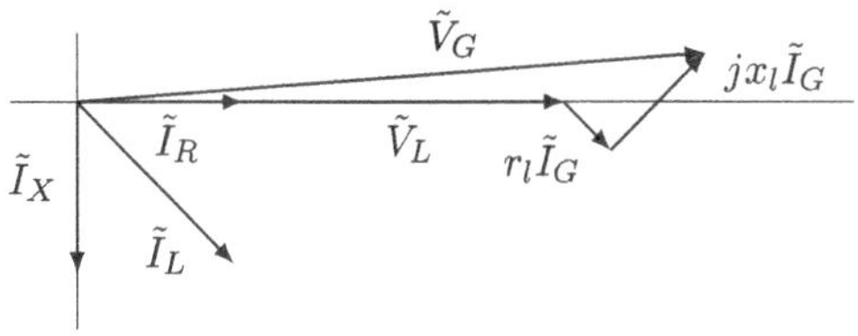


Figure 5.4-4 Phasor diagram for Fig. 5.4-3.

Figure 5.4-4 depicts the phasor diagram for the elementary system given in Fig. 5.4-3. With only the generator supplying the reactive power the transmission line must carry the reactive component, $\tilde{I}_X$, of the load current, $\tilde{I}_L$.

If now we place a capacitor at or near the load, the power system becomes that shown in Fig. 5.4-5 and if we select $X_C$ equal to $X_p$ the phasor diagram becomes Fig. 5.4-6. Since the transmission line carries only the real component of the load current, the voltage drop across the transmission line is reduced and thus $\tilde{V}_L$ is increased. Some three-phase capacitor banks are connected permanently near the load center and others are automatically "switched in" during heavy load conditions when the load voltage has dropped below a preset level. This common means of load voltage support is referred to as power factor correction.

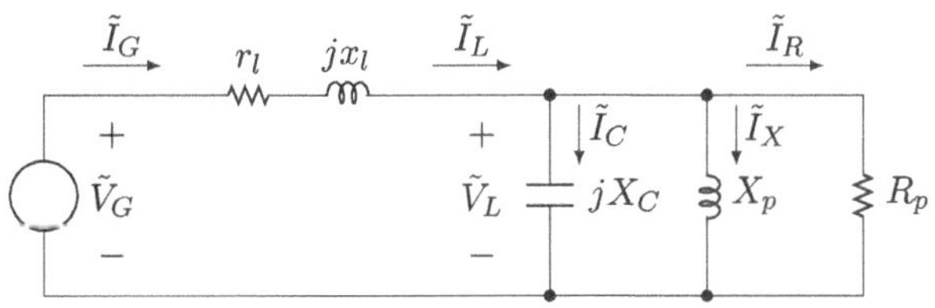


Figure 5.4-5 One-line diagram of an elementary power system with capacitor in parallel with the load – power factor correction.

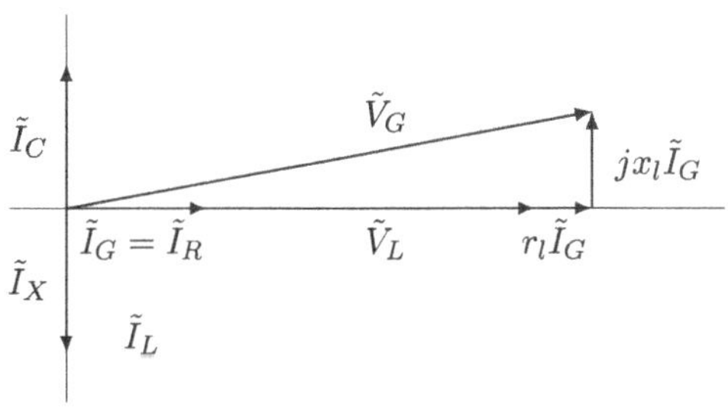


Figure 5.4-6 Phasor diagram for Fig. 5.4-5 with $X_C = X_p$.

**Example 5C. Power factor correction.** The 60 Hz current supplied to a single-phase load is $\tilde{I}_{L1} = 10\underline{/-30^\circ}$ A and the voltage is $110\underline{/0^\circ}$ V. A second load is connected in parallel and the current supplied to this load is $\tilde{I}_{L2} = 20\underline{/-60^\circ}$ A. (a) Determine the parallel equivalent $R_{p1}$, $X_{p1}$, $R_{p2}$, and $X_{p2}$. (b) Determine the capacitance that when connected in parallel with the loads will return the power factor angle to 30°. (c) Calculate the percent decrease in total load current from (a) to (b).

For (a), let us first write the load currents as

$$\tilde{I}_{L1} = 8.66 - j5\text{ A} \tag{5C-1}$$

$$\tilde{I}_{L2} = 10 - j17.3\text{ A} \tag{5C-2}$$

Now,

$$R_{p1} = \frac{110}{8.66} = 12.7\,\Omega \tag{5C-3}$$

$$X_{p1} = \frac{110}{5} = 22\,\Omega\text{ (inductive)} \tag{5C-4}$$

$$R_{p2} = \frac{110}{10} = 11\,\Omega \tag{5C-5}$$

$$X_{p2} = \frac{110}{17.3} = 6.3\,\Omega\text{ (inductive)} \tag{5C-6}$$

We could have used (5.4-6) and (5.4-7) to calculate $R_p$ and $X_p$. For example,

$$\begin{aligned} Z_{L1} &= \frac{\tilde{V}}{\tilde{I}_{L1}} \\ &= \frac{110\underline{/0^\circ}}{10\underline{/-30^\circ}} = 11\underline{/30^\circ} = 9.5 + j5.5\,\Omega \end{aligned} \tag{5C-7}$$

Substituting into (5.4-6)

$$\begin{aligned} R_{p1} &= \frac{R_s^2 + X_s^2}{R_s} \\ &= \frac{(9.5)^2 + (5.5)^2}{9.5} = 12.7\,\Omega \end{aligned} \tag{5C-8}$$

From (5C-1) and (5C-2) the total load current is

$$\begin{aligned} \tilde{I}_{LTa} &= \tilde{I}_{L1} + \tilde{I}_{L2} \\ &= (8.66 + 10) - j(5 + 17.3) \\ &= 18.66 - j22.3\text{ A} \end{aligned} \tag{5C-9}$$

For (b) we want the total load current, $I_{LTb}$, to have a phase angle of −30°, thus

$$|\tilde{I}_{LTb}| = \frac{18.66}{\cos 30^\circ} = 21.55\text{ A} \tag{5C-10}$$

and

$$\begin{aligned} \tilde{I}_{LTb} &= 18.66 - j21.55\sin 30^\circ \\ &= 18.66 - j10.8\text{ A} \end{aligned} \tag{5C-11}$$

Therefore, the parallel capacitor must "supply"

$$|\tilde{I}_{cap}| = 22.3 - 10.8 = 11.5\text{ A} \tag{5C-12}$$

Whereupon,

$$Z_{cap} = \frac{|\tilde{V}|}{|\tilde{I}_{cap}|} = \frac{110}{11.5} = 9.57\,\Omega \tag{5C-13}$$

Therefore, the capacitance is

$$C = \frac{1}{Z\omega_e} = \frac{1}{(9.57)(377)} = 277\,\mu\text{F} \tag{5C-14}$$

For (c), the magnitude of the load current for (a) from (5C-9) is

$$|\tilde{I}_{LTa}| = (18.66^2 + 22.3^2)^{\frac{1}{2}} = 29.1\,\text{A} \tag{5C-15}$$

From (5C-11)

$$|\tilde{I}_{LTb}| = (18.66^2 + 10.8^2)^{\frac{1}{2}} = 21.6\,\text{A} \tag{5C-16}$$

The percent decrease in the magnitude of the load (line) current is

$$\%\ \text{decrease} = \left(\frac{29.1 - 21.6}{29.1}\right)100 = 25.8\% \tag{5C-17}$$

Although this example is perhaps academic, a similar situation occurred when in the mid-1950s, the eldest author lived with his young family in veteran's housing at a distance from the power source. Window air conditioners were being installed by many residents which were cheaply made with a large power factor. The large reactive load current caused the voltage at the housing area (World War II converted hospital barracks) to drop to where the motor in the air conditioners would stop; unable to handle the load of the fan and compressor. It was a hot Nebraska summer. Finally, banks of capacitors were installed by the utility. Later, restrictions were placed on air conditioner manufacturers to improve the power factor.

---

**SP5.4-1.** Show that (5.4-1) is $|\tilde{I}|^2X$ for a series resistive and reactive circuit.

**SP5.4-2.** Derive (5.4-6) and (5.4-7).

**SP5.4-3.** Repeat Example 5C(b) if $I_{LTb} = 18.66\underline{/0^\circ}$ A.[C = 538 μF].

**SP5.4-4.** Locate the stator and rotor poles in Fig. 5.4-1. [$N^r$ at $-\frac{\pi}{2}$ in (b) and (c). No rotor poles in (a); (a) $S^s$ at $-\frac{\pi}{2}$; (b) $I_{as}$ assumed zero; (c) $S^s$ at $\frac{\pi}{2}$]

## 5.5 PER UNIT SYSTEM

The per unit system is a means of normalizing the system parameters and variables to a specified base power for the entire system being considered. The base voltage is generally selected as the rated voltage of each system component: generator, transmission line, and load. The power system engineers are especially fond of the per unit system since it is far more convenient to express the voltage of the synchronous machine in Example 5B as one per unit rather than 26,000 V or 26 kV. More importantly, it allows a means of dealing with different voltage levels with all transformers considered ideal. Base power for a power system would generally be much larger than that of a single machine and it is generally given in volt-amperes rather than watts. If we were to per unitize just the synchronous machine in Example 5B, the power base would be 835 MVA and the line-to-line voltage base would be 26 kV. The base phase voltage would be $\frac{1}{\sqrt{3}}(26\text{ kV})$. Since the 26 kV is an rms value, base peak voltage would be $\sqrt{2}$ times the rms value. Once the base power is specified, the base current will depend upon the voltage base selected for each system component. This can be confusing if the base voltage is not defined as either a phase or line-to-line voltage; and either rms or peak value.

The parameters of the power system and of the machine are commonly per-phase quantities. With the phase rms voltage and current as base quantities, base power may be expressed as

$$P_{\text{base}} = 3V_{\text{base}}I_{\text{base}} \tag{5.5-1}$$

where $P_{\text{base}}$ is base system power and $V_{\text{base}}$ is the rated phase voltage of the system component being considered. The impedance base is then

$$Z_{\text{base}} = \frac{V_{\text{base}}}{I_{\text{base}}} = \frac{3V_{\text{base}}^2}{P_{\text{base}}} \tag{5.5-2}$$

where $I_{\text{base}}$ is determined from (5.5-1) once $V_{\text{base}}$ is established. Although $P_{\text{base}}$ is the same throughout the system being considered, $V_{\text{base}}$ and $I_{\text{base}}$ will change for each system voltage level.

Base torque is

$$T_{\text{base}} = \frac{P_{\text{base}}}{\left(\frac{2}{P}\right)\omega_{\text{base}}} \tag{5.5-3}$$

where $\omega_{\text{base}}$ is the electrical angular velocity corresponding to rated frequency. If the interest is only in per unitizing the power system and not the synchronous generator, we can stop here and skip to Example 5D and follow through part (a).

## Per Unitizing the Synchronous Generator

We will consider the synchronous machine and play power engineer by selecting positive current out of the machine. With positive power out of the machine, $T_e$ is now positive for generator action. When using the per unit system, it is common to express the flux linkage equations as flux linkages per second by multiplying (5.3-29) and (5.3-30) by $\omega_{\text{base}}$ ($\omega_e$). Thus,

$$\begin{aligned}\psi_{qs}^r &= \omega_e \lambda_{qs}^r \\ &= -X_{ls} i_{qs}^r - \frac{3}{2} X_{ms} i_{qs}^r \end{aligned} \tag{5.5-4}$$

$$\begin{aligned}\psi_{ds}^r &= \omega_e \lambda_{ds}^r \\ &= -X_{ls} i_{ds}^r + \frac{3}{2} X_{ms}\left(-i_{ds}^r + i_{fd}^r\right)\end{aligned} \tag{5.5-5}$$

For the power system engineer, the voltage equations (5.3-27) and (5.3-28) become

$$v_{qs}^r = -r_s i_{qs}^r + \frac{\omega_r}{\omega_{\text{base}}} \psi_{ds}^r + \frac{p}{\omega_{\text{base}}} \psi_{qs}^r \tag{5.5-6}$$

$$v_{ds}^r = -r_s i_{ds}^r - \frac{\omega_r}{\omega_{\text{base}}} \psi_{qs}^r + \frac{p}{\omega_{\text{base}}} \psi_{ds}^r \tag{5.5-7}$$

We are ready to per unitize these equations; for $\psi_{qs}^r$, we will normalize to $I_{\text{base}} Z_{\text{base}}$. Although $P_{\text{base}}$ is the same as before, the base voltage and current become

$$V_{\text{base}} = \sqrt{2}\, V_{\text{rms}} \tag{5.5-8}$$

$$I_{\text{base}} = \sqrt{2}\, I_{\text{rms}} \tag{5.5-9}$$

Thus, in per unit,

$$\frac{\psi_{qs}^r}{I_{\text{base}} Z_{\text{base}}} = -\frac{X_{ls}}{Z_{\text{base}}} \frac{i_{qs}^r}{\sqrt{2} I_{\text{rms}}} - \frac{\frac{3}{2} X_{ms}}{Z_{\text{base}}} \frac{i_{qs}^r}{\sqrt{2} I_{\text{rms}}} \tag{5.5-10}$$

We now see why we are using inductive reactances rather than inductances in the flux linkage equations; however, there could be confusion here. We would like to simply rewrite (5.5-4) and (5.5-5) and state that they are now in per unit. We cannot do that since it is customary to per unitize $\frac{3}{2} X_{ms}$ rather than $X_{ms}$. To get around this let us define

$$X_{MS} = \frac{3}{2} X_{ms} \tag{5.5-11}$$

then we will say that the per unitized equations of (5.5-4) and (5.5-5) are

$$\psi_{qs}^r = -X_{ls} i_{qs}^r - X_{MS} i_{qs}^r \tag{5.5-12}$$

$$\psi_{ds}^r = -X_{ls} i_{ds}^r + X_{MS}\left(-i_{ds}^r + i_{fd}^r\right) \tag{5.5-13}$$

where we understand that all reactances and currents have been per unitized. Perhaps the upper case *MS* subscript will help to remind us that we are in per unit.

Since we have per unitized $\psi_{qs}^r$ and $\psi_{ds}^r$, we need only to per unitize the *ri* term. For $v_{qs}^r$ given by (5.5-6),

$$\frac{v_{qs}^r}{\sqrt{2} V_{\text{rms}}} = -\frac{r_s}{Z_{\text{base}}} \frac{i_{qs}^r}{\sqrt{2} I_{\text{rms}}} + \frac{\omega_r}{\omega_{\text{base}}} \psi_{ds}^r + \frac{p}{\omega_{\text{base}}} \psi_{qs}^r \tag{5.5-14}$$

Since the per unit voltage equations are of the same form as (5.5-6) and (5.5-7), we will give notice when we are dealing in per unit which is nearly always when we are power system engineers.

We have yet to per unitize the torque and the torque-rotor angle relation. From (5.3-33) per unit torque, positive for generator action (positive currents out of machine), becomes

$$\frac{T_e}{T_{\text{base}}} = \left(\frac{3}{2}\right)\left(\frac{P}{2}\right)\left(\lambda^r_{ds}i^r_{qs} - \lambda^r_{qs}i^r_{ds}\right)\left(\frac{2}{P}\right)\frac{\omega_{\text{base}}}{P_{\text{base}}}$$

$$= \left(\frac{3}{2}\right)\frac{\omega_{\text{base}}\lambda^r_{ds}i^r_{qs}}{\left(\frac{3}{2}\right)V_{\text{base}}I_{\text{base}}} - \left(\frac{3}{2}\right)\frac{\omega_{\text{base}}\lambda^r_{qs}i^r_{ds}}{\left(\frac{3}{2}\right)V_{\text{base}}I_{\text{base}}} \tag{5.5-15}$$

Now, since we are dealing with $qs^r$- and $ds^r$-variables, base voltage and current are given by (5.5-8) and (5.5-9), respectively. This explains the $\frac{3}{2}$ factor in base power. Let us rewrite (5.5-15) as

$$\frac{T_e}{T_{\text{base}}} = \frac{\omega_{\text{base}}\lambda^r_{ds}}{Z_{\text{base}}I_{\text{base}}}\frac{i^r_{qs}}{I_{\text{base}}} - \frac{\omega_{\text{base}}\lambda^r_{qs}}{Z_{\text{base}}I_{\text{base}}}\frac{i^r_{ds}}{I_{\text{base}}} \tag{5.5-16}$$

Therefore, per unit torque, positive for generator action, is

$$T_e = \psi^r_{ds}i^r_{qs} - \psi^r_{qs}i^r_{ds} \tag{5.5-17}$$

The steady-state power output $P_e$ may be expressed in per unit by starting with (5.3-51) and realizing that for positive current out of the machine the negative sign on the right-hand side of (5.3-51) becomes positive and stable generator operation occurs for $0 < \delta < \frac{\pi}{2}$. Also, we must recall that steady-state power is torque times the actual rotor speed $\omega_{rm}$. The per unit electrical power output, positive for generator action, becomes

$$P_e = \frac{|\tilde{E}_a|\,V_s}{X_{ss}}\sin\delta \tag{5.5-18}$$

You are asked to show this in a problem at the end of this chapter.

The torque-rotor angle relationship is given by (5.3-55), where we have neglected the term containing $B_m$ since this term is very small in comparison to the power level of the power system. Thus, for generator action, (5.3-55) becomes

$$T_e = -J\left(\frac{2}{P}\right)p^2\delta + T_I \tag{5.5-19}$$

where $p^2 = \frac{d^2}{dt^2}$ and $T_e$ is positive for generator action and $T_I$ is positive for an input torque to the generator shaft from the prime mover (steam turbine). Now, to per unitize (5.5-19) we can write

$$\frac{T_e}{T_{\text{base}}} = -\frac{J\left(\frac{2}{P}\right)p^2\delta}{T_{\text{base}}} + \frac{T_I}{T_{\text{base}}} \tag{5.5-20}$$

The power system engineer prefers to express (5.5-20) in per unit as

$$T_e = -\frac{2H}{\omega_{\text{base}}} p^2 \delta + T_I \tag{5.5-21}$$

where, in seconds

$$H = \left(\frac{1}{2}\right)\left(\frac{2}{P}\right)\frac{J\omega_{\text{base}}}{T_{\text{base}}} \tag{5.5-22}$$

**Example 5D.** (a) Consider the elementary power system shown in Fig. 5D-1. Use $P_{\text{base}} = 1000$ MVA and determine $V_{\text{base}}$, $I_{\text{base}}$, and $Z_{\text{base}}$ for each of the voltage levels. (b) Per unitize the parameters of the steam turbine generator given in Example 5B. Use $P_{\text{base}} = 835$ MVA with the rms phase voltage as the base voltage.

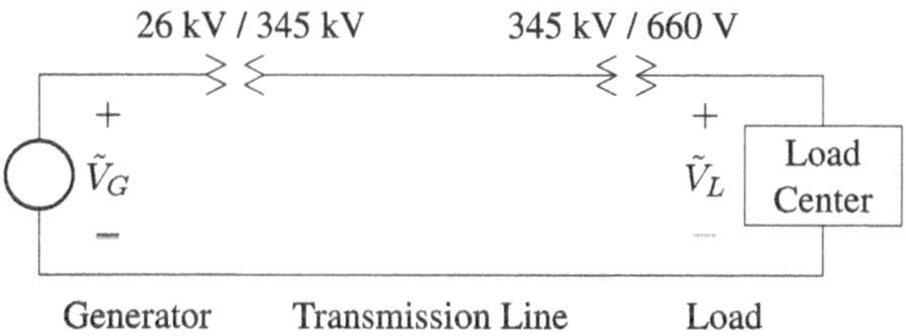


Figure 5D-1 One phase of an elementary power system for part (a). Voltage levels are given in line-to-line values.

For part (a), since the ratings are given in line-to-line, the phase voltage ratings ($V_{\text{base}}$) are obtained by dividing by $\sqrt{3}$. The power base per phase is $\frac{1000}{3}$ MVA or 333 MVA.

For the generator given in Example 5B,

$$V_{\text{base}} = \frac{26\text{ kV}}{\sqrt{3}} = 15\text{ kV} \tag{5D-1}$$

$$I_{\text{base}} = \frac{333 \times 10^6}{(15 \times 10^3)} = 22\text{ kA} \tag{5D-2}$$

$$Z_{\text{base}} = \frac{V_{\text{base}}}{I_{\text{base}}} = \frac{15\text{ kV}}{22\text{ kA}} = 0.682\ \Omega \tag{5D-3}$$

Note that we are using $P_{\text{base}}$ different than the machine rating. For the transmission line,

$$V_{\text{base}} = \frac{345\text{ kV}}{\sqrt{3}} = 199\text{ kV} \tag{5D-4}$$

$$I_{\text{base}} = \frac{333 \times 10^6}{199 \times 10^3} = 1.673\text{ kA} \tag{5D-5}$$

$$Z_{\text{base}} = \frac{199\text{ kV}}{1.673\text{ kA}} = 118.9\ \Omega \tag{5D-6}$$

For the load,

$$V_{\text{base}} = \frac{660}{\sqrt{3}} = 381\,\text{V} \tag{5D-7}$$

$$I_{\text{base}} = \frac{333 \times 10^6}{382} = 874\,\text{kA} \tag{5D-8}$$

$$Z_{\text{base}} = \frac{382}{872 \times 10^3} = 0.437 \times 10^{-3}\,\Omega \tag{5D-9}$$

For (b)

$$V_{\text{base}} = \frac{1}{\sqrt{3}} 26\,\text{kV} = 15.01\,\text{kV} \tag{5D-10}$$

$$\begin{aligned} I_{\text{base}} &= \frac{P_{\text{base}}}{3V_{\text{base}}} \\ &= \frac{835\,\text{MVA}}{(3)15.01\,\text{kV}} = 18.5\,\text{kA} \end{aligned} \tag{5D-11}$$

$$\begin{aligned} Z_{\text{base}} &= \frac{V_{\text{base}}}{I_{\text{base}}} \\ &= \frac{15.01\,\text{kV}}{18.3\,\text{kA}} = 0.812\,\Omega \end{aligned} \tag{5D-12}$$

$$\begin{aligned} T_{\text{base}} &= \frac{P_{\text{base}}}{\left(\frac{P}{2}\right)\omega_{\text{base}}} \\ &= \frac{835\,\text{MVA}}{\left(\frac{2}{2}\right)(377)} = 2.215 \times 10^6\,\text{N}\cdot\text{m} \end{aligned} \tag{5D-13}$$

The per unit parameters are obtained by dividing the parameters given in ohms in Example 5B by $Z_{\text{base}}$. The following listing gives these parameters in ohms and in per unit (pu)

$$r_s = 2.43 \times 10^{-3}\,\Omega, 3 \times 10^{-3}\,\text{pu} \qquad X_{ls} = 0.1538\,\Omega,\ 0.189\,\text{pu}$$

$$\frac{3}{2}X_{ms} = 1.3032\,\Omega, 1.6\,\text{pu} \qquad X_{ss} = 1.457\,\Omega, 1.797\,\text{pu}$$

$$r'_{fd} = 7.5 \times 10^{-4}\,\Omega, 9.248 \times 10^{-4}\,\text{pu} \quad X'_{lfd} = 0.1145\,\Omega, 0.141\,\text{pu}$$

In part (a) we used $P_{\text{base}} = 1000\,\text{MVA}$, the above parameters are per unitized on a $P_{\text{base}}$ of 835 MVA. To convert the above per unit values to a 1000 MVA base, we would multiply by 1000/835 if $V_{\text{base}}$ remains the same. Now from (5.5-22) with $J = 0.0658 \times 10^6\,\text{J}\cdot\text{s}^2$

$$\begin{aligned} H &= \left(\frac{1}{2}\right)\left(\frac{2}{2}\right)\frac{J\omega_{\text{base}}}{T_{\text{base}}} \\ &= \left(\frac{1}{2}\right)\frac{0.0658 \times 10^6(377)}{2.215 \times 10^6} = 5.6\,\text{s} \end{aligned} \tag{5D-14}$$

**SP5.5-1.** Some prefer to use per unit inductance rather than per unit reactance. Determine base inductance for Example 5D part (b). [$L_{base} = 2.15$ mH]

**SP5.5-2.** If the rms phase voltage is used to per unitize $qs^r$- and $ds^r$-voltage, determine the per unit value of maximum $v^r_{qs}$ or $v^r_{ds}$. [$\sqrt{2}$ pu]

## 5.6 DISCUSSION OF TRANSIENT STABILITY

Large-excursion stability, which is commonly referred to as transient stability, is of major concern to the power system engineer. Although we will not get too involved, we can use the material in Section 5.3 to describe some of the major aspects of this phenomenon. As we have mentioned, the vast majority of electric power is generated by synchronous generators and in order for multiple generators to supply power to an electric system all of the generators must be synchronized with the electric system; in other words, all synchronous generators connected to an electric system must rotate at the same electrical angular velocity. Large areas of the power grid of the United States are "synchronously tied together." Since there are thousands of synchronous generators connected to a large power grid, we can understand the concern of the power system engineer.

### Three-Phase Fault

Our purpose is to describe large-excursion stability. To do this we will consider the system in Fig. 5.6-1. Although this example may be considered academic, it serves our need without becoming overly involved. Initially, the steam turbine generator that we considered in Example 5B is delivering rated MVA (mega volt-ampere) at rated power factor with essentially all of the power being delivered to the power grid. A three-phase fault occurs on the line feeding the negligibly small local load near the generator. This is simulated by stepping the three-phase voltages at the bus to zero. These voltages are held at zero until the circuit breaker, CB, opens, clearing the fault from the bus. Whereupon it is assumed that the voltage at the bus steps back to normal magnitude, phase, and frequency due to the robustness or "stiffness" of the large system.

Figures 5.6-2 and 5.6-3 illustrate the dynamic performance of the steam turbine generator considered in Example 5B during and following the three-phase fault.

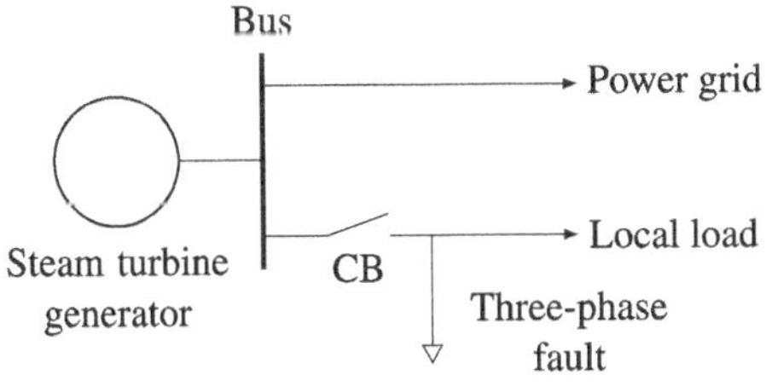


Figure 5.6-1 One-line diagram of system configuration for three-phase fault.

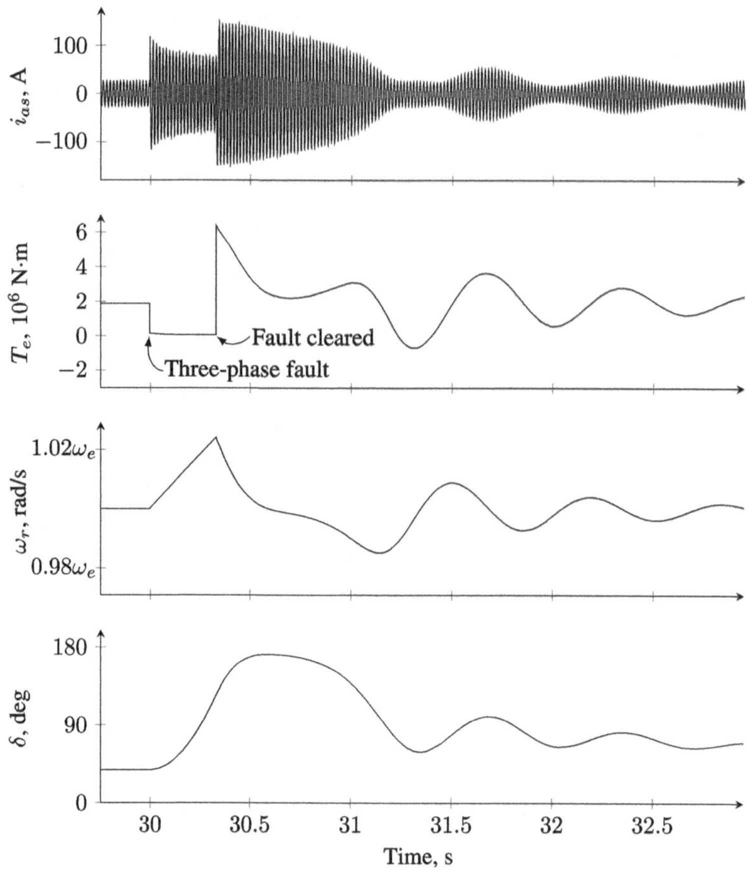


Figure 5.6-2 Dynamic performance of the steam turbine generator during a three-phase fault at the terminals predicted with stator electric transients neglected.

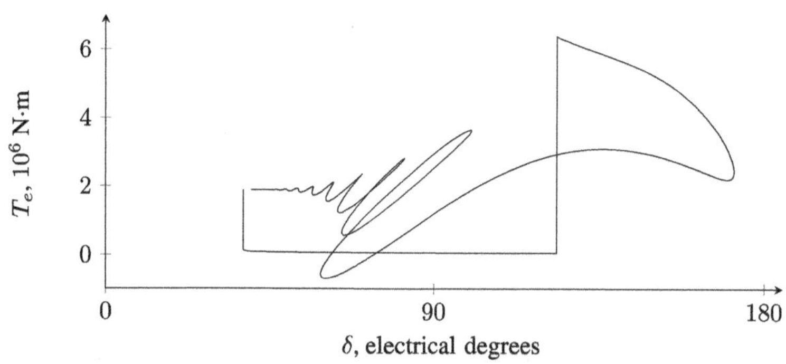


Figure 5.6-3 Torque versus angle characteristics for the study shown in Fig. 5.6-2.

There are a host of things that we need to explain regarding these computer traces. In Fig. 5.6-2, $i_{as}$, $T_e$, $\omega_r$, and the power angle $\delta$ are plotted. From Example 5B, the initial $\delta$ is $\delta_0 = 38.1°$.

The dynamic torque-angle characteristics during and following the fault are illustrated in Fig. 5.6-3. The generator is initially operating in the steady state with the torque input $T_I = (0.85)2.22 \times 10^6$ N · m and $E'_x = (2.48)\sqrt{\frac{2}{3}}26$ kV. Recall we are now power system engineers and we changed the assumed direction of positive current; thus, the input torque is $T_I$ and it is positive for generator action. With this change, stable steady-state operation occurs on the positive-slope part of the $T_e$ versus $\delta$ plot ($-90° < \delta < 90°$); however, positive $T_e$ occurs for $0 < \delta < 90°$.

In the more sophisticated present-day transient stability programs, the $q$- and $d$-circuits are used for the machines and the transmission network with the machines represented in their rotor reference frame and the network represented in the synchronously rotating reference frame. The electric transients are neglected in the stator windings of the machines and in the network, thereby, eliminating the need to integrate for each inductance in the stator and network circuit equations while maintaining sufficient accuracy for transient stability studies [2]. If the electric transients were taken into account a rapidly decaying dc offset would be present in stator and network currents following an electrical disturbance. The decaying dc offset in the stator currents gives rise to a short term 60 Hz decaying pulsation in $T_e$.

As we mentioned in Section 5.3, damper windings, which are short-circuited rotor windings, are needed to damp rotor oscillations after a disturbance. Although we included these windings in the simulation of the steam turbine generator for studies shown in Figs. 5.6-2 and 5.6-3, we will not include these windings in our analysis here since the effects of damper windings on transient stability is minimal.

Now, (5.3-38) defines $\delta$ and as we have mentioned, the power system is large and is often considered to have infinite electrical and mechanical inertia. Although the generator connected to the power system bus can deliver power to the system, the voltage and frequency of a large power system is essentially independent of the generator operation. This large power system bus is often referred to as an "infinite bus." Now, it can be argued that no such thing exists; clearly, a micro-grid is not an infinite inertia system; nevertheless, we will assume an infinite bus in this discussion whereupon $\theta_{esv}$ is constant and since we are now power system engineers, (5.3-55) becomes

$$T_I - T_e = J\left(\frac{2}{P}\right)\frac{d^2\delta}{dt^2} \tag{5.6-1}$$

When the input torque, $T_I$, from the steam turbine is greater than the output torque, $T_e$, or power output, the machine accelerates; when $T_e > T_I$, it slows. Equation (5.6-1) is often referred to as the "swing equation." It is the dynamic relationship between the "swing" of the rotor of a generator relative to a large power system. That is, each generator connected to the power system would have its own swing equation made up of its own $T_e$, $T_I$, inertia, and rotor angle $\delta$.

Let us now get back to Figs. 5.6-2 and 5.6-3. As mentioned, the three-phase fault is simulated by stepping the terminal voltages to zero; therefore, power cannot be transferred from the generator to the power grid. Thus, the input torque $T_I$

accelerates the rotor as shown in the plot of $\omega_r$ in Fig. 5.6-2 where $\omega_e$, in the plot of $\omega_r$, is the electrical angular velocity of the power system. The fault is removed by opening the circuit breakers to the local load in 0.334 s whereupon the machine is reconnected to the power system and the oscillations in rotor speed subside due to the action of the damper windings. Had the fault been allowed to remain slightly longer and $T_e$ had become less than $T_I$, the machine would again accelerate and would not have returned to synchronous operation with the grid without going through a time-consuming resynchronization, or by what is called "slipping poles," which is not a common practice since system and/or machine damage can result.

Now, we should mention that a sudden three-phase fault is rare and a "bolted" three-phase fault near the machine is even rarer; however, it is often used in transient stability studies since it is a measure of the robustness of the system to faults. The dynamic torque-angle characteristics due to the three-phase fault shown in Fig. 5.6-3 are shown in the plot of $T_e$ versus $\delta$ from Fig. 5.6-2. Now, we understand that the torque or power from the machine is essentially zero during the fault since the voltage is zero. The goal is to clear the fault, reconnect the machine to the grid, and maintain synchronization of the machine with the grid. As we mentioned, synchronism is maintained if the fault is isolated by opening the circuit breaker feeding the fault in 0.334 s. Thus, the so-called "critical-clearing time" is 0.334 s and the "critical-clearing angle" which is the rotor displacement at the critical-clearing angle is approximately 128 electrical degrees.

Let us "fast-backwards" to the early twentieth century when power systems were being built and the electric industry was growing rapidly and yet computers were years away. It appears that even at the start of World War II, there were less than twenty integrators available in the United States and these were mechanical. Nevertheless, with only slide rules, the engineers were faced with the problem of predicting transient stability. Fortunately, R.E. Doherty and C.A. Nickle derived an approximate transient torque-angle characteristic that was sufficiently accurate when used with "equal-area criterion" to predict transient stability [1]. We are not going to dwell on a technique that has long since given way to modern-day computation; however, it is interesting that Doherty and Nickle's method is based on the fact that during the "first swing" of the rotor, which determines stable or unstable operation, the flux linkages of circuits that are largely inductive with a relative small resistance will tend to remain constant. Starting with this principle, they derived a transient torque-angle curve for the first rotor swing that was easy to determine and use. We should also mention in passing that during the 1930s and 1940s, point-to-point integration for transient stability studies was conducted using a "network calculator" which was a physical mimic of the main high voltage part of the power system. This was an inconvenient time-consuming device that rapidly gave way to the digital computer in the 1960s.

**Example 5E. Calculation of $\omega_r$ and $\delta$ at the time of clearing the fault.** Consider Figs. 5.6-2 and 5.6-3. Calculate the rotor speed $\omega_r$, and rotor angle $\delta$, at the clearing time $t_c$ of 0.334 s after the occurrence of the three-phase bolted fault when the synchronous generator is reconnected to the system. The input torque is

$$T_I = (0.85) \times 2.22 \times 10^6 = 1.89 \times 10^6 \text{ N} \cdot \text{m} \tag{5E-1}$$

This input shaft torque remains on the generator during the duration of the fault, 0.334 s. For generator action, the power engineers write (5.3-34) as

$$T_I - T_e = J\left(\frac{2}{P}\right)\frac{d\omega_r}{dt} \tag{5E-2}$$

where $T_L$ has become $-T_I$, $T_e$ has become positive for generation action, and $B_m$ is neglected. Solving for $\omega_r$ yields

$$\begin{aligned}\omega_r &= \frac{P}{2J}\int (T_I - T_e)dt \\ &= \frac{P}{2J}\int_0^{t_c}(T_I - T_e)d\xi\end{aligned} \tag{5E-3}$$

where $t_c$ is the clearing time, $J = 0.0658 \times 10^6$, and $P = 2$. Now, $T_e$ is essentially zero during the time the fault is on the system; thus, (5E-3) becomes

$$\begin{aligned}\omega_r &= \frac{(2)(1.89 \times 10^6)}{(2)(0.0658 \times 10^6)}\xi|_0^{0.334} + 377 \\ &= 9.59 + 377 = 386.6\,\text{rad/s}\end{aligned} \tag{5E-4}$$

This is 1.025 $\omega_e$.

Now, (5.6-1) may be written as

$$\frac{d^2\delta}{dt^2} = \frac{P}{2J}(T_I - T_e) \tag{5E-5}$$

From Example 5B, the initial rotor angle is $\delta_0 = 38.1°$ and during the fault $T_e = 0$; thus

$$\begin{aligned}\delta &= \frac{P}{2J}\int_0^{t_c}\int_0^{t} T_I d\tau dt + \delta_0 \\ &= \left(\frac{P}{2J}\right)(T_I)\int_0^{t_c} t dt + \delta_0 \\ &= \left[\left(\frac{P}{2J}\right)\frac{T_I}{2}\xi^2|_0^{0.334}\right]\frac{360°}{2\pi} + \delta_0 \\ &= \left[\frac{2(1.89 \times 10^6)(0.334)^2}{(2)(0.0658 \times 10^6)(2)}\right]\frac{360°}{2\pi} + 38.1° = 129.9°\end{aligned} \tag{5E-6}$$

Note that $t_c$ is the clearing time and that the $\frac{360°}{2\pi}$ multiplier is to convert radians to degrees.

---

**SP5.6-1.** Why is $T_e$ during the three-phase fault not exactly zero? $[3|\tilde{I}_{as}|^2 r_s]$

**SP5.6-2.** Assume $|\tilde{I}_{as}| = 40\,\text{kA}$ during the fault. Calculate $T_e$. $[0.031 \times 10^6\ \text{N} \cdot \text{m}]$

## REFERENCES

[1] R. E. Doherty and C. A. Nickle, "Synchronous machines—III, Torque-angle characteristics under transient conditions," *AIEE Transactions*, vol. 46, January 1927, pp. 1–8.

[2] P. C. Krause, F Nozari, T. L. Skvarenina, and D. W. Olive, "The theory of neglecting stator transients," *IEEE Transactions on Power Apparatus and Systems*, vol. 98, January/February 1979, pp. 141–148.

## PROBLEMS

1. Consider the steam turbine generator in Example 5B. Assume it is operating as a motor with rated power input and rated leading power factor. Draw the phasor diagram with all the quantities that are shown in Fig. 5B-1.

2. Repeat Problem 1 for motor action with rated power factor lagging.

3. Consider Example 5C. Rework part (a) in per unit with the power base of 5.5 kVA. Check your work by comparing with the ohmic values given by (5C-3) through (5C-6) and current value given by (5C-9).

4. The transmission line parameters in Fig. 5D-1 are $r_t = 0.5\,\Omega$ and $x_t = 10\,\Omega$ (inductive). The load current is $755 - j436$ kA at rated phase voltage of $382\underline{/0^\circ}$ V. Per unitize the elementary power system using the base values determined in Example 5D part (a). Use the system turbine generator considered Example 5D, part (b) as the source.

5. The elementary power system shown in Fig. 5D-1 is shown in per unit on a 835 MVA three-phase power base in Fig. 5.8-1 below. (a) Calculate $\tilde{V}_L$. (b) Capacitors are distributed throughout the load center so that the load appears only resistive. Recalculate $\tilde{V}_L$.

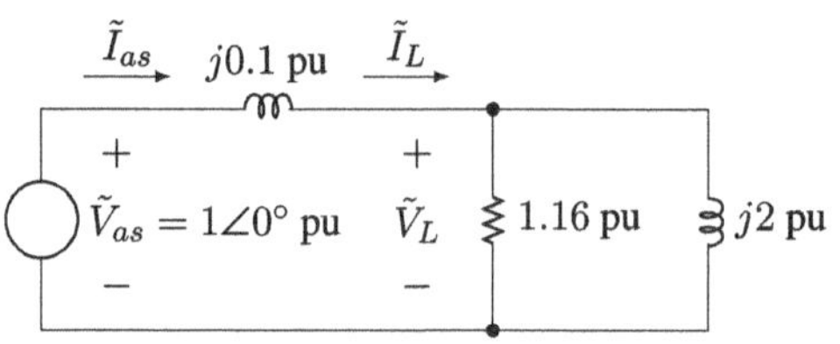


Figure 5.8-1 Per unit of Fig. 5D-1.

6. Assume that in part (a) of Problem 5 $\tilde{I}_{as} = 1\underline{/-30^\circ}$ A and in part (b), $\tilde{I}_{as} = 0.866\underline{/0^\circ}$ A. Calculate per unit $\tilde{E}_a$ and $P_e$ for both cases.

7. Start with (5.3-51) and obtain the expression for steady-state electrical output power in per unit given by (5.5-18).

APPENDIX A

# TRIGONOMETRIC RELATIONS, CONSTANTS AND CONVERSION FACTORS, AND ABBREVIATIONS

## A.1 BASIC TRIGONOMETRIC RELATIONS

$$e^{jx} = \cos x + j\sin x$$

$$a\cos x + b\sin x = \sqrt{a^2+b^2}\cos(x+\phi) \quad \phi = \text{angle}(a-jb)$$

$$\cos^2 x + \sin^2 x = 1$$

$$\sin 2x = 2\sin x\cos x$$

$$\cos 2x = \cos^2 x - \sin^2 x = 2\cos^2 x - 1 = 1 - 2\sin^2 x$$

$$\cos x\cos y = \frac{1}{2}\cos(x+y) + \frac{1}{2}\cos(x-y)$$

$$\sin x\sin y = \frac{1}{2}\cos(x-y) - \frac{1}{2}\cos(x+y)$$

$$\sin x\cos y = \frac{1}{2}\sin(x+y) + \frac{1}{2}\sin(x-y)$$

$$\cos(x\pm y) = \cos x\cos y \mp \sin x\sin y$$

$$\sin(x\pm y) = \sin x\cos y \pm \cos x\sin y$$

## A.2 THREE-PHASE TRIGONOMETRIC RELATIONS

$$\cos x + \cos\left(x - \frac{2\pi}{3}\right) + \cos\left(x + \frac{2\pi}{3}\right) = 0$$

$$\sin x + \sin\left(x - \frac{2\pi}{3}\right) + \sin\left(x + \frac{2\pi}{3}\right) = 0$$

$$\cos^2 x + \cos^2\left(x - \frac{2\pi}{3}\right) + \cos^2\left(x + \frac{2\pi}{3}\right) = \frac{3}{2}$$

$$\sin^2 x + \sin^2\left(x - \frac{2\pi}{3}\right) + \sin^2\left(x + \frac{2\pi}{3}\right) = \frac{3}{2}$$

*Introduction to Electric Power and Drive Systems*, First Edition.
Paul Krause, Oleg Wasynczuk, Timothy O'Connell and Maher Hasan.
© 2017 by The Institute of Electrical and Electronics Engineers, Inc. Published 2017 by John Wiley & Sons, Inc.
Companion website: www.wiley.com/go/krause/electricpower

$$\sin x \cos x + \sin\left(x - \frac{2\pi}{3}\right)\cos\left(x - \frac{2\pi}{3}\right) + \sin\left(x + \frac{2\pi}{3}\right)\cos\left(x + \frac{2\pi}{3}\right) = 0$$

$$\sin x \cos y + \sin\left(x - \frac{2\pi}{3}\right)\cos\left(y - \frac{2\pi}{3}\right) + \sin\left(x + \frac{2\pi}{3}\right)\cos\left(y + \frac{2\pi}{3}\right) = \frac{3}{2}\sin(x - y)$$

$$\sin x \sin y + \sin\left(x - \frac{2\pi}{3}\right)\sin\left(y - \frac{2\pi}{3}\right) + \sin\left(x + \frac{2\pi}{3}\right)\sin\left(y + \frac{2\pi}{3}\right) = \frac{3}{2}\cos(x - y)$$

$$\cos x \cos y + \cos\left(x - \frac{2\pi}{3}\right)\cos\left(y - \frac{2\pi}{3}\right) + \cos\left(x + \frac{2\pi}{3}\right)\cos\left(y + \frac{2\pi}{3}\right) = \frac{3}{2}\cos(x - y)$$

$$\sin x \cos y + \sin\left(x + \frac{2\pi}{3}\right)\cos\left(y - \frac{2\pi}{3}\right) + \sin\left(x - \frac{2\pi}{3}\right)\cos\left(y + \frac{2\pi}{3}\right) = \frac{3}{2}\sin(x + y)$$

$$\sin x \sin y + \sin\left(x + \frac{2\pi}{3}\right)\sin\left(y - \frac{2\pi}{3}\right) + \sin\left(x - \frac{2\pi}{3}\right)\sin\left(y + \frac{2\pi}{3}\right) = -\frac{3}{2}\cos(x + y)$$

$$\cos x \cos y + \cos\left(x + \frac{2\pi}{3}\right)\cos\left(y - \frac{2\pi}{3}\right) + \cos\left(x - \frac{2\pi}{3}\right)\cos\left(y + \frac{2\pi}{3}\right) = \frac{3}{2}\cos(x + y)$$

## A.3 ABBREVIATIONS

| | | | |
|---|---|---|---|
| Alternating current | ac | Megawatt | MW |
| Ampere | A | Meter | m |
| Ampere-turn | At | Microfarad | μF |
| Coulomb | C | Millihenry | mH |
| Direct current | dc | Newton | N |
| Foot | ft | Newton meter | N·m |
| Gram | g | Oersted | Oe |
| Henry | H | Pound | lb |
| Hertz | Hz | Poundal | pdl |
| Horsepower | hp | Power factor | pf |
| Inch | in. | Pulse width modulation | PWM |
| Joule | J | Radian | rad |
| Kilogram | kg | Revolution per minute | r/min (rpm) |
| Kilovar | kvar | Second | s |
| Kilovolt | kV | Volt-ampere reactive | var |
| Kilovolt-ampere | kVA | Volt | V |
| Kilowatt | kW | Volt-ampere | VA |
| Magnetomotive force | mmf | Watt | W |
| Maxwell | Mx | Weber | Wb |

## A.4 CONSTANTS AND CONVERSION FACTORS

| | |
|---|---|
| Permeability of free space | $\mu_0 = 4\pi \times 10^{-7}$ Wb/A·m |
| Permittivity (capacitivity) of free space | $\varepsilon_0 = 8.854 \times 10^{-12}$ C$^2$/N·m$^2$ |
| Acceleration of gravity | $g = 9.807$ m/s$^2$ |
| Length | 1 m = 3.281 ft = 39.37 in. |

| | |
|---|---|
| Mass | 1 kg = 0.0685 slug = 2.205 lb (mass) |
| Force | 1 N = 0.225 lb = 7.23 pdl |
| Torque | 1 N·m = 0.738 lb·ft |
| Energy | 1 J (W·s) = 0.738 lb·ft |
| Power | 1 W = $1.341 \times 10^{-3}$ hp |
| Moment of inertia | 1 kg·m$^2$ = 0.738 slug·ft$^2$ = 23.7 lb·ft$^2$ |
| Magnetic flux | 1 Wb = $10^8$ Mx (lines) |
| Magnetic flux density | 1 Wb/m$^2$ = 10,000 G = 64.5 klines/in.$^2$ |
| Magnetizing force | 1 At/m = 0.0254 At/in. = 0.0126 Oe |

APPENDIX **B**

# *WINDING INDUCTANCES*

**IN THE EARLY CHAPTERS**, we did not spend time deriving the inductances of the machine; instead, we more or less intuitively set forth the inductances in order to focus on machine analysis. Calculating machine inductances is often an appropriate introduction to machine design; however, we can set the stage for those who wish to pursue this area by deriving expressions for inductances of elementary windings in relative motion. In Chapter 2 we expressed the sinusoidally distributed *as*-winding, thus, for a two-pole machine, (2.2-4) and (2.2-5) are

$$N_{as} = \frac{N_s}{2}\sin\phi_s \quad \text{for } 0 < \phi_s < \pi \tag{B-1}$$

$$N_{as} = -\frac{N_s}{2}\sin\phi_s \quad \text{for } \pi < \phi_s < 2\pi \tag{B-2}$$

where $N_s$ is the number of turns per phase.

Our goal is to derive expressions for the magnetizing inductances $L_{ms}$ and $L_{mr}$ and the mutual inductance $L_{sr}$ of the windings shown in Fig. B-1. The *as*-winding is defined by (B-1) and (B-2), the *ar*-winding may be expressed as

$$N_{ar} = \frac{N_r}{2}\sin\phi_r \quad \text{for } 0 < \phi_r < \pi \tag{B-3}$$

$$N_{ar} = -\frac{N_r}{2}\sin\phi_r \quad \text{for } \pi < \phi_r < 2\pi \tag{B-4}$$

where $N_r$ is the number of turns per phase of the *ar*-winding and we know that there must be an even number of poles and an equal number on the rotor and the stator. Now, $\phi_s$, $\phi_r$, and $\theta_r$ are related as given by (2.4-2).

From Chapter 1, the flux $\Phi$, mmf, and the reluctance of the air gap $\mathcal{R}_g$ (neglecting the reluctance of the iron) are related as

$$\Phi = \frac{\text{mmf}}{\mathcal{R}_g} \tag{B-5}$$

Although we are using $\phi_s$ and $\phi_r$ as angular displacements and $\Phi$ (uppercase $\phi$) as flux, we will make every effort to keep things straight. Since we are considering a magnetically linear system, the self-inductances of the *as*- and *ar*-windings are

$$L_{asas} = L_{ls} + L_{ms} \tag{B-6}$$

$$L_{arar} = L_{lr} + L_{mr} \tag{B-7}$$

*Introduction to Electric Power and Drive Systems*, First Edition.
Paul Krause, Oleg Wasynczuk, Timothy O'Connell and Maher Hasan.
© 2017 by The Institute of Electrical and Electronics Engineers, Inc. Published 2017 by John Wiley & Sons, Inc.
Companion website: www.wiley.com/go/krause/electricpower

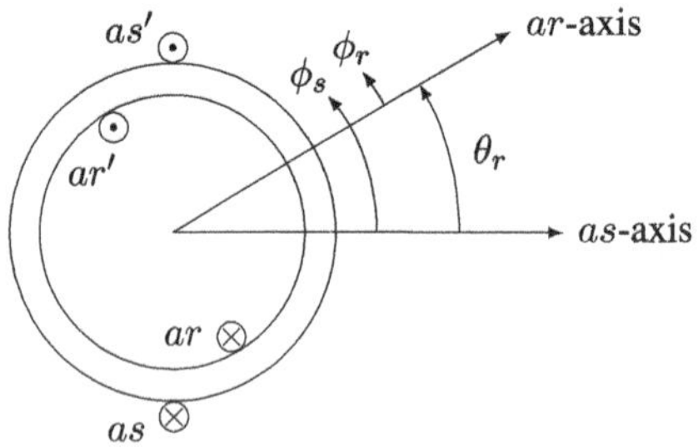


Figure B-1 Windings in relative motion: two-pole equivalent.

As we know, the leakage inductances $L_{ls}$ and $L_{lr}$ make up 5–10% of the self-inductances depending on the design of the machine. We will not derive expressions for the leakage inductances. This derivation is given in Reference [1].

To express the self-inductance of a winding, we need to determine the flux linking the winding due to current flowing in the winding. Let us consider the flux linkages of a single turn of the *as*-winding which spans $\pi$ radians and is located at an angle $\phi_s$. Thus, from (B-1),

$$\text{mmf}_{as} = \frac{N_s}{2} i_{as} \cos \phi_s \tag{B-8}$$

Now, the flux $\Phi$ is determined by performing a surface integral over the open surface of a single turn. Thus, from (B-5) and (B-8),

$$\Phi_{as} = \int_{\phi_s}^{\phi_s+\pi} \frac{N_s}{2} i_{as} \frac{\mu_0 lr}{g} \cos \xi d\xi \tag{B-9}$$

where, in determining the incremental reluctance of the air gap, $r$ is the radius of the rotor, $l$ is the axial length of the air gap of the machine, $g$ is the air-gap length, $d\phi_s$ is the incremental displacement, and $\xi$ is a dummy variable of integration. We will obtain the total flux linkages by summing the flux linking all turns; thus, from (B-2) and (B-9),

$$\lambda_{as} = L_{ls} i_{as} - \int_{\pi}^{2\pi} \left(\frac{N_s}{2}\right) \sin \phi_s \int_{\phi_s}^{\phi_s+\pi} \left(\frac{N_s}{2}\right) i_{as} \frac{\mu_0 lr}{g} \cos \xi d\xi d\phi_s \tag{B-10}$$

Hold on, why did we use (B-2)? Well, we are integrating over the part of the winding that, by the right-hand rule, produces a positive mmf drop from rotor to stator in the direction of the positive *as*-axis at $\phi_s = 0$. In other words, this choice of integration produces a positive $L_{ms}$. Now, since

$$\lambda_{as} = L_{ls} i_{as} + L_{ms} i_{as} \tag{B-11}$$

we can write $L_{ms}$ from (B-10) as

$$L_{ms} = -\left(\frac{N_s}{2}\right)^2 \frac{\mu_0 lr}{g} \int_{\pi}^{2\pi} \sin \phi_s \left| \sin \xi \right|_{\phi_s}^{\phi_s+\pi}$$

$$= 2\left(\frac{N_s}{2}\right)^2 \frac{\mu_0 lr}{g} \int_{\pi}^{2\pi} \sin^2\tau d\tau$$

$$= \left(\frac{N_s}{2}\right)^2 \frac{\mu_0 lr\pi}{g} \tag{B-12}$$

where we used $\tau$ as the dummy variable of integration. It follows that if we substitute subscript $r$ for subscript $s$ in (B-9) and (B-10), we will obtain

$$L_{mr} = \left(\frac{N_r}{2}\right)^2 \frac{\mu_0 lr\pi}{g} \tag{B-13}$$

The mutual inductance between the *as*- and *ar*-winding may be determined by following a similar procedure. Here we will determine the flux linking the *as*-winding due to current flowing only in the *ar*-winding by substituting subscript $r$ for subscript $s$ in (B-9); whereupon, (B-10) becomes

$$\lambda_{as} = -\int_{\pi}^{2\pi} \left(\frac{N_s}{2}\right) \sin\phi_s \int_{\phi_r}^{\phi_r+\pi} \left(\frac{N_r}{2}\right) i_{ar} \frac{\mu_0 lr}{g} \cos\xi d\xi d\phi_s \tag{B-14}$$

Therefore, since $\phi_r$ may be expressed in terms of $\phi_s$ and $\theta_r$ by (2.4-2), we can determine $L_{asar}$ by

$$L_{asar} = -\left(\frac{N_s}{2}\right)\left(\frac{N_r}{2}\right) \frac{\mu_0 lr}{g} \int_{\pi}^{2\pi} \sin\phi_s \left|\sin\xi\right|_{\phi_r}^{\phi_r+\pi} d\phi_s$$

$$= 2\left(\frac{N_s}{2}\right)\left(\frac{N_r}{2}\right) \frac{\mu_0 lr}{g} \int_{\pi}^{2\pi} \sin\phi_s \sin(\phi_s - \theta_r) d\phi_s$$

$$= 2\left(\frac{N_s}{2}\right)\left(\frac{N_r}{2}\right) \frac{\mu_0 lr}{g} \int_{\pi}^{2\pi} (\sin^2\tau\cos\theta_r - \sin\tau\cos\tau\sin\theta_r) d\tau$$

$$= \left(\frac{N_s}{2}\right)\left(\frac{N_r}{2}\right) \frac{\mu_0 lr\pi}{g} \cos\theta_r \tag{B-15}$$

For compactness we have used

$$L_{sr} = \left(\frac{N_s}{2}\right)\left(\frac{N_r}{2}\right) \frac{\mu_0 lr\pi}{g} \tag{B-16}$$

whereupon

$$L_{asar} = L_{sr}\cos\theta_r \tag{B-17}$$

**Example.** Assume that a *bs*-winding with $N_s$ turns is placed on the stator of the device given in Fig. B-1 with the *bs*-axis at $\phi_s = \frac{2}{3}\pi$. Determine the mutual inductance $L_{asbs}$. First, let us express $\text{mmf}_{bs}$ since we will determine $L_{asbs}$ by expressing the flux linking the *as*-winding due to current flowing only in the *bs*-winding. Thus,

$$\text{mmf}_{bs} = \left(\frac{N_s}{2}\right) i_{bs} \cos\left(\phi_s - \frac{2}{3}\pi\right) \tag{B-18}$$

Rewriting (B-14) to fit our situation

$$\lambda_{as} = -\int_{\pi}^{2\pi}\left(\frac{N_s}{2}\right)\sin\phi_s\int_{\phi_s}^{\phi_s+\pi}\left(\frac{N_s}{2}\right)i_{bs}\frac{\mu_0 lr}{g}\cos\left(\xi-\frac{2}{3}\pi\right)d\xi d\phi_s$$

$$= -\left(\frac{N_s}{2}\right)^2 i_{bs}\frac{\mu_0 lr}{g}\int_{\pi}^{2\pi}\sin\phi_s\int_{\phi_s}^{\phi_s+\pi}\cos\left(\xi-\frac{2}{3}\pi\right)d\xi d\phi_s \tag{B-19}$$

Therefore, $L_{asbs}$ becomes

$$L_{asbs} = -\left(\frac{N_s}{2}\right)^2\frac{\mu_0 lr}{g}\int_{\pi}^{2\pi}\sin\phi_s\left|\sin\left(\xi-\frac{2}{3}\pi\right)\right|_{\phi_s}^{\phi_s+\pi}d\phi_s$$

$$= -\left(\frac{N_s}{2}\right)^2\frac{\mu_0 lr}{g}\int_{\pi}^{2\pi}\sin\phi_s\left[\sin\left(\phi_s+\frac{\pi}{3}\right)-\sin\left(\phi_s-\frac{2}{3}\pi\right)\right]d\phi_s$$

$$= -\left(\frac{N_s}{2}\right)^2\frac{\mu_0 lr}{g}\int_{\pi}^{2\pi}(\sin^2\tau+\sqrt{3}\sin\tau\cos\tau)d\tau$$

$$= -\frac{1}{2}\left(\frac{N_s}{2}\right)^2\frac{\mu_0 lr\pi}{g} \tag{B-20}$$

Thus, from (B.12)

$$L_{asbs} = -\frac{1}{2}L_{ms} \tag{B-21}$$

Schmitz and Novotny introduced winding functions in the mid-1960s [2]. Their derivation was to divide the air-gap mmf by the current to obtain what they referred to as the winding function. More recently, Sudhoff set forth a comprehensive derivation and extension of the application of winding functions [1, 3]. Those interested in machine design are well-advised to become familiar with this work. In this Appendix, we have used the approach in Reference [4].

## REFERENCES

[1] S. D. Sudhoff, *Power Magnetic Devices*, Wiley-IEEE Press, Hoboken, NJ, 2013.

[2] N. L. Schmitz and D. W. Novonty, *Introductory Electromechanics*, Roland Press, New York, 1965.

[3] P. C. Krause, O. Wasynczuk, S. D. Sudhoff, and S. D. Pekarek, *Analysis of Electric Machinery and Drive Systems*, 3rd edition, Wiley-IEEE Press, Hoboken, NJ, 2013.

[4] P. C. Krause, O. Wasynczuk, and S. D. Sudhoff, *Analysis of Electric Machinery*, 1st edition, IEEE Press, 1994.

# APPENDIX C

# ANIMATIONS

**ANIMATIONS can be accessed at** www.wiley.com/go/krause/electricpower.

**ANIMATION A. ROTATING air-gap mmf; Tesla's rotating magnetic field.** In this animation, the rotating air-gap mmf of an elementary two-pole, two-phase, symmetrical stator is viewed as a stationary and synchronously rotating observer. This is illustrated with balanced voltages applied to stator windings which are symmetrical $rL$ circuits. The transient and steady-state currents, $i_{as}$ and $i_{bs}$, are shown along with $\text{mmf}_s^s$ and $\text{mmf}_s^e$. The fields are shown for $\text{mmf}_s^s$ (stationary observer) and $\text{mmf}_s^e$ (observer rotating at $\omega_e$).

**ANIMATION B. DYNAMIC magnetic field orientation of a brushless dc motor during operation with $\boldsymbol{\phi_v = 0}$, $\boldsymbol{\phi_v = \phi_{vMT/V}}$, and $\boldsymbol{\phi_v = \phi_{vMT/A}}$.** The dynamic orientations of the magnetic fields of a permanent-magnet ac motor, as viewed by an observer rotating with the rotor, are portrayed for free acceleration and load torque switching. Normal brushless dc motor operation, $\phi_v = 0$, maximum torque operation, $\phi_v = \phi_{vMT/V}$, and $\phi_{vMT/A}$ are shown. The modes of operation illustrated are those depicted in machine variables in Figs. 4.2-1 and 4.2-2 for $\phi_v = 0$ and Figs. 4.2-4 and 4.2-5 for $\phi_v = \phi_{vMT/V}$ and Figs. 4.2-6 and 4.2-7 for $\phi_{vMT/A}$.

---

*Introduction to Electric Power and Drive Systems*, First Edition.
Paul Krause, Oleg Wasynczuk, Timothy O'Connell and Maher Hasan.
© 2017 by The Institute of Electrical and Electronics Engineers, Inc. Published 2017 by John Wiley & Sons, Inc.
Companion website: www.wiley.com/go/krause/electricpower

# INDEX

abbreviations, 220
ac-ac conversion, 144
ac-dc conversion, *see* rectifier
air gap, 12
air-gap mmf, 49, 53
Ampere's Law, 11, 49
animations, 229
arbitrary reference frame, 76

balanced systems
  three-phase, 37–40
  two-phase, 35–37
battery, 92–93, 110, 128
bipolar junction transistor (BJT), 97
boost converter
  circuit diagram, 113
  circuit waveforms, 116
  ideal voltage conversion ratio, 114
  lossy voltage conversion ratio, 114, 117
  power balance, 115
brushless dc drive
  constant torque, constant power operation, 177–179
  inverter-machine drive, 164–166
  pulse width modulation, 176
  simulation of, 180–185
  without phase shift, $\phi_v = 0$
    drive operation
      control mode, 174–175
      free-acceleration, 166–168
      load switching, 169
      sixth harmonic, 170
    two-phase operation
      efficiency, 155
      free-acceleration, 150–152
      load switching, 153
      phasor diagram, 154
      phasor voltage equation, 152
      torque equation, 152
  with phase shift, $\phi_{vMT/A}$
    drive operation
      control mode, 175–176
      free-acceleration, 172–174
    two-phase operation
      calculation of $\phi_{vMT/A}$, 160–162
      efficiency, 163
      free-acceleration, 161–162
      phasor diagram, 163
      phasor voltage equation, 162
      torque equation, 162
  with phase shift, $\phi_{vMT/V}$
    drive operation
      control mode, 175
      free acceleration, 171–172
  two-phase operation
    calculation of $\phi_{vMT/V}$, 156
    efficiency, 160
    free-acceleration, 156–158
    phasor diagram, 159
    phasor voltage equation, 158
    torque equation, 159
buck converter
  circuit diagram, 104
  circuit waveforms, 112
  design example, 110–111
  power balance, 106
  voltage conversion ratio, 105

capacitor
  average current for energy balance, 102
  in classical rectifier, 128
  diode capacitor circuits, 122

*Introduction to Electric Power and Drive Systems*, First Edition.
Paul Krause, Oleg Wasynczuk, Timothy O'Connell and Maher Hasan.
© 2017 by The Institute of Electrical and Electronics Engineers, Inc. Published 2017 by John Wiley & Sons, Inc.
Companion website: www.wiley.com/go/krause/electricpower

capacitor (*continued*)
  "large" definition ($C_{\min}$) in buck converter, 109
  switched capacitor circuit, 101–102
classical rectifier, *see* rectifier
coenergy
  definition, 13
  electromagnetic torque evaluation, 32–35
  multiple electrical inputs, 21
coil, 48
coil side, 48
compact Fourier series, *see* Fourier analysis
conservation of energy, 101–103
constants and conversion factors, 220–221
coupled circuits
  relative motion, 22–25
  stationary, 16–22

damping coefficient, 68
dc-ac conversion, *see* inverter
dc-dc conversion
  advanced circuit topologies, 118
  boost converter, *see* boost converter
  buck converter, *see* buck converter
  ideal, 103–104
delta-wye impedance conversion, 193
diode, 98
diode trial method
  in full-wave rectifier, 125–126
  in half-wave rectifier, 119
direct axis, 73
distortion, *see* harmonics
duty ratio, *D*
  of boost converter, 114–115, 117
  of buck converter, 105–106
  of dc-dc converter, 103
  definition of, 98
  harmonics and, 101
  of switching function, 99

electromagnetic torque, 21, 61
electromechanical systems
  energy in, 32
  force calculation, 71
  windings in relative motion, 22–25
energy
  balance, 27
  electromechanical, 27–28
  field, 57–64
  storage elements, 101–103
even function, 96

field energy, 13–14
field intensity, 10
field strength, 10
field winding, synchronous machine, 194
flux
  definition, 8–9
  density, 10
  leakage, 9
  magnetizing, 9
flux linkages, definition of, 12
force calculation, 30–31
Fourier analysis, 95–97

half-wave symmetry, 97
harmonics
  in power electronic circuits, 144–146
  total harmonic distortion (THD), 145
hysteresis, 13

ideal transformer, 17
inductance(s), definition of
  leakage, 12
  magnetizing, 12
  mutual, 18
  self, 12
inductor
  average voltage for energy balance, 102
  current in buck converter, 108
  "large" definition ($L_{\min}$) in buck converter, 108
  lossy, in boost converter, 115
  switched inductor circuit, 101–102
infinite bus, 217
insulated-gate bipolar transistor (IGBT), 97–98
inverter
  single-phase
    circuit diagram, 133
    definition of, 133
    power calculation, 135–136
    with series resonant *rLC* load, 134–135
    with series *rL* load, 134
  three-phase/"six-step"
    circuit diagram, 137
    current calculation, 143–144
    definition of, 136

description of, 138, 164–166
Fourier expansion of phase voltage, 138
with series $rL$ load, 140–142
simulation of, 183

leakage fluxes, definition, 9
line-to-line voltage, 189
line-to-neutral voltage, 189
load, definition of, 92

magnetic circuit
electric analog, 9
leakage flux, 9
reluctance, 9
magnetic field strength, 10
magnetic flux density, 10
magnetically coupled circuits
in relative motion, 22–26
stationary, 16–22
magnetization curve, 12
magnetomotive force (mmf)
definition of, 10
rotating air-gap, 52
sinusoidally distributed air-gap, 44
Tesla's rotating magnetic field, 52
mean square, 136
metal-oxide-semiconductor field-effect transistor (MOSFET), 97
mutual inductance, 18

odd function, 96

$P$-pole machines, 58–63
Park, R. H., 43, 85
permanent-magnet ac machine
Tesla's rotating magnetic field
stationary observer, 69
synchronously rotating observer, 80
three-phase
brushless dc motor-inverter, 164–166
brushless dc motor operation, 139–163
comparison with two-phase, 86–87
machine variables, 82–84
rotor reference frame variables, 84–87
torque
machine variables, 67
maximum per ampere, 160–163
maximum per volt, 155–160
rotor reference frame variables, 74
two-phase
change of variables, 69–72
equivalent circuit, 73
flux linkages, 66–72
inductances, 62, 66–67
machine variables, 66–67
phasor diagram, 79–80
power balance, 75–76
rotor reference frame variables, 72
steady-state equations
phasor, 78
rotor reference frame, 77
voltage equations, instantaneous
machine variables, 65
rotor reference frame variables, 73–74
voltage phase shifting, 155–163
permeability, 10
phase sequence
three-phase, 37
two-phase, 33
phasor analysis, 1–7
power balance, 75
power conversion principles, 92–95
power electronics
definition of, 91
devices, 92
power factor angle, 6
power, steady-state, 6
pulse width modulation (PWM), 133–134, 176–177

quadrature axis, 69

reactive power, 6
rectifier
classical, 128
full-wave/full-bridge
average power, 132
circuit diagram, 125
Fourier series of voltage, $v_{rec}$, 128
with $LC$ output filter, 131–132
with parallel $rC$ load, 128–129
with resistive load, 127–128
with series $rL$ load, 128, 130
half-wave/half-bridge
average power, 123–124
circuit diagram, 118
with parallel $rC$ load, 120–122
with resistive load, 119–120
with series $rL$ load, 122–125

reference frame, rotor, 76, 84
reluctance, 9
reluctance machines, 64
resonant filter, 135
right-hand rule, 9
ripple
  of capacitor voltage, $\Delta V$, 109, 121, 128
  as a design parameter, 110–111, 118
  of inductor current, $\Delta I$, 107–108
  relationship to switching frequency, 110
rotating magnetic field
  three-phase, 53–58
  two-phase
    air-gap mmf, 53–56
    Tesla's, 53–56
  viewed by
    arbitrary rotating observer, 54
    stationary observer, 54
    synchronously rotating, 54–56
rotating magnetic poles, *see* rotating magnetic field

saturation, 13–14
sinusoidally distributed winding, 49
source, definition of, 92
space sinusoid, 44
switch
  active, 97
  binary variable, $q$, 98
  definition of, 97
  ideal, 98
  passive, 97
  real, 100
switching
  "fast", definition of, 110
  frequency, $f_{sw}$, 98
  function, $q(t)$
    definition of, 98–99
    Fourier series of, 99–101
  loss, 100
  period, $T_{sw}$, 98
synchronous condenser, 206
synchronous machine
  damper windings, 198
  equivalent circuit, rotor reference frame, 107
  excitation voltage, 201
  fault, three-phase, 215–219
  flux linkage equations
    machine variables, 195
    rotor reference frame variables, 196–197
  inductances, 195–197
  per-unit equations, 210–213
  phasor diagram, 204
  rotor angle, 201–202
  steady-state analysis, 203–204
  Tesla's rotating magnetic field, 202
  torque-angle characteristics, 216
  torque equations, 198, 201–202
  transient stability, 215–219
  voltage equations
    machine variables, 194, 200
    rotor reference frame variables, 197
    steady-state, 200
synchronous speed (synchronism), 193–194

torque-angle (rotor angle), 201–202
torque equation, 61, 152, 198
total harmonic distortion (THD), *see* harmonics
transformer
  equivalent circuit, 19
  field energy in, 20
  ideal, three-phase, 192
  leakage fluxes in, 17
  magnetizing flux, 17
  open-circuit test, 21–22
  short-circuit test, 21–22
  three-phase connections
    delta-delta, 190–191
    delta-wye, 191
    wye-delta, 191
    wye-wye, 188–190
trigonometric identities, 219–220

voltage
  conversion ratio, $\alpha$, 94, 105, 114–115, 117
  divider, 93–95

windings
  coil, 48
  coil side, 48
  concentrated, 44–47
  distributed, 47–51
  field, synchronous machine, 194
  functions, 226
  inductance, calculation of, 223–226
  pitch, 48
  sinusoidally distributed, 48–49
  uniformly distributed, 48, 51

# IEEE Press Series on Power Engineering

**Series Editor: M. E. El-Hawary,** Dalhousie University, Halifax, Nova Scotia, Canada

The mission of IEEE Press Series on Power Engineering is to publish leading-edge books that cover the broad spectrum of current and forward-looking technologies in this fast-moving area. The series attracts highly acclaimed authors from industry/academia to provide accessible coverage of current and emerging topics in power engineering and allied fields. Our target audience includes the power engineering professional who is interested in enhancing their knowledge and perspective in their areas of interest.

1. *Principles of Electric Machines with Power Electronic Applications, Second Edition*
M. E. El-Hawary

2. *Pulse Width Modulation for Power Converters: Principles and Practice*
D. Grahame Holmes and Thomas Lipo

3. *Analysis of Electric Machinery and Drive Systems, Second Edition*
Paul C. Krause, Oleg Wasynczuk, and Scott D. Sudhoff

4. *Risk Assessment for Power Systems: Models, Methods, and Applications*
Wenyuan Li

5. *Optimization Principles: Practical Applications to the Operations of Markets of the Electric Power Industry*
Narayan S. Rau

6. *Electric Economics: Regulation and Deregulation*
Geoffrey Rothwell and Tomas Gomez

7. *Electric Power Systems: Analysis and Control*
Fabio Saccomanno

8. *Electrical Insulation for Rotating Machines: Design, Evaluation, Aging, Testing, and Repair, Second Edition*
Greg Stone, Edward A. Boulter, Ian Culbert, and Hussein Dhirani

9. *Signal Processing of Power Quality Disturbances*
Math H. J. Bollen and Irene Y. H. Gu

10. *Instantaneous Power Theory and Applications to Power Conditioning*
Hirofumi Akagi, Edson H. Watanabe, and Mauricio Aredes

11. *Maintaining Mission Critical Systems in a 24/7 Environment*
Peter M. Curtis

12. *Elements of Tidal-Electric Engineering*
Robert H. Clark

13. *Handbook of Large Turbo-Generator Operation and Maintenance, Second Edition*
Geoff Klempner and Isidor Kerszenbaum

14. *Introduction to Electrical Power Systems*
Mohamed E. El-Hawary

15. *Modeling and Control of Fuel Cells: Distributed Generation Applications*
M. Hashem Nehrir and Caisheng Wang

16. *Power Distribution System Reliability: Practical Methods and Applications*
Ali A. Chowdhury and Don O. Koval

17. *Introduction to FACTS Controllers: Theory, Modeling, and Applications*
Kalyan K. Sen and Mey Ling Sen

18. *Economic Market Design and Planning for Electric Power Systems*
James Momoh and Lamine Mili

19. *Operation and Control of Electric Energy Processing Systems*
James Momoh and Lamine Mili

20. *Restructured Electric Power Systems: Analysis of Electricity Markets with Equilibrium Models*
Xiao-Ping Zhang

21. *An Introduction to Wavelet Modulated Inverters*
S.A. Saleh and M.A. Rahman

22. *Control of Electric Machine Drive Systems*
Seung-Ki Sul

23. *Probabilistic Transmission System Planning*
Wenyuan Li

24. *Electricity Power Generation: The Changing Dimensions*
Digambar M. Tigare

25. *Electric Distribution Systems*
Abdelhay A. Sallam and Om P. Malik

26. *Practical Lighting Design with LEDs*
Ron Lenk and Carol Lenk

27. *High Voltage and Electrical Insulation Engineering*
Ravindra Arora and Wolfgang Mosch

28. *Maintaining Mission Critical Systems in a 24/7 Environment, Second Edition*
Peter Curtis

29. *Power Conversion and Control of Wind Energy Systems*
Bin Wu, Yongqiang Lang, Navid Zargari, and Samir Kouro

30. *Integration of Distributed Generation in the Power System*
Math H. Bollen and Fainan Hassan

31. *Doubly Fed Induction Machine: Modeling and Control for Wind Energy Generation Applications*
Gonzalo Abad, Jesus Lopez, Miguel Rodrigues, Luis Marroyo, and Grzegorz Iwanski

32. *High Voltage Protection for Telecommunications*
Steven W. Blume

33. *Smart Grid: Fundamentals of Design and Analysis*
James Momoh

34. *Electromechanical Motion Devices, Second Edition*
Paul C. Krause, Oleg Wasynczuk, and Steven D. Pekarek

35. *Electrical Energy Conversion and Transport: An Interactive Computer-Based Approach, Second Edition*
George G. Karady and Keith E. Holbert

36. *ARC Flash Hazard and Analysis and Mitigation*
J. C. Das

37. *Handbook of Electrical Power System Dynamics: Modeling, Stability, and Control*
Mircea Eremia and Mohammad Shahidehpour

38. *Analysis of Electric Machinery and Drive Systems, Third Edition*
Paul Krause, Oleg Wasynczuk, S. D. Sudhoff, and Steven D. Pekarek

39. *Extruded Cables for High-Voltage Direct-Current Transmission: Advances in Research and Development*
Giovanni Mazzanti and Massimo Marzinotto

40. *Power Magnetic Devices: A Multi-Objective Design Approach*
S. D. Sudhoff

41. *Risk Assessment of Power Systems: Models, Methods, and Applications, Second Edition*
Wenyuan Li

42. *Practical Power System Operation*
Ebrahim Vaahedi

43. *The Selection Process of Biomass Materials for the Production of Bio-Fuels and Co-Firing*
Najib Altawell

44. *Electrical Insulation for Rotating Machines: Design, Evaluation, Aging, Testing, and Repair, Second Edition*
Greg C. Stone, Ian Culbert, Edward A. Boulter, and Hussein Dhirani

45. *Principles of Electrical Safety*
Peter E. Sutherland

46. *Advanced Power Electronics Converters: PWM Converters Processing AC Voltages*
Euzeli Cipriano dos Santos Jr. and Edison Roberto Cabral da Silva

47. *Optimization of Power System Operation, Second Edition*
Jizhong Zhu

48. *Power System Harmonics and Passive Filter Designs*
J. C. Das

49. *Digital Control of High-Frequency Switched-Mode Power Converters*
Luca Corradini, Dragan Maksimovic, Paolo Mattavelli, and Regan Zane

50. *Industrial Power Distribution, Second Edition*
Ralph E. Fehr, III

51. *HVDC Grids: For Offshore and Supergrid of the Future*
Dirk Van Hertem, Oriol Gomis-Bellmunt, and Jun Liang

52. *Advanced Solutions in Power Systems: HVDC, FACTS, and Artificial Intelligence*
Mircea Eremia, Chen-Ching Liu, and Abdel-Aty Edris

53. *Power Grid Operation in a Market Environment: Economic Efficiency and Risk Mitigation*
Hong Chen

54. *Current Signature Analysis for Condition Monitoring of Cage Induction Motors: Industrial Application and Case Histories*
William T. Thomson and Ian Culburt

55. *Model Predictive Control of Wind Energy Conversion Systems*
Venkata Yaramasu and Bin Wu

56. *Electric Power System Basics for the Nonelectrical Professional, Second Edition*
Steven W. Blume

57. *Energy Production Systems Engineering*
Thomas H. Blair

58. *Understanding Symmetrical Components for Power System Modeling*
J. C. Das

59. *High-Power Converters and AC Drives, Second Edition*
Bin Wu and Mehdi Narimani

60. *Practical Lighting Design with LEDs, Second Edition*
Ron Lenk and Carol Lenk

61. *Introduction to Electric Power and Drive Systems*
Paul Krause, Oleg Wasynczuk, Timothy O'Connell, and Maher Hasan